Automorphisms and Derivations of Associative Rings

Mathematics and Its Applications (*Soviet Series*)

Volume 69

Automorphisms and Derivations of Associative Rings

by

V. K. Kharchenko

Institute for Mathematics,
Novosibirsk, U.S.S.R.

KLUWER ACADEMIC PUBLISHERS

DORDRECHT / BOSTON / LONDON

Library of Congress Cataloging-in-Publication Data

Knarchenko, V. K.
Automorphisms and derivations of associative rings / by V.K. Kharchenko.
p. cm. -- (Mathematics and its applications. Soviet series ; 69)
Translated from the Russian.
Includes bibliographical references and index.
ISBN 0-7923-1382-8 (acid free paper)
1. Associative rings. 2. Automorphisms. I. Title. II. Series: Mathematics and its applications (Kluwer Academic Publishers). Soviet series ; 69.
QA251.5.K48 1991
512'.24--dc20 91-25930

ISBN 0-7923-1382-8

Published by Kluwer Academic Publishers,
P.O. Box 17, 3300 AA Dordrecht, The Netherlands.

Kluwer Academic Publishers incorporates
the publishing programmes of
D. Reidel, Martinus Nijhoff, Dr W. Junk and MTP Press.

Sold and distributed in the U.S.A. and Canada
by Kluwer Academic Publishers,
101 Philip Drive, Norwell, MA 02061, U.S.A.

In all other countries, sold and distributed
by Kluwer Academic Publishers Group,
P.O. Box 322, 3300 AH Dordrecht, The Netherlands.

Printed on acid-free paper

Translated from the Russian by
L. Yuzina

Printed in the Netherlands

In memory of my teachers

AKILOV Gleb Pavlovich
KARGAPOLOV Mikhail Ivanovich
SHIRSHOV Anatolii Illarionovich

SERIES EDITOR'S PREFACE

'Et moi, ..., si j'avait su comment en revenir, je n'y serais point allé.'

Jules Verne

The series is divergent; therefore we may be able to do something with it.

O. Heaviside

One service mathematics has rendered the human race. It has put common sense back where it belongs, on the topmost shelf next to the dusty canister labelled 'discarded non-sense'.

Eric T. Bell

Mathematics is a tool for thought. A highly necessary tool in a world where both feedback and non-linearities abound. Similarly, all kinds of parts of mathematics serve as tools for other parts and for other sciences.

Applying a simple rewriting rule to the quote on the right above one finds such statements as: 'One service topology has rendered mathematical physics ...'; 'One service logic has rendered computer science ...'; 'One service category theory has rendered mathematics ...'. All arguably true. And all statements obtainable this way form part of the raison d'être of this series.

This series, *Mathematics and Its Applications*, started in 1977. Now that over one hundred volumes have appeared it seems opportune to reexamine its scope. At the time I wrote

> "Growing specialization and diversification have brought a host of monographs and textbooks on increasingly specialized topics. However, the 'tree' of knowledge of mathematics and related fields does not grow only by putting forth new branches. It also happens, quite often in fact, that branches which were thought to be completely disparate are suddenly seen to be related. Further, the kind and level of sophistication of mathematics applied in various sciences has changed drastically in recent years: measure theory is used (non-trivially) in regional and theoretical economics; algebraic geometry interacts with physics; the Minkowsky lemma, coding theory and the structure of water meet one another in packing and covering theory; quantum fields, crystal defects and mathematical programming profit from homotopy theory; Lie algebras are relevant to filtering; and prediction and electrical engineering can use Stein spaces. And in addition to this there are such new emerging subdisciplines as 'experimental mathematics', 'CFD', 'completely integrable systems', 'chaos, synergetics and large-scale order', which are almost impossible to fit into the existing classification schemes. They draw upon widely different sections of mathematics."

By and large, all this still applies today. It is still true that at first sight mathematics seems rather fragmented and that to find, see, and exploit the deeper underlying interrelations more effort is needed and so are books that can help mathematicians and scientists do so. Accordingly MIA will continue to try to make such books available.

If anything, the description I gave in 1977 is now an understatement. To the examples of interaction areas one should add string theory where Riemann surfaces, algebraic geometry, modular functions, knots, quantum field theory, Kac-Moody algebras, monstrous moonshine (and more) all come together. And to the examples of things which can be usefully applied let me add the topic 'finite geometry'; a combination of words which sounds like it might not even exist, let alone be applicable. And yet it is being applied: to statistics via designs, to radar/sonar detection arrays (via finite projective planes), and to bus connections of VLSI chips (via difference sets). There seems to be no part of (so-called pure) mathematics that is not in immediate danger of being applied. And, accordingly, the applied mathematician needs to be aware of much more. Besides analysis and numerics, the traditional workhorses, he may need all kinds of combinatorics, algebra, probability, and so on.

In addition, the applied scientist needs to cope increasingly with the nonlinear world and the extra mathematical sophistication that this requires. For that is where the rewards are. Linear models are honest and a bit sad and depressing: proportional efforts and results. It is in the non-linear world that infinitesimal inputs may result in macroscopic outputs (or vice versa). To appreciate what I am hinting at: if electronics were linear we would have no fun with transistors and computers; we would have no TV; in fact you would not be reading these lines.

There is also no safety in ignoring such outlandish things as nonstandard analysis, superspace and anticommuting integration, p-adic and ultrametric space. All three have applications in both electrical engineering and physics. Once, complex numbers were equally outlandish, but they frequently proved the shortest path between 'real' results. Similarly, the first two topics named have already provided a number of 'wormhole' paths. There is no telling where all this is leading - fortunately.

Thus the original scope of the series, which for various (sound) reasons now comprises five sub-series: white (Japan), yellow (China), red (USSR), blue (Eastern Europe), and green (everything else), still applies. It has been enlarged a bit to include books treating of the tools from one subdiscipline which are used in others. Thus the series still aims at books dealing with:

- a central concept which plays an important role in several different mathematical and/or scientific specialization areas;
- new applications of the results and ideas from one area of scientific endeavour into another;
- influences which the results, problems and concepts of one field of enquiry have, and have had on the development of another.

Galois theory deals with the action of a finite group of automorphisms on a field. The power and the usefulness of the theory is enormous, and hence, for a long time, there has been great interest in more general Galois theory. This has turned out to be a long and involved quest. Finally, it has led to the modern theory of automorphisms and derivations of associative rings and algebras.

A great many technical and conceptual advances were needed to bring the theory to its present rich state. This includes 'rings with generalized identities', 'non-standard algebra' and a powerful logic-algebraic metatheorem.

The author is one of the main, currently active contributors to the subject and I am greatly pleased that he has written this first, comprehensive authoritative monograph on the topic.

The shortest path between two truths in the real domain passes through the complex domain.

J. Hadamard

La physique ne nous donne pas seulement l'occasion de résoudre des problèmes ... elle nous fait pressentir la solution.

H. Poincaré

Never lend books, for no one ever returns them; the only books I have in my library are books that other folk have lent me.

Anatole France

The function of an expert is not to be more right than other people, but to be wrong for more sophisticated reasons.

David Butler

Amsterdam, August 1991

Michiel Hazewinkel

TABLE OF CONTENTS

INTRODUCTION

Starting from the Bergman-Isaacs work on regular actions of finite groups, a great number of papers have been devoted to studying the automorphisms of associative rings. For some time the central aim of these studies was to find out and investigate those ring properties which are preserved under passage to a fixed ring under finite group actions and, conversely, from the fixed ring to the initial ring. The methods which emerged in these studies proved their value for investigating arbitrary automorphisms and derivations, basically from the point of view of their algebraic dependences. These methods proved so effective that they made it possible to prove Galois correspondence theorems in the class of semiprime rings both for automorphism groups and for Lie algebras of derivations.

A rapid development of the theory was facilitated by the data accumulated by that time on the structural theory of rings with generalized identities (Amitsur, Martindale theorems), and it profited greatly from the concepts formed in noncommutative Galois theory as it was transferred from skew fields to complete rings of continuous linear transformations of spaces in work by E.Noether, N.Jacobson, G.Cartan, T.Nakayama, G.Azumaya, J.Diedonnè, A.Rosenberg and D.Zelinsky.

In the process of studying automorphisms, involutions and derivations the authors started noting that the basic obstacles in proving the theorems disappeared as soon as one considers prime rings. However, for these problems, it turned out that the direct reduction from semiprime rings to prime rings was impossible and one had to do things directly at the semiprime level, which quite frequently resulted in monotonous considerations. This effect was investigated in an interesting paper by K.I.Beidar and A.V.Mikhalev [16]. The essence of their result (from now on called "a metatheorem") is that all theorems written in terms of Horn formulas of the elementary language can be transferring from the class of prime rings to that of orthogonally complete semiprime rings. If we add to what was said above that in the class of prime rings the formulas of the elementary language are equivalent to Horn ones, then the "applied" value of the metatheorem becomes evident.

At approximately the same time Burris and Werner [25] obtained similar results on sheaves of algebraic systems independent of applications in the theory of rings. Additionally, in studies by E.I.Gordon and V.A.Lyubetzky [100] based on non-standard analysis, similar problems were considered. Within these frameworks a semiprime orthogonal complete ring can be viewed as a non-standard prime ring. The formulation and proof of the metatheorem presented in this book take, more or less, into account the background intuition of all three of the approaches just mentioned.

The material is arranged in the following way. The first chapter is an introductory part presenting the current state of the art of the structural theory of rings for the Baer radical. We, naturally, tried to keep the presentation as close to the basic subject as possible and as early as in the third paragraph one can find the Bergman-Isaacs theorem on the nilpotency of any ring having a regular finite group of automorphisms (under natural limitation on the characteristics), as well as the Quinn theorem on the integrality of a ring over the ring of invariants.

These theorems together with Martindale theorem and the metatheorem are the key results of the chapter.

In the second chapter problems of algebraic dependences of automorphisms and derivations are considered. The results presented show, essentially, the algebraic independence of automorphisms and derivations of semiprime rings. To grasp this idea certain refinements are necessary since some dependences of a certain type (trivial ones) still exist. For instance, the sum of any two derivations is equal to a suitable third one. In a similar way, the product of any two automorphisms will be an automorphism. To be more exact, the results of the second chapter show that all the algebraic dependences between automorphisms and derivations are consequences of the simplest ones, i.e., of those resulting from the algebraic structure of a set of derivations and automorphisms, as well as of those relations

$\mu = r_a - l_a, g = r_a l_{a^{-1}}$ which determine the inner derivations and automorphisms for a Martindale ring of quotients.

Associated with the above refinement, there arises a problem of obtaining criteria allowing one to distinguish between trivial generalized polynomials with automorphisms and derivations from non-trivial ones. In the sixth section we give such a criterion for the multilinear case in the spirit of the L.Rowen's approach and prove that if the ideal generated by the values of generalized monomials of an identity with automorphisms contains a unit, then the ring satisfies a usual polynomial identity. This statement is also valid for differential identities with automorphisms in the case when the ring has no additive torsion.

The seventh chapter presents certain immediate corollaries of independence. Let us cite two of them: any algebraic derivation of semiprime ring with zero characteristic is an inner one for a Martindale ring of quotients; if the fixed ring of a finite group G of automorphisms of the ring R satisfies a polynomial identity and R has no additive torsion, then R also satisfies a polynomial identity.

In the next two chapters the Galois theory of prime rings is developed and then, in the fifth chapter, using the metatheorem, it is transferred to semiprime rings. The results presented in these chapters are, basically, due to the author; however, the formulations used and the proofs incorporate the ideas from a paper by S.Montgomery and D.Passman [117], as well as the results of studies presented in a series of works by J.Goursaud, J.Osterburg, J.Pascaud and J.Valette [50, 51, 52]. In transferring the results to semiprime rings, the ideas by A.V.Yakovlev [155] concerning the Galois theory of sheaves of sets also proved useful.

In any event there arises the question of relating the theory developed in this book with the classical Galois theory of fields, sfields and rings of continuous transformations. The fact that the classical results follow from those cited in the book is not enough information, since any true statement follows from other true statement (as well as from a false one). Therefore, from our viewpoint, the problem should be posed at another level, i.e., to what extent is acquaintance with the general results useful for understanding the classical ones? As to the Galois theory of fields, the answer is unambiguous; namely, it is the ABC of modern algebra and it is, therefore, doubtful that anyone who wishes to get acquainted with it, will start with studying noncommuta-tive rings. In the case of the automorphisms of skew fields a direct construction of the Galois theory is much simpler, so that it would be reasonable to get acquainted with this theory by means of the seventh chapter of the monograph by N.Jacobson [65], which, however, does not cover the case of derivations given in the fourth chapter of the present monograph. The latter results for skew fields were obtained by M.Weisfeld [153].

As to the case of rings of continuous linear transformations of vector spaces over skew fields (or simple artinian rings), the classical proof is based on the fact that the automorphisms turn out to be conjugate under semilinear transformations and the problems are, therefore, reduced to a great extent to skew fields. In the general case there is no description of automorphisms, so, as applied to rings of transformations, the new proofs reveal other reasons for the Galois correspondence theorems. It is a *posteriori* obvious that for a ring of linear transformations the correspondence described in 3.10.6 reduces to the classical one, and thus the BM and RF conditions, as well as the centralizer condition, can be viewed as an internal characterization of the homogenous rings of endomorphisms.

Therefore, the Galois theory of prime rings sheds additional light on rings of continuous linear transformations as well. Of much greater importance, however, is the application of this theory to other classes of rings which have not been studied in terms of the Galois theory so far.

For instance, the correspondence theorem for domains is rather far from the analogous theorem for skew fields; there exist domains which cannot be embedded into skew fields at all. The application of this theory to free algebras also proved unexpected, since there is nothing of this kind for rings of polynomials.

This application leads into the concluding chapter of the book which is mainly intended for the reader trained in the theory of rings. The chapter treats the A.N.Koryukin theorems on noncommutative invariants of linear groups which gave a complete solution of an analogue of Hilbert's 14-th problem for a ring of noncommutative polynomials. The chapter considers a wide range of problems on the transfer of the properties from a ring of invariants (a ring of constants) to the initial ring and conversely, it clarifies the relations between the spectra of prime ideals of a ring and those of a fixed ring (the Montgomery equivalence), it proves the Fisher and Grzesczuk-Puczylowsky's theorems on finite groups acting on modular lattices, and it gives a proof of the Goursaud-Pascaud-Valette and Piers Dos Santos theorems on maximal rings of quotients. The last section presents a general concept of Hopf algebra actions embracing both the case of automorphisms and that of derivations. The well-known Kostant-Sweedler theorem (Theorem **6.5.1**) demonstrates that the study of the action of cocommutative Hopf algebras over algebraically closed fields of characteristic zero reduces to those of actions of the corresponding groups and Lie algebras. Therefore, it is the study of non-commutative Hopf algebras that is of prime interest. In the last section we consider some approaches to studying skew derivations from this viewpoint.

The author wishes to express his gratitude to the participants of Shirshov seminars on the ring theory held at the Institute of Mathematics of the Siberian Branch of the USSR Academy of Sciences for their interest in his work.

INTRODUCTORY PART

CHAPTER 1. STRUCTURE OF RINGS

1.1 Baer Radical and Semiprimeness

We shall start constructing Baer radicals with a simple remark that any abelian group can be turned into a ring by way of defining multiplication by means of a formula $xy = 0$. Such a ring is called a *zero-multiplication ring* and is trivial in terms of the theory of rings. Now, starting from an arbitrary ring R, we can try to "kill off" the trivial part in it. If the ring R contains a nonzero ideal I_1 with a zero multiplication, $I_1^2 = 0$, then it would be natural to go over to considering a factor-ring R / I_1, since the whole ring R is presented as an extension of the "trivial" ring I_1 by R / I_1. The same approach can be used for the ring R / I_1, i.e., we can try to find in it a non-zero ideal I_1 / I_2 with a trivial multiplication, and focus then our attention on a factor-ring $R / I_2 \cong (R / I_1) / (I_2 / I_1)$. A continuation of this process is equivalent to finding in the ring R an ascending chain of ideals

$$(0) \subset I_1 \subset I_2 \subset \ldots \subset I_n \subset \ldots,$$

such that the factors I_n / I_{n-1} have a zero multiplications, i.e., $I_n^2 \subseteq I_{n-1}$. It goes without saying that such a chain can prove infinite. In this case we can introduce into the consideration an ideal $I_\omega = \cap I_n$ which is, as has been shown above, obtained by extending the "trivial" rings (using the "trivial" ones), and nothing can prevent us from using the same approach for killing off "trivial" parts in the ring R / I_ω.

Therefore, we get a transfinite chain of the ideals of the ring R.

$$(0) \subset I_1 \subset I_2 \subset \ldots \subset I_\alpha \subset \ldots,$$

where $I^2_{\alpha+1} \subseteq I_\alpha$, and the equality $I_\alpha = \bigcup_{\beta<\alpha} I_\beta$ holding for limit ordinals α. Obviously, in its cardinality the length of such a chain is not greater than that of the set of elements of the ring R, so that we can consider a union $\boldsymbol{B}(R) = \bigcup I_\alpha$. An ideal $\boldsymbol{B}(R)$ is called a *Baer radical* (or a *lower nil-radical*) of the ring R.

By construction, we can no longer "kill off" any trivial part of the ring $R/\boldsymbol{B}(R)$, since this ring contains no nonzero ideals with a zero multiplication.

1.1.1 Definition. A ring is called *semiprime* if all its nonzero ideals have a nonzero multiplication, i.e., for an ideal I the equality $I^2 = 0$ implies $I = 0$.

1.1.2 Definition. An ideal I of the ring R is called *semiprime* if the factor-ring R/I is semiprime.

We see, therefore, that the Baer radical of a ring is a semiprime ideal.

From the viewpoint of the construction process above, it is also possible (as a limiting case) that the chain being constructed reaches the whole ring R in a finite number of steps. It is easily seen that in this case there exists a natural number n, such that the product of any n elements of the ring R equals zero, i.e., $R^n = 0$. A ring with such a property is called *nilpotent* and, accordingly, the ideal I is called nilpotent if for a certain natural n we have $I^n = 0$.

1.1.3 Note. A semiprime ring contains no nilpotent ideals since one of the powers of a nilpotent ideal has a zero multiplication. Accordingly, for a semiprime ideal J and and ideal I the condition $I^n \subseteq J$ implies $I \subseteq J$.

The above definition of a Baer radical of a ring clearly demonstrates that the radical is "composed" of rings with a zero multiplication, its form, however, depending on a certain arbitrariness of the choice of the ideals I_α, which is a disadvantage. It is, therefore, useful to prove the following theorem.

1.1.4 Theorem.(a) *The Baer radical of a ring can be obtained as a union of the transfinite chain of ideals*

$$(0) = N_0 \subseteq N_2 \subseteq \ldots \subseteq N_\alpha \subseteq \ldots$$

wherein the ideal $N_{\alpha+1}$ *is chosen in such a way that the sum of all the nilpotent ideals of the ring* R/N_α *is* $N_{\alpha+1}/N_\alpha$ *and the equality* $N_\alpha = \bigcup_{\beta<\alpha} N_\beta$ *holds for the limit ordinals* α.

(b) *The Baer radical of a ring is the smallest semiprime ideal.*

Proof.(a) Let N be a union of the chain given in the formulation of the theorem. Using transfinite induction, let us prove that $I_\alpha \subseteq N_\alpha$ for all α. Because the ideal with a zero multiplication is nilpotent, $I_1 \subseteq N_1$. If α is a limit ordinal and $I_\beta \subseteq N_\beta$ for all $\beta < \alpha$, then $\bigcup I_\alpha = \bigcup_\beta I_\beta = N_\alpha$. While if α is not a limit ordinal and $I_{\alpha-1} \subseteq N_{\alpha-1}$, then $I_\alpha^2 \subseteq I_{\alpha-1} \subseteq N_{\alpha-1}$, so that $I_\alpha^2 \subseteq I_{\alpha-1} \subseteq N_{\alpha-1}$ is a nilpotent ideal of $R/N_{\alpha-1}$, and, hence, $I_\alpha \subseteq N_\alpha$. Therefore, $\mathbf{B}(R) \subseteq N$.

Let us, conversely, assume that $N \not\subset \mathbf{B}(R)$ and choose the smallest α, such that $N_\alpha \not\subset \mathbf{B}(R)$. In this case α cannot be a finite ordinal, i.e., there exists an ordinal $\alpha - 1$ for which $N_{\alpha-1} \subseteq \mathbf{B}(R)$. Because the ideal N_α is, by definition, equal to the sum of all the ideals of the ring R which are nilpotent modulo $N_{\alpha-1}$, then for a certain ideal N a nilpotent modulo $N_{\alpha-1}$ (i.e., such that $N^n \subseteq N_{\alpha-1}$, $N \not\subset \mathbf{B}(R)$ holds. The condition $N^n \subseteq N_{\alpha-1} \subseteq \mathbf{B}(R)$, however, implies that the ideal $N + \mathbf{B}(R)$ is nilpotent modulo $\mathbf{B}(R)$, which contradicts the semiprimitivity of the factor-ring $R/\mathbf{B}(R)$.

(b) Let J be an arbitrary semiprime ideal. Since $I_1^2 = (0) \subseteq J$, then $I_1 \subseteq J$. If $I_\alpha \subseteq J$, then $I_{\alpha+1}^2 \subseteq I_\alpha \subseteq J$, and, therefore, $I_{\alpha+1} \subseteq J$. Finally, for a finite ordinal α we see that the conditions $I_\beta \subseteq J(\beta < \alpha)$ yield $I_\alpha = \bigcup_\beta I_\beta \subseteq J$. Therefore, according to the principle of transfinite induction, $\mathbf{B}(R) \subseteq J$. The theorem is proved.

1.1.5 Definition. A ring R is called *prime* if the product of any two of its nonzero ideals is different from zero. Accordingly, the ideal I is called *prime* if the factor-ring R/I is prime. Therefore, the ring R is prime iff (0) is a prime ideal.

The following characterization of prime and semiprime ideals in terms of the ring elements often proves to be useful.

1.1.6 Lemma.(a)*The ideal* I *of the ring* R *is prime iff for any* $a, b \in R$ *the inclusion* $aRb \in I$ *implies* $a \in I$ *or* $b \in I$. *In particular,* *the ring* R *is prime iff for any nonzero* a, b *there exists* $x \in R$, *such that* $axb \neq 0$.

(b)*The ideal* I *of the ring* R *is semiprime iff for any element* $a \in R$ *the inclusion* $aRa \in I$ *implies* $a \in R$. *In particular, the ring* R *is semiprime iff for any nonzero element* $a \in R$ *there exists* $x \in R$, *such that* $axa \neq 0$.

Proof. (a) Let I be a prime ideal and , and suppose that $b \notin I$. Then the ideal $(b) = b\mathbf{Z} + bR + Rb + RbR$ which is generated by the element b is not contained in I, but the product of the ideals $(aR + RaR) \cdot (b)$ is contained in I, which implies that $aR + RaR \subseteq I$. Let (a) be an ideal generated by the element a, then $(a) \cdot (a) \subseteq aR + RaR$ and,by the definition of a prime ideal, $(a) \subseteq I$ i.e., $a \in I$, which is the required proof. Conversely, if condition (a) is valid for the ideal I, and the product PQ of the ideals P and Q is contained in I, then for any $p \in P$, $q \in Q$ we have $pRq \subseteq PRQ \subseteq PQ \subseteq I$. If now $q \notin I$, then $p \in I$ and, since p is an arbitrary element of the ideal P, then $P \subseteq I$. If there exists no element $q \in Q$ with the properties $q \notin I$, then $Q \subseteq I$, which is the required proof.

(b) If I is a semiprime ideal, $aRa \subseteq I$, then $(a)^3 \subseteq I$. That implies that the ideal $(a) + I / I$ is nilpotent in the factor-ring R / I, i.e., $(a) \subseteq I$ and, in particular, $a \in I$.Let condition (b) of the lemma be fulfilled and $P^2 \subseteq I$, where P is an ideal of the ring R. Then for any $p \in P$ we have $pRp \subseteq PRP \subseteq P^2 \subseteq I$, i.e., $p \in I$, which is the required proof.

1.1.7 Lemma. *Any semiprime ring is a subdirect product of prime rings.*

It should be recalled that a ring R is called a *subdirect product* of rings R_α, $\alpha \in A$, if there exists an imbedding π of the ring R to the direct product $\prod R_\alpha$, such that its composition with the projections $p_\alpha : \prod R_\alpha \to R_\alpha$, results in epimorphisms (it is implied here that the subdirect product is not uniquely defined by the rings R_α, $\alpha \in A$). If we consider the kernels I_α of the compositions πp_α, then we can easily see that the condition of decomposability of the ring R into a subdirect product of the rings R_α, $\alpha \in A$ is equivalent to the existence of a set of the ideals I_α, such that $\bigcap_\alpha I_\alpha = 0$ and $R / I_\alpha \cong R_\alpha$. Under these conditions the ring R is also said to be *approximated* by the rings R_α.

Proof of lemma 1.1.7. Let $a = a_0$ be a nonzero element of the ring R. According to Lemma **1.1.6**, we can find an element $x_1 \in R$, such that $a_1 = ax_1a \neq 0$. Using the element a_1 we find an element x_2 such that

$a_2 = a_1x_2a_1 \neq 0$. Continuing this process, we can construct a countable sequence of nonzero elements $a_0, a_1,.., a_n,..., a_{n+1} = a_nx_{n+1}a_n$ for certain elements $x_1, x_2,.. x_n,..$ of the ring R.

Let us consider a set M of all the ideals of the ring R containing no elements of the sequence constructed and ordered by inclusion. This set is not empty since it contains the zero ideal as an element. Moreover, the set is directed, i.e., according to the Zorn lemma, the set M contains maximal elements. Let P_a be one of them. In this case the ideal P_a does not intersect the sequence $a_0, a_1,.., a_n,...,$ but any ideal strictly containing P_a has a nonempty intersection with this sequence. Since $a \notin P_\alpha$, then $\bigcap_{0 \neq a \in R} P_a = 0$ and it remains to show that P_a is a prime ideal.

Let A and B be ideals of the ring R not contained in P_a. Then $A_1 = A + P_a \supset P_a$, $B_1 = B + P_a \supset P_a$ and, since P_a is maximal in the set M, the ideals A_1 and B_1 do not belong to M, i.e., there can be found natural n, m, such that $a_n \in A_1$, $a_m \in B_1$. Let, for instance, $n \geq m$. In this case, since $a_{k+1} = a_kx_{k+1}a_k \in (a_k)$, then $a_n \in (a_m) \subseteq B_1$. In this case we get:

$$a_{n+1} = a_nx_{n+1}a_n \in A_1B_1 \subseteq AB + P_\alpha .$$

Therefore, the ideal $AB + P_\alpha$ is not contained in P_a and, hence, $AB \not\subset P_a$, which is the required proof.

The reader could ask a natural question as to how far the Baer radical can be away from being nilpotent because of the transfinite process of its construction. To a certain extent this question can be answered by the following theorem.

1.1.8 Theorem *The Baer radical of a ring is locally nilpotent, i.e., any finite set of its elements generates a nilpotent subring.*

Proof. Let us consider the inductive construction. We have already noticed that at any finite step nilpotent (so *afortiori* locally nilpotent) ideals are obtained. Let α be a limit ordinal, $I_\alpha = \bigcup_{\beta < \alpha} I_\beta$ and let us suppose that all the ideals I_β be locally nilpotent. If $a_1,..., a_n$ are elements from I_α, then there exist transfinite numbers $\beta_1,.., \beta_n$ which are less than α and such that $a_1 \in I_{\beta_1},..., a_n \in I_{\beta_n}$. Let β be the largest from the numbers $\beta_1,..., \beta_n$, then, since the ideals $\{I_\gamma\}$ form a chain, we get

$a_1, a_2, \dots, a_n \in I_\beta$ and, due to the local nilpotency of I_β these elements generate a nilpotent subring.

If α is not a limit ordinal, $\alpha = \beta + 1$ and the ideal I_β is locally nilpotent, then any finite set $s_1, \dots, s_n$ of the elements of the ideal I_α generate a subring S, such that $S^2 \subseteq I_\beta$. However, the subring S^2 is generated by a finite set of elements $\{s_i s_j \mid 1 \leq i, j \leq n\}$ and, hence, S^2 and, consequently, also S, are nilpotent. The theorem is proved.

No further details pertaining to the Baer radical are going to be discussed here; an interested reader is referred to the thorough study by V.A.Andrunakievich and Yu.M.Ryabukhin "Radicals of Algebras and Structural Theory" (Moscow, "Nauka", 1979).

1.2 Automorphism Groups and Lie Differential Algebras

1.2.1. An *automorphism* of the ring R is an isomorphism onto itself, i.e., the one-to-one correspondence $g: R \to R$, which preserves the operations

$$(xy)^g = x^g y^g, \quad (x \pm y)^g = x^g \pm y^g.$$

One can easily see that a composition of two automorphisms as well as the mapping which is inverse to an automorphism will be an automorphism. This fact implies that a set of all automorphisms forms a group designated by $Aut\, R$.

In the case of non-commutative rings of importance is the notion of *inner* automorphism. Let R be a ring with a unit and b be its invertible element, i.e., $bb^{-1} = b^{-1}b = 1$ for a certain $b^{-1} \in R$. In this case the mapping

$$\tilde{b}: x \to b^{-1} x b$$

is an automorphism of ring R which is called *inner*. A set of all inner automorphisms $Int\, R$ forms an invariant subgroup in $Aut\, R$, since

$$\tilde{b} \cdot \tilde{d} = \widetilde{bd}, \quad g^{-1}\tilde{b}g = \widetilde{b^g}$$

where $b, d \in R$, $g \in Aut\, R$.

Let us denote through R^* a group of all invertible elements of the ring R. Then the mapping $b \to \tilde{b}$ will be a homomorphism of this group onto the group $Int\, R$, the kernel of this homomorphism consisting of all central invertible elements. An element z of the ring R is called *central* if it commutes with all the elements of the ring R: $zx = xz$ for all $x \in R$. The set $Z(R)$ of all central elements is called a *center* of the ring R.

Therefore, we have an exact sequence of the homomorphisms of the groups

$$\{1\} \to Z(R)^* \to R^* \to IutR.$$

1.2.2. A *derivation* of the ring R is a mapping $\mu: R \to R$ which obeys the following properties:

$$(x \pm y)^{\mu} = x^{\mu} \pm y^{\mu};\ (xy)^{\mu} = x^{\mu} y + xy^{\mu}.$$

A set of all derivations of the ring R is denoted through $Der R$. This set is closed relative to the commutator operation: if μ, ν are derivations, then $[\mu, \nu] \overset{df}{=} \mu\nu - \nu\mu$ is also a derivation:

$$(xy)^{[\mu,\nu]} = ((xy)^{\mu})^{\nu} - ((xy)^{\nu})^{\mu} = (x^{\mu} y + xy^{\mu})^{\nu} -$$
$$-(x^{\nu} y + xy^{\nu})^{\mu} = x^{\mu\nu} y + x^{\mu} y^{\nu} + x^{\nu} y^{\mu} + xy^{\mu\nu} -$$
$$x^{\nu\mu} y - x^{\nu} y^{\mu} - x^{\mu} y^{\nu} - xy^{\nu\mu} = x^{[\mu,\nu]} y + xy^{[\mu,\nu]}.$$

Therefore, $Der R$ is a Lie ring. This ring is simultaneously a right module over the center of the ring R, provided the multiplication by the central elements is determined by the formula $x^{(\mu z)} \overset{df}{=} x^{\mu} \cdot z$. Indeed,

$$(xy)^{(\mu z)} = (x^{\mu} y + xy^{\mu}) z = x^{\mu} zy + xy^{\mu} z = x^{(\mu z)} y + xy^{(\mu z)}.$$

At the same time, one cannot claim that $Der R$ is an algebra over the ring $Z(R)$ since the definition of an algebra includes the identity

$$[\mu z, \nu] = [\mu, \nu] z,$$

while in the ring $Der R$ the following identity holds instead:

$$[\mu z, \nu] = [\mu, \nu] z + \mu \cdot z^{\nu} \qquad (1)$$

This peculiarity results in the following definition.

1.2.3.Definition. Let Z be a commutative ring, and D be a certain Lie ring which is simultaneously a right module over Z, and let us assume that there is a certain homomorphism of Lie rings given: $u: D \to Der\ R$. Then D is called a *differential Lie-Z-algebra* (or, for short, a *Lie-∂-algebra*), if identity (1) holds in D, where the z^{ν} value is determined as $z^{u(\nu)}$.

Therefore, we come to the conclusion that the set $Der\,R$ forms a differential Lie $Z(R)$-algebra, where the homomorphism $u: Der\,R \to Der\ Z$ is determined as a restriction of the derivations onto the center. It should, naturally, be noted that all the derivations transform central elements into central ones: if $zx = xz$ for all x, then $z^{\mu}x + zx^{\mu} = x^{\mu}z + xz^{\mu}$, which results in $z^{\mu}x + xz^{\mu}$, i.e., $z^{\mu} \in Z(R)$.

Let us now assume that the ring R has a simple characteristic $p > 0$, $pR = 0$. In this case the p-th power of the derivation appears to be another derivation

$$(xy)^{\mu^p} = \sum C_p^k x^{\mu^k} y^{\mu^{p-k}} = x^{\mu^p} y + xy^{\mu^p},$$

since the binomial coefficients C_p^k are divisible by p, and, hence, a set of derivations forms a restricted Lie ring.

It should be recalled that a Lie ring is called *restricted* if a unary operation $\mu \to \mu^{[p]}$ is given on it, which is related to addition and Lie multiplication with the following identities

$$\begin{aligned} &[\mu, \nu^{[p]}] = [\ldots[\mu, \nu]\nu]\ldots \nu], \\ &(\mu + \nu)^{[p]} = \mu^p + \nu^p + W(\mu, \nu), \end{aligned} \tag{2}$$

where $W(x, y)$ is a commutator presentation of the noncommutative associative polynomial $(x + y)^p - x^p - y^p$. Well known is the fact that such a presentation exists. For instance, at $p = 3$:

$$\begin{aligned} &(x + y)^3 - x^3 - y^3 = x^2y + xyx + yx^2 + xy^2 + \\ &yxy + y^2x \equiv [[x, y]y] + [[y, x]x] \pmod 3. \end{aligned}$$

The Lie algebra over a commutative ring Z is called a *restricted Lie algebra* (or sometimes a *Lie p-algebra*) if it is a restricted Lie ring and the operation $\mu \to \mu^{[p]}$ is related to the module structure through the formula

$$(\mu c)^{[p]} = \mu^{[p]} c^p .$$

In the ring of derivations the operation of bringing to the power p does not obey the above identity, and the following identity holds instead:

$$(\mu c)^p = \mu^p c^p + \mu \cdot (\ldots((c^{\mu} c)^{\mu} c)^{\mu} \ldots)^{\mu} c . \qquad (3)$$

We thus naturally come to the following definition.

1.2.4. Definition. A differential Lie Z-algebra D is called *restricted* if the operation $\mu \rightarrow \mu^{[p]}$ is given on it, which transforms it into a restricted Lie ring and is associated with the module structure through the formula

$$(\mu c)^{[p]} = \mu^{[p]} c^p + \mu \cdot (\ldots((\overbrace{c^{\mu} c)^{\mu})^{\mu} c \ldots)^{\mu}}^{p-1} c . \qquad (4)$$

It is clear that if D acts trivially on Z (i.e., if $u = 0$), then this definition turns into that of a Lie p-algebra over Z.

Hence, if the ring R has a simple characteristic, then the set of derivations forms a restricted differential Lie $Z(R)$-algebra.

Among derivations, as well as among automorphisms, one can single out an inner part. Let b be an element of ring R, in which case the mapping adb, which operates by the formula

$$ad\, b\colon x \rightarrow [x, b] \overset{df}{=} xb - bx,$$

will be a derivation, since

$$[xy, b] = [x, b]\, y + x[y, b] .$$

This derivation is called an *inner* one.

Let us consider R as a (restricted) Lie ring with a multiplication operation $[xy] \rightarrow xy - yx$ (and the operation $x^{[p]} = x^p$ at $p > 0$; the addition remaining the same). This Lie ring is called *adjoined* to R and is designated by $R^{(-)}$.

In this case the mapping $x \rightarrow adx$ sets a homomorphism of the (restricted) Lie rings $ad\colon R^{(-)} \rightarrow Der\, R$, which can be easily seen to preserve the $Z(R)$-module structure:

$$[ad\ b, ad\ r] = ad[b, r]; ad\,(bz) = (ad\ b)\,z\,.$$

The kernel of this homomorphism coincides with the center of the ring R, so that we have an exact sequence of the homomorphisms of Lie rings

$$(0) \to Z(R)^{(-)} \to R^{(-)} \to Der\ R.$$

It should be also remarked that $[adb, \mu] = adb^{\mu}$ and, hence, the image adR forms an ideal of the (restricted) Lie ring $Der\,R$, which makes it possible to consider the *outer* part of the set of derivations as a factor-ring as well:

$$D_{ext}(R) = Der\ R\,/\,adR.$$

1.2.5. Adjoined Lie ∂-algebras and universal envelopings. By analogy with adjoined Lie rings, one can also determine adjoined Lie ∂-algebras. Let A be an associative ring with a unit, Z be its subring, and let us suppose that $Z^{(-)}$ is an ideal with a zero multiplication in the Lie ring $A^{(-)}$. This implies that Z is a commutative subring in A, which is invariant relative to all inner derivations: $az - za = z^{a} \in Z$. Viewing $A^{(-)}$ as a right module over the commutative (associative) ring Z, we see $A^{(-)}$ converting into a (restricted) differential Lie Z-algebra, where the action of $A^{(-)}$ on Z is governed by the formula $z^{a} = az - za$. This algebra will hereafter be designated by $A_Z^{(-)}$ and called an *adjoined (restricted) Lie ∂-algebra.*

The ring A will be referred to as an associative enveloping of the (restricted) differential Lie ∂-algebra D if there exists a Z-linear homomorphism of the (restricted) Lie rings $\xi_A: D \to A_Z^{(-)}$, such that the action of $d \in D$ on Z coincides with that of $\xi_A(d)$, and the set $\xi_A(D)$ generates A as an associative algebra with a unit.

The associative enveloping A will be called a *universal enveloping for D* if for any associative enveloping B there exists a Z-linear homomorphism of the rings $\varphi: A \to B$, such that $\xi_A \varphi = \xi_B$.

From the theory of categories one can easily deduce that the universal enveloping for D exists in all cases. It is much more difficult to elucidate under what conditions the homomorphism ξ_A proves to be an embedding into the universal enveloping.

If D is presented by the derivations of ring R with a unit and center Z, i.e., $D \subseteq \mathrm{Der}\, R$, then D is embedded into the ring of endomorphisms of the abelian group $< R, + >$, in which case the subring $\Phi(D)$ generated by D and the (right) multiplication by the elements of Z will be an associative enveloping for D (the element $z \in Z$ is associated with the multiplication $x \to xz$). It is clear that $\xi_\Phi : D \to \Phi(D)_Z^{(-)}$ will be an embedding and, hence, the universal homomorphism ξ_A will be also an embedding.

In a general case, if D is a free module over Z, one can prove that ξ_A is an embedding, which fact will be proved below only for the case when Z is a field (see **6.1.12**) and we shall see that any Lie ∂-algebra over a field can be presented by the derivations of a free algebra.

1.2.6. By way of concluding this paragraph we should note that the automorphisms of the ring R act in a natural way on its derivations: if $g \in \mathrm{Aut}\, R$, $\mu \in \mathrm{Der}\, R$, then the superposition $g^{-1}\mu g \overset{df}{=} \mu^g$ will be a derivation

$$(xy)^{g^{-1}\mu g} = (x^{g^{-1}\mu} y^{g^{-1}} + x^{g^{-1}} y^{g^{-1}\mu})^g = x^{g^{-1}\mu g} y + xy^{g^{-1}\mu g}.$$

1.3 Bergman-Isaacs Theorem, Shelter Integrality

In this paragraph we shall prove some results by Bergman and Isaacs on fixed rings of finite groups of automorphisms, as well as the Queen's theorem.

1.3.1. Definition. Let G be a group of automorphism of the ring R. *A fixed ring of group* G is a subring $R^G = \{r \in R \mid \forall\, g \in G,\ r^g = r\}$ of all elements which are fixed relative G.

Our nearest aim is to find out the conditions under which the ring R^G proves to be sufficiently "small" (zero or nilpotent), which makes further studies of invariants of finite groups possible.

1.3.2. Example. Let F be a field with a characteristic $p \neq 0$ and an element $\omega \neq 0, 1$ of a finite multiplicative order $\omega^n = 1$. Let $F < x, y >$ be a ring of all noncommutative polynomials of all variables without free terms. Let us denote by R a ring of all 2×2 matrices with the elements from $F < x, y >$. The matrices

$$A_1 = \begin{pmatrix} 1 & x \\ 0 & 1 \end{pmatrix}, A_2 = \begin{pmatrix} 1 & y \\ 0 & 1 \end{pmatrix}, A_3 = \begin{pmatrix} \omega & 0 \\ 0 & 1 \end{pmatrix}$$

though not belonging to R, do determine the automorphisms $g_i: M \to A_i^{-1} MA_i$ of the ring R. The order of a group G generated by these automorphisms is np^2. Direct calculations also show that $R^G = \{0\}$. (It is sufficient to remark that the matrix which is commutative with A_1 and A_2 has the form $\begin{pmatrix} 0 & f \\ 0 & 0 \end{pmatrix}$, while the matrix of this form commutes with A_3 only if $f = 0$.)

This example shows that under most general conditions there is no tight association between the rings R and R^G. Nonetheless, when in R there is no additive n-torsion, where n is the order of the group, such associations can be established.

1.3.3. Definitions. Let $X = \{g_1, \ldots, g_n\}$ be a finite set (not obligatory a group) of automorphisms of the ring R. Let us denote by $U(X)$ a set of all sequences $u = (u_g)_{g \in X}$ of the elements of the ring R. Let τ be a mapping from $U(X)$ in R, which sets the sequences $u = (u_{g_1}, \ldots, u_{g_n})$ in correspondence with the element $u_{g_1} + u_{g_2} + \ldots + u_{g_n}$.

1.3.4. Lemma. *Let $V(X)$ be a subset of the sequences $(u_{g_1}, \ldots, u_{g_n})$ from $U(X)$, such that*

$$u_{g_1} x^{g_1} + u_{g_2} x^{g_2} + \ldots + u_{g_n} x^{g_n} = 0$$

for all $x \in R$. In this case the left ideal $\tau V(X)$ is nilpotent to the power not greater than $\prod_{k=0}^{n} (C_n^k + 1)$.

Proof. For any $a \in R$ and $u \in U(X)$ let us determine the products $au = (au_g) \in U(X)$ and $u.\, a = (u_g a^g) \in U(X)$. Then $V(X) = \{u \in U(X) : \tau(u.\, a) = 0$ for all $a \in R\}$. For any $Y \subseteq X$ let us identify a set $V(Y)$ with a subset of sequences $(u_g) \in V(X)$, for which $u_g = 0$ at any $g \in X \setminus Y$.

Let $u, v \in V(Y)$, $g \in Y$. We have

$$u_g v - u.(v_g^{g^{-1}}) \in V(Y\setminus\{g\}).$$

Indeed, this element lies in $V(Y)$, and its g–component is

$$u_g v_g - u_g (v_g^{g^{-1}})^g = 0.$$

If we apply τ to this element, then the second term disappears by the definition of $V(X)$, and we get $u_g(\tau v) \in \tau V(Y\setminus\{g\})$, obtaining, after summing it up over all $g \in Y$,

$$(\tau u)(\tau v) \in \sum_{g \in Y} \tau V(Y\setminus\{g\}).$$

Hence,

$$(\tau V(Y))^2 \subseteq \sum_{g \in Y} \tau V(Y\setminus\{g\}). \tag{5}$$

Let now Y run through all $C_n^m = k$ of the subsets X of m elements. Then we get

$$\Big(\sum_{|Y|=m} \tau V(Y)\Big)^{k+1} \subseteq \sum_{|Z|=m-1} \tau V(Z) R^{\#}.$$

(Here $R^{\#}$ is a ring obtained from R by joining the unit). In order to visualize it, the left part should be presented as a sum of products $\tau V(Y_0)\cdot \ldots \cdot \tau V(Y_k)$. According to the Dirichlet principle, one of the factors is encountered twice, i.e., each product can be presented as $A\tau V(Y)B\tau V(Y)C \subseteq (R^{\#}\tau V(Y))(R^{\#}\tau(V(Y))R^{\#} = (\tau V(Y))^2 R^{\#}$, i.e., according to (5), it is contained in the left part.

By induction over m we now get $(\tau V(X))^{\prod_{i=0}^{n}(C_n^i+1)} = 0$. The lemma is proved.

1.3.5. Proposition. *Let* G *be a finite group of automorphisms of the ring* R, n *be its order,* $h = 1 + \prod_{i=0}^{n}(C_n^i+1)$. *In this case for any natural* d *the following inclusion is valid:* $(nR)^{h^d} \subseteq nR^{\#} t(R)^d R$, *where*

$t(r) = \sum_{g\in G} r^g$ *is a trace of element* r.

Proof. Let $L = \{l \in R,\ lt(R) = 0\}$ be a left annihilator of all traces $t(R)$. If $l \in L$, then $\sum_{g\in G} lx^g = 0$, which implies that $u = (l, l, ..., l) \in V(G)$ and, hence, $nl = \tau(u) \in \tau V(G)$. According to Lemma **1.3.4**, $(nl)^{h-1} = \{0\}$.

Let now J be an arbitrary G-invariant left ideal of the ring R, $RJ \subseteq J$, $J^g = J$ for all $g \in G$. Let us consider a factor-ring $\bar{R} = R\,/\,Jt(R)R^{\#}$. Since the two-sided ideal $T = Jt(R)R^{\#}$ is G-invariant, then the action of the group $G: (r+T)^g = r^g + T$ will be naturally transferred onto $\bar{R}$. For the element $\bar{r} = r + T \in \bar{R}$ we have $t(\bar{r}) = t(r) + T = \overline{t(r)}$ and hence, $\bar{J}\,\overline{t(R)} = \bar{0}$, i.e. $\bar{J}$ is contained in the left annihilator of all the traces of the ring R. Therefore, $(n\bar{J})^{h-1} = \{\bar{0}\}$. For the ring R this equality implies that $(nJ)^{h-1} \subseteq Jt(R)R^{\#}$. Multiplying from the right by nJ, we see that

$$(nJ)^h \subseteq nJt(R)R^{\#}. \tag{6}$$

Stemming from this conclusion, let us carry out induction over d. Let

$$(nJ)^{h^{d-1}} \subseteq nJt(R)^{d-1}R^{\#}.$$

Raising both parts to the power h and using inclusion (6) for the ideal $J_0 = Jt(R)^{d-1}J$, we get

$$(nJ)^{h^d} \subseteq (nJt(R)^{d-1})^h R^{\#} = (nJ_0)^h R^{\#} \subseteq$$

$$J_0 t(R)R^{\#} = Jt(R)^d R^{\#}.$$

Setting $J = R$, we come to the statement of the proposition.

1.3.6.Theorem *Let* G *be a finite group of the automorphisms of the ring* R, n *be its order. In this case, if* R *has no additive* n-*torsion, the* d-*power of the nilpotency of the fixed ring implies that of the ring* R *of the power not greater than* n^d. *In particular, if* $R^G = \{0\}$, *then* $R^h = \{0\}$, *where*

$$h = 1 + \prod_{i=0}^{n} (c_n^i + 1).$$

The proof follows immediately from proposition **1.3.5**. The estimate of nilpotency in this theorem seems to be extremely rough. For instance, the degree of nilpotency for solvable groups is known not to exceed nd. A detailed study, however, has not yet been carried out, and it would be of interest first to find the boundaries for estimating nilpotency from below, using a series of examples.

The theorem obtained above is an extremely important instrument for studying invariants of finite groups in the case when the ring has no additive n-torsion (where n is the order of the group).

In this paragraph we shall consider only one application.

1.3.7. Corollary. *Let G be a finite group of the automorphisms of a semiprime ring R which has no additive n-torsion, where n is the order of the group G. In this case the fixed ring R^G is also semi-prime.*

Proof. Let I be a non-zero ideal of the ring R^G, such that $I^2 = (0)$. Let us consider a right ideal nIR in R. If $ns = n\sum_m i_m r_m$ is a fixed element of this ideal, then

$$ns = t(s) = t(\sum_m i_m r_m) = \sum_m i_m t(r_m) \in I.$$

If we apply theorem **1.3.6** to the ring nIR, we get $(nIR)^G$ is a ring with a zero multiplication and, hence, nIR is a nilpotent ring of the power h^2. Since the ring R has no additive n-torsion, the right ideal IR will also be nilpotent. And, finally, the two-sided ideal J generated by IR will also be nilpotent:

$$(R^{\#}IR)^{h^2+1} \subseteq R^{\#}(IR)^{h^2}R^{\#} = 0,$$

which does not agree with the semiprimeness of the ring R. The corollary is proved.

If the initial ring R is commutative, then all the coefficients of the polynomial $\prod_{g \in G}(x - a^g)$ at $a \in R$ will be fixed relative the action of the finite group G. Opening the brackets, we find an integer polynomial over R^G, a root of which is a, i.e., R is an integer extension of R^G of the degree $|G|$. This fact has an analog for non-commutative rings as well (under natural limitations on the additive structure), which has recently been discovered by D.Quinn.

1.3.8. Definitions. Let T be a subring of the ring R. The *integer T-polynomial of the degree* n will be expressed as

$$x^n + \sum_{\substack{i \\ k \leq n}} t_{i1} x t_{i2} x \dots x t_{ik}$$

where t_{ij} are either elements from T or formal units; in both cases in every term under the sign of the sum there is at least one factor from T (if T contains the unit, the latter amendment is excessive).

The ring R is called *Shelter integer over* T (of *the degree* n) if for every $a \in R$ there exists an integer T-polynomial $f(x)$ (of the *degree* n), such that $f(a) = 0$.

A linearized variation of the above notion is somewhat deeper and more useful.

A *quasi-integer T-polynomial of the degree* n is an expression of the following kind:

$$x_1 x_2 \dots x_n + \sum_{\substack{k \leq n \\ i,\, j}} t_{i1} x_{j1} t_{i2} x_{j2} \dots x_{j\,k-1} t_{ik},$$

where, as above, t_{ij} is either an element from T or the unit, in which cases not all t_{ij} under the sign of the sum are equal to the formal unit.

The ring R is termed *quite integer* over T of the degree n if for every set $a_1, \dots, a_n$ of elements from R there exists a quasi-integer T-polynomial $f(x_1, \dots, x_n)$ of the degree n, such that $f(a_1, \dots, a_n) = 0$.

1.3.9. Theorem (D.Quinn). *Let G be a finite group of automorphisms of the ring R. If $|G| R = R$; $|G| = 0 \to x = 0$, then the ring R is quite integer over R^G of a certain degree m.*

The proof of this theorem is based on the following statement obtained by R.Paré and W.Schelter [131].

1.3.10. Theorem. *Let a group G of an order n act on a ring R. In the ring of $n \times n$ matrices R_n let us consider a subring of diagonal matrices $T = \{diag(a^{g_1}, \dots, a^{g_n}) \mid a \in R\} \cong R$ where $(g_1, \dots, g_n) = G$. In this case the ring R_n will be quite integer over T of a certain degree m.*

Proof. For every $k = 1, 2, \dots, n$ we shall denote by R_k a subring in

R_n, which consists of all the matrices the last $n-k$ rows and $n-k$ columns of which are zero. By induction over k we will show R to be quite integer over T of a certain degree m.

Let $k=1$ and $a_1 = diag(r_1, 0, ..0) \in R_1$. Let us find an element $t \in T$, such that $t = diad(r_1, ...)$. Then for any $a_2 \in R_1$ we have $(a_1 - t)a_2 = 0$, and, hence, $a_1 a_2 = t a_2$ which is the required proof.

Let $k > 1$. Let us present an arbitrary matrix $a \in R_k$ as

$$a = \begin{pmatrix} a' & r'' & 0 \\ * & * & 0 \\ 0 & 0 & 0 \end{pmatrix} \tag{7}$$

where a' is the $(k-1) \times (k-1)$ matrix, r'' is a column of the height $k-1$. Let us set

$$\tilde{a} = \begin{pmatrix} a' & 0 \\ 0 & 0 \end{pmatrix} \in R_{k-1}.$$

Then for all the elements $a_1, a_2 \in R_k$ we shall introduce the relation $a_1 \equiv a_2$ iff the columns r''_1 and r''_2 corresponding to the matrices a_1 and a_2 in presentation (7) coincide, $r''_1 = r''_2$. It should be remarked that this relation obeys the following properties.

(1). Let $a_1, a_2, a \in R_k$ and $t \in T$. If $a_1 \equiv a_2$ then $\tilde{a}\, a_1 \equiv \tilde{a}\, a_2$ and $t\, a_1 \equiv t\, a_2$.

(2). If $a_1, a \in R_k$, then there exists an element $t \in T$, such that

$$\tilde{a}_1\, a \equiv a_1\, a - a_1\, t \tag{8}$$

Indeed, one can find $t \in T$, such that

$$a - t = \begin{pmatrix} * & r'' & 0 \\ * & 0 & 0 \\ 0 & 0 & * \end{pmatrix}$$

and, hence,

$$a_1(a - t) = \begin{pmatrix} * & a'_1 r'' & 0 \\ * & * & 0 \\ 0 & 0 & 0 \end{pmatrix} \equiv \tilde{a}_1 a.$$

Let $a_1,\dots, a_m$ be arbitrary elements from R_k, where the number m equals the degree of integrality of R_{k-1} over T. In this case we get the following dependence:

$$\tilde{a}_1 \dots \tilde{a}_m + \sum t_{i1} \tilde{a}_{j_1} \dots \tilde{a}_{j_{m-1}} t_{ik} = 0 .$$

Let us multiply the above equality from the right by an arbitrary element $a \in R_k$. Then, due to the first property of relation $\equiv$, a multiple application of relation (8) enables one to eliminate the wavy lines over the letters. Since relation (8) is homogenous both over a_1 and the totality of the variables comprising it, the senior form is transformed into a quasi-integer polynomial with the senior term $a_1 \dots a_m, a$, while the remaining terms are transformed into a T-polynomial f of degree $m+1$, such that $f(a_1,\dots, a_m, a) \equiv 0$.

This fact implies that there is an element $t \in T$, such that the difference $s = f(a_1,\dots) - t$ has the form

$$s = \begin{pmatrix} * & 0 & 0 \\ * & 0 & 0 \\ 0 & 0 & * \end{pmatrix} \qquad (9)$$

In line with the proposition on induction for any elements $s_1,\dots, s_m$ of type (9), there exists a quasi-integer polynomial φ of degree m, such that the matrix $\varphi(s_1,\dots, s_m)$ has the form: $\begin{pmatrix} 0 & 0 & 0 \\ * & 0 & 0 \\ 0 & 0 & * \end{pmatrix}$ and, hence, $\varphi(s_1,\dots, s_m)\varphi'(s_1',\dots, s_m')a = 0$, where φ' is a quasi-integer polynomial for other elements $s'_1,\dots, s_m'$ of type (9). If we now assume that the elements s'_i, s_i are constructed by the arbitrary sets $a_1^{(i)},\dots, a_m^{(i)}, a^{(i)}$, $(1 \subseteq i \subseteq 2m)$ in the same way as the element s by the set $a_1,\dots, a_m, a$, then we will get the required dependence between the elements $a_1^{(i)},\dots, a_m^{(i)}, a^{(i)}$, $(1 \subseteq i \subseteq 2m)$, a of the degree $2(m+1)^2+1$. The theorem is proved.

1.3.11.Corollary. *The ring of matrices R_n is Shelter integer over the ring of scalar matrices* $\{diag(r,\dots, r) \setminus r \in R\}$.

Proof. It suffices to assume that in the preceding theorem the group acts trivially.

We are now ready to prove Theorem **1.3.9.** Let us consider a formal $n\times n$ matrix $e=\left\|\frac{1}{n}\right\|$, with all its points occupied by $\frac{1}{n}$, where n is the order of the group G. Since $nR=R$ and $nx=0\Rightarrow x=0$ in the ring R, then the multiplication of matrices from R_n by e is defined, in which case $e^2=e$ both as a matrix and an operator. Let $a_1,\dots, a_m$ be arbitrary elements from R_n. Let us consider an integer dependence of the elements $ea_1e,\dots, ea_me$ over $T=\{diag(r^{g_1},\dots, r^{g_n})\mid r\in R\}$:

$$ea_1\, ea_2\dots ea_me=\sum t_{i1}\, ea_{j_1}\, e\dots a_{j_{m-1}}\, et_{im}.$$

Multiplying this equality from the right and from the left by e, we see that the ring eR_ne is quite integer over the subring eTe of a certain degree m. The ring eR_ne, however, consists of the matrices, all the coefficients of which are the same, while the ring eTe consists of the matrices, all the coefficients of which are fixed relative G and equal to each other. Bearing in mind the fact that the product of the matrices $\|a\|\cdot\|b\|$ has the form $\|nab\|$, and the possibility of cancelling by n, we come to the conclusion that R is quite integer over R^G. The theorem is proved.

1.4. Martindale Ring of Quotients

Let R be a semiprime ring. Let us denote by F a set of all (two-sided) ideals of the ring R which have zero annihilators in R. It should be recalled that the *right annihilator* $r_R(A)$ of a set A in R is a totality of all $x\in R$, such that $Ax=0$. Accordingly, the *left annihilator* $l_R(A)$ is a set of all $x\in R$, such that $xA=0$. The intersection $ann_R A= r_R(A)\cap l_R(A)$ is called an *annihilator* of A in R.

The lemma below demonstrates that all the three notions coincide for ideals of a semiprime ring.

1.4.1.Lemma. *Let I be an ideal of a semiprime ring R. Then $ann_R(I)= r_R(I)= l_R(I)$.*

Let us first make the following remark.

1.4.2.Remark. *A semiprime ring has no nonzero nilpotent one-sided ideals.*

Indeed, let, for instance, B be the right nilpotent ideal, $B^n=0$. Let us consider an ideal $B+RB$ generated by the set B. Then, allowing for

$BR \subseteq B$, we have $(B+RB)^n = 0$, i.e., due to the semipriness, $B+RB=0$ and, moreover, $B=0$.

Proof (of the lemma). It is evident that it would suffice to establish the second equality. Let $Ix=0$. Then $(xI)^2 = xIxI = 0$, i.e., xI is a nilpotent right ideal. According to the remark, $xI=0$. Thus, $r_R(I) \subseteq l_R(I)$. The inverse inclusion is proved in an analogous way. The lemma is proved.

Lemma **1.4.1** shows, in particular, that the set F is multiplicatively closed, i.e., together with the ideals I_1, I_2. it contains their product $I_1 I_2$, because the equality $I_1 I_2 x = 0$ yields $I_2 x \subseteq ann_R I_1 = 0$, i.e., $I_2 x = 0$ and $x \in ann_R I_2 = 0$.

1.4.3. Lemma. *Any ideal* I *of a semiprime ring* R *has a zero intersection with its annihilator ann* I. *The direct sum* $I + annI$ *belongs to* F.

Proof. The intersection $I \cap annI$ has a zero multiplication and, hence, equals zero. If $(I + annI)x = 0$, then $Ix = 0$ and, hence, x belongs to the ideal $A = annI$. On the other hand, $Ax = 0$ and, therefore, x also belongs to the annihilator of the ideal A, i.e., $x \in A \cap annA = 0$, which is the required proof.

1.4.4. Definition. An ideal of the ring R is called *essential* if it has a zero intersection with any nonzero ideal of the ring R.

1.4.5. Lemma. *The ideal* I *of a semiprime ring* R *belongs to* F *iff it is essential.*

Proof. If $I \in F$, then $0 \neq IJ \subseteq I \cap J$. Inversely, according to the previous lemma, $I \cap annI = 0$ and, hence, $annI = 0$ if I is essential, which is the required proof.

1.4.6. For every $I \in F$ let us denote by $Hom({}_R I, R)$ an abelian group of all homomorphisms of the left R-module I to R. When considering the union V of all these abelian groups, $V = \bigcup_{I \in F} Hom({}_R I, R)$, let us introduce the equivalence relation $\varphi_1 \approx \varphi_2$ iff there exists an ideal $I \in F$ lying in the intersection of the domains of the definitions of φ_1 and φ_2, such that $\varphi_1(a) = \varphi_2(a)$ for all $a \in I$. On the factor-set $V/\approx$ let us introduce ring operations: if $\varphi_1(a) \in Hom(I_1, R), \varphi_2 \in Hom(I_2, R)$, then we assume that

$$\varphi_1 \pm \varphi_2,\ \varphi_1 \varphi_2 \in Hom(I_1 I_2,\ R),$$

in which case

$$(\varphi_1 \pm \varphi_2)(a) = \varphi_1(a) \pm \varphi_2(a); \ (\varphi_1\varphi_2)(a) = \varphi_2(\varphi_1(a)).$$

1.4.7. Lemma. *The set* $V/\approx$ *with the above operations introduced is a ring.*

Proof. We should, first of all, check the correctness of the operations. Let $\varphi'_1 \approx \varphi_1$, $\varphi'_2 \approx \varphi_2$, and let the action of φ_1 coincide with that of φ'_1 on the ideal $I'_1 \in F$, while the action of φ_2 coincides with that of φ'_2 on the ideal $I'_2 \in F$. In this case $I'_2 I'_1 \subseteq I'_2 \cap I'_1$ and $I'_2 I'_1 \in F$, and, hence, $\varphi'_1 \pm \varphi'_2 \approx \varphi_1 \pm \varphi_2$; $\varphi'_1 \varphi'_2 \approx \varphi_1 \varphi_2$. Moreover, the homomorphism φ_2 is determined on $\varphi_1(I'_2 I'_1)$, since φ_1 is the homomorphism of the left R-modules and, hence, $\varphi_1(I'_2 I'_1) \subseteq I'_2 \varphi_1(I'_1) \subseteq I'_2 R \subseteq I'_2$. Verification of the laws of associativity, distributivity, etc. is also trivial.

1.4.8. The constructed ring $V/\approx$ is called a *left Martindale ring of quotients,* and will be denoted by R_F. The reader familiar with the notion of a direct limit can easily see that

$$R_F = \varinjlim_{I \in F} Hom_R({}_R I, R).$$

If a is a certain element of the ring R, then we can determine a homomorphism $\tilde{a} \in Hom({}_R R, R)$ of the left R-modules by the formula $\tilde{a}(x) = xa$. The mapping i which puts the class of the equivalence defined by the homomorphism $\tilde{a}$ in $V/\approx$ into correspondence to the element a, will be an embedding of the ring R into the ring R_F. Indeed, if $\tilde{a} \approx 0$, then, by the definition, $\tilde{a}(I) = 0$. for a suitable $I \in F$. The latter equality, however, implies that $Ia = 0$, i.e., $a = 0$. Consequently, if $a_1, a_2 \in R$, then

$$\widetilde{a_1 a_2}(x) = xa_1 a_2 = (xa_1)a_2 = \tilde{a}_2(\tilde{a}_1(x)),$$

$$\widetilde{a_1 \pm a_2}(x) = x(a_1 \pm a_2) = xa_1 \pm xa_2 = \tilde{a}_1(x) \pm \tilde{a}_2(x),$$

and, hence, i preserves the ring operations. Let us now identify R with its image $i(R)$ and assume R to be a subring of R_F.

1.4.9. Theorem. *The ring* R_F *satisfies the following properties:*

(1) *For any element* $a \in R_F$. *there exists an ideal* $I_a \in F$, *such that* $I_a a \subseteq R$.

(2) *If* $a \in R_F$ *and* $Ia=0$ *for a certain* $I \in F$, *then* $a=0$.

(3) *If* $I \in F$ *and* $\varphi: I \rightarrow R$ *is a homomorphism of left* $R-$*modules, then there exists an element* $r \in R_F$, *such that* $\varphi(i)=ir$ *for all* $i \in I$.

(4) *Let* W *be an* (R, R)-*subbimodule in* R_F *and* $\varphi: W \rightarrow R_F$ *be a homomorphism of left* R-*modules. If* W *contains the ideal* I *of the ring* R, *such that* $\varphi(I) \subseteq R$ *and* $\mathrm{ann}_R I = l_R(W)$, *then there is an element* $r \in R_F$, *such that* $\varphi(b)=br$ *for any* $b \in W$ *and* $lr=0$ *for any* $l \in l_R(W)$.

Properties (1)-(3) result immediately from the definition. Moreover, these properties define the ring of quotients R_F uniquely, to the accuracy of an isomorphism over R. Let us show the way properties (1)-(3) yield property (4).

1.4.10. Lemma. *If* W *is a nonzero submodule of the left (right)* R-*module* R_F, *then* $W^2 \neq 0$. *In particular,* R_F *is a semiprime ring.*

Indeed, if $0 \neq v \in W$ and I_v is an essential ideal of the ring R, such that $I_v v \subseteq R$, then $0 \neq (I_v v)^2 \subseteq W^2$.

Therefore, the left and right annihilators of the (R, R)-submodule W in R_F coincide, since the equality $lW=0$ implies $(Wl)^2=0$ and, vice versa, $Wl=0$ implies $(lW)^2=0$. This peculiarity makes it possible to substitute right or two-sided annihilators for left ones in the definition of property (4).

Let W obey condition (4). Let us denote by L an annihilator of the ideal I in R. Let us also extend φ onto $L+I$, assuming $\varphi(L)=0$. Since the annihilator of the ideal $L+I$ equals zero, then, according to property (3), we find an element $r \in R_F$, such that $Lr=0, tr=\varphi(t)$, where $t \in I$. Then, if $v \in W$, $a \in II_v$, then $av \in I$ and, hence, $a\varphi(v)=\varphi(av)=avr$. Therefore, $I(\varphi(v)-vr)=0$. Besides, according to the condition, $L \cdot W=0$, i.e., $L(\varphi(v)-vr)=0$, which implies $\varphi(v)=vr$. If $lW=0$, then $I_1 \cdot lW=0$ and, hence, $I_1 \cdot l \subseteq L$. Consequently, $I_1 lr=0$, which, according to property (2), affords $1r = 0$, which is the required proof.

1.4.11. By analogy with R_F one can define the right ring of quotients.

$$R_F^0 = \varinjlim_{I \in F} Hom(I_R, R).$$

The difference in the determination of the operations is in the only fact that the product in R_F^0 is determined by the formula $(\varphi_1 \varphi_2)(x) = \varphi_1(\varphi_2(x))$. This formula yields the correspondence $a \to \bar{a}$, where $\bar{a}(x) \to ax$ is an embedding of R in R_F. In R_F let us consider a subring $Q(R)$ which consists of all elements a for which there exists ideals $I'_a \in F$, such that $aI'_a \subseteq R$. In this case the mapping $a \to ax$, $x \in I'_a$ determines a certain element $\bar{a} \in R_F^0$. Analogously, if $b \in R_F^0$ and $Ib \in R$, then the mapping $x \to xb$, $x \in I$ determines the element $\bar{b} \in R_F$. It is evident that the mappings $a \to \bar{a}$, $b \to \bar{b}$ are mutually inverse isomorphisms, and the rings $Q(R)$ and $\overset{\circ}{Q}(R) = \{r \in R_F^0 \mid E\ I \in F, I_r \subseteq R$ can be identifies. The ring $Q(R)$ is called a *two-sided Martindale ring of quotients* (or a *symmetrical ring of quotients*).

1.4.12. Proposition. *Let the ring R be expanded into a direct sum of ideals $R = T \oplus S$. Then $R_F = T_F \oplus S_F$*

Proof. It should be remarked that the rings T_F and S_F are naturally embedded into R_F: if $t \in T_F$ and $It \subseteq T$, where I is an essential ideal of T, then $TIT + S$ is an essential ideal of R and the mapping $a + s \to at$ defined on this ideal determines the element $t \in R_F$ which is naturally identical to the element t.

Since $ST_F = 0$ and $TS_F = 0$ by construction, then we have $(S + T)(T_F \cap S_F) = 0$, i.e., the intersection $(T_F \cap S_F)$ equals zero.

If $q \in R_F$ and $Iq \subseteq R, I \in F$, then $TIq \subseteq T$, $SIq \subseteq S$, in which case TI and SI are essential ideals in T and S, respectively. Hence, right multiplications by q determine the elements $t \in T_F$, $s \in S_F$. Let us show that $q = t + s$. Indeed, if $a \in TI$, then $aq = at$, $as = 0$, if $a \in SI$, then $aq = as$, $at = 0$ and, hence, for $a \in TI + SI$ the equality $aq = at + as$ is valid, i.e., $RI(q - t - s) = 0$, and property (2) of the ring R_F yields $q = s + t$.

Let us show that $T_F \cdot S_F = 0$. Indeed, if $t \in T_F$, then $(TIT + S)t \subseteq T$ for a certain essential ideal I of the ring T, since by the construction of S_F we have $TS_F = 0$, then $(TIT + S)tS_F = 0$, i.e., by property (2) we get $tS_F = 0$.

Therefore, the subrings T_F, S_F annihilate each other and as a sum they present the whole ring R_F, i.e., they are ideals and we have $R_F = T_F \oplus S_F$.

For the ring of quotients Q the considerations are the same, the only requirement is to check when constructing t if $t \in Q(R)$ when $t \in Q(T)$. The proposition is proved.

1.4.13. Proposition. *If* R *is a semiprime ring and* $I \in F$ *then the left Martindale ring of quotients of* I *can naturally be identified with* R_F, *in which case* $Q(I) = Q(R)$.

Proof. If A is an essential ideal of the ring I, then $IAI \in F$, in which case $IAI \subseteq A$ and, hence, if $q: A \to I$ is a homomorphism of the left I-modules, then the restriction of q on IAI will be a homomorphism of the left R-modules, and we thus get the embedding $I_F \subseteq R_F$.

The inverse inclusion is also evident: if J is an essential ideal of R, then JIJ is simultaneously an essential ideal of the ring R and the ring I. In this case the image of the restriction $\bar{\varphi}$ of the homomorphism $\varphi: J \to R$ onto JIJ is contained in $I\varphi(J) \subseteq I$, which allows one to identify $\varphi/\approx$ with the element $\bar{\varphi}/\approx\, \in I_F$. The equality $Q(I) = Q(R)$ is obvious under such an identification. The proposition is proved.

1.5 The Generalized Centroid of a Semiprime Ring

1.5.1. Definition. *A generalized centroid* of a semiprime ring R is a center C of its Martindale ring of quotients R_F. Is is evident that the center of a ring is a commutative subring. It should be recalled that the *centralizer* of the set S in the ring R is a subring $Z_A(S) = \{a \in A \mid \forall\, s \in S\ (as = sa)\}$.

1.5.2. Lemma. *The centralizer of any ideal* $I \in F$ *in the ring* R_F *is equal to the generalized centroid of the ring* R.

Proof. As usual, let us denote $[x, y] = xy - yx$. In this case the identity $[xz, y] - [x, y]z = x[z, y]$ is valid. Let an element a belong to the centralizer of I, i.e., $[I, a] = 0$. If $r \in R_F$, then, in conformity with the above identity, we have

$$II_r[r, a] \subseteq [II_r r, a] + [II_r, a]r = 0$$

Herefrom, by property (2) of the ring of quotients (see **1.4.9**), we get $[r, a] = 0$, i.e., $a \in C$. The lemma is proved.

1.5.3. Definition. A ring is called *regular* if for any element x there exists an element x', such that $xx x' = x$.

1.5.4. Lemma. *The generalized centroid of a semiprime ring is a regular ring.*

Proof. Let a be a central element of a Martindale ring of quotients. Let us show that the mapping $\varphi: xa^2 \to xa$, where x runs through the set $I_{a^2} \cap I_a$, is correctly determined. Let $xa^2 = 0$. Then $(xa)R_F(xa) = 0$. Therefore, φ is a homomorphism of the left R-module $J = I_{a^2} \cap I_a \in F$ to R, i.e., by property (3) of the ring of quotients (see theorem **1.4.9**), there is an element $a_1 \in R$ such that $xa^2a_1 = xa$ for all $x \in J$. According to property (2) of the ring of quotients we have $a^2a_1 = a$. Let us show that the element a_1 lies in the center. Let x be an arbitrary element of the ring of quotients. Then $[x, (a^2a_1)^2] = [x, a^2] = 0$, and, hence, $a^4[x, a_1^2] = 0$. Multiplying the obtained equality from the left by a_1^3, we get $0 = a[x, a_1^2] = [x, a_1]$. The lemma is proved.

As any regular ring, a generalized centroid has a sufficient number of idempotents. Indeed, if $a^2a' = a$, then $(aa')^2 = a^2a' \cdot a' = aa'$, i.e., $e = aa'$ is an idempotent, in which case $ea = a$. The latter condition implies, in particular, that the C-module generated by the element a is generated by the idempotent e, i.e., $eC = aC$. In the set E of all the central idempotents the following relation of a partial order naturally arises:

$$e_1 \le e_2 \overset{df}{\Leftrightarrow} e_1 e_2 = e_1.$$

1.5.5. Definition. The *support* of the set $S \subseteq R_F$ is the least idempotent $\mathbf{e}(S) = e \in C$, such that $se = s$ for all $s \in S$.

1.5.6. Lemma. *Any set* $S \subseteq R_F$ *has a support.* $\mathbf{e}(S) = e \in C$. *In this case the equality* $xRS = 0$ *for an element* $x \in R_F$ *(as well as the equality* $SRx = 0$*) is equivalent to* $x\,\mathbf{e}(S) = 0$.

Proof. Let V be a two-sided R-submodule in R_F, generated by the set S. In this case $I = V \cap R$ is a two-sided ideal of the ring R. Let us prove that its annihilator in the ring R coincides with the annihilator of V in R. Let $Ix = 0$. If $v \in V$, then (see **1.4.9**) $I_v \cdot v \in R \cap V = I$ and, hence, $I_v vx = 0$. By property (2), we get $vx = 0$.

By property (4) of the ring of quotients, for the identical mapping $\varphi\colon V \to V$, there exists an element $e \in R_F$, such that $ve = v$ for all $v \in V$ (in particular, $se = s$ for all $s \in S$) and e annihilates the annihilator L of the set V in the ring R_F. This fact implies that for any $l \in L, v \in V, x \in R_F$ the following equalities are valid: $(l + v)[x, e] = 0, (l + v)(e^2 - e) = 0$. Since the annihilator of the sum $L + V$ has a zero multiplication, then, by lemma **1.4.10**, the element e is a central idempotent.

Further on, if e_1 is a central idempotent, such that $se_1 = s$ for all $s \in S$, then, in accordance with the centrality, $ve_1 = v$ for all $v \in V$ and, hence, $1 - e_1 \in L$, i.e., $0 = (1 - e_1)e = e - e_1 e$, which implies $e_1 \ge e$ by the definition.

Let, finally, $SRx = 0$. Then Rx lies in the annihilator of V and, hence, $eRx = 0$, which affords $Rex = 0$ and $ex = 0$. The lemma is proved.

1.5.7. Lemma. *If* $0 \ne e_1 \le \mathbf{e}(S)$ *then* $e_1 S \ne 0$.

Proof. If $e_1 S = 0$, then the idempotent $f = 1 - e_1$ obeys the equality $sf = s$ for all $s \in S$. Therefore, $f \ge \mathbf{e}(S) \ge e_1$, i.e., $e_1 = fe_1 = 0$, which is a contradiction.

The essence of proportionality over C in terms of the initial ring R is well demonstrated by the following lemma.

1.5.8. Lemma. *The elements* $a, b \in R$ *are proportional over* C, *i.e.,* $a = cb$ *for a certain* $c \in C$ *iff* $e(a) \geq e(b)$ *and* $axb = bxa$ *for all* $x \in R$.

This lemma will later be singled out from the general statement **1.7.14,** and meanwhile the reader is offered the idea to prove it as an exercise.

1.5.9. Proposition. *A generalized centroid of a prime ring is a field. Inversely, if a generated centroid of a semiprime ring is a field, then the ring is prime.*

Proof. Let R be a prime ring. As the ring C is regular, it is sufficient to prove that it has no zero divisors. If $c_1 \cdot c_2 = 0$, $c_1, c_2 \in C$, then, choosing nonzero ideals I_1, I_2 in R, such that $c_1 I_1 \subseteq R, c_2 I_2 \subseteq R$ we see that the product of the two-sided ideals $(c_1 I_1)(c_2 I_2)$ is equal to zero, which affords $c_1 = 0$ or $c_2 = 0$.

Inversely, if $aRb = 0$, then by lemma **1.5.6** we have $e(a)b = 0$. From here we get $e(a)Rb = 0$, i.e., in C there are zero divisors. The proposition is proved.

1.6 Modules Over a Generalized Centroid

In this paragraph we in fact shall be speaking about modules over an arbitrary regular self-injective commutative ring, though it becomes evident only at the very end. The approach for presenting such a classical material is clear to be quite different. We prefer a topological approach by an important reason that all module homomorphisms, ring automorphisms and derivations, as well as termal unary operations prove to be continuous and, hence, the problem of transferring many properties onto an injective hull (or, in our terms, onto a topological closure) proves to be absolutely evident.

In the case of a prime ring all the modules considered here are converted into linear spaces, and the topology becomes discrete. Therefore, the reader mainly interested in prime rings can omit this paragraph.

1.6.1. Let R be a semiprime ring, C be its generalized centroid. Let us determine the topology in an arbitrary C-module M with the help of idempotents from C.

It should be recalled that a partially ordered set A is called *directed* if for any two elements there exist their upper boundary. The set E of all the central idempotents is partially ordered with respect to the order introduced above (see **1.5.4**). In this case any subset $E_1 \subseteq E$ has an exact upper boundary sup E_1 which, as can be easily seen, is equal to the support $e(E_1)$ of the set E_1.

1.6.2. Definitions. Let A be a directed partially ordered set. Let us call an element m of module M over ring C *a limit* of the *family* $\{m_\alpha \in M, \alpha \in A\}$, if there exists a directed family of idempotents $\{e_\alpha, \alpha \in A\}$, such that $e_\alpha \leq e_\beta$ for $\alpha \leq \beta$; $\sup\{e_\alpha\} = 1$, and for all $\alpha \in A$ the equalities

$$me_\alpha = m_\alpha e_\alpha \tag{10}$$

are valid.

In this case we shall write $m = \lim\limits_A m_\alpha$.

Accordingly, the set $T \subseteq M$ is called *closed* if any limit of the family of elements from T lies in T. *A closure* of the set is the least closed set containing the given one. Therefore, the operation of closure determines a certain topology on the C-module M.

Let now $\varphi: M_1 \to M$ be a certain mapping of the C-modules. We shall call φ *quite continuous* if the equality $m = \lim\limits_A m_\alpha$ yields $\varphi(m) = \lim\limits_A \varphi(m_\alpha)$. It should be remarked that any quite continuous mapping is continuous. Indeed, in this case the total preimage of a closed set is closed and, hence, φ is a continuous mapping. It is evident that if φ preserves the operators of multiplication by the idempotents from C: $\varphi(me) = \varphi(m)e$, then φ will be quite continuous. We, in particular, can conclude that any module homomorphism is continuous, while an isomorphism is a homeomorphism.

1.6.3. Proposition. *The operators of left and right multiplication, those of the shift* $\Pi_a(x) = x + a$, *as well as all the derivations and automorphisms of a ring* R_F *are quite continuous. Moreover, if* $r = \lim\limits_A r_\alpha$, $s = \lim\limits_A s_\alpha$ *then* $r \pm s = \lim\limits_A (r_\alpha \pm s_\alpha)$, $r \cdot s = \lim\limits_A (r_\alpha s_\alpha)$.

Proof. The operators of the left and right multiplications are quite continuous, since they are module homomorphisms.

If $r = \lim\limits_A r_\alpha$, then

$$(r + a)e_\alpha = re_\alpha + ae_\alpha = r_\alpha e_\alpha + ae_\alpha = (r_\alpha + a)e_\alpha,$$

i.e., $r+a=\lim_A (r_\alpha + a)$.

In order to prove the continuity of derivations, let us first remark that the central idempotents are constants for any derivation. If $e^2=e$, then $2ee^\mu = e^\mu$. Multiplying by e, we get $e^\mu = 2ee^\mu = ee^\mu$ and, hence, $ee^\mu = 0$, i.e., $e^\mu = 0$.

Now we have $r^\mu_\alpha e_\alpha = (r_\alpha e_\alpha)^\mu = r^\mu e_\alpha$, i.e., $r^\mu = \lim_A r^\mu_\alpha$.

If g is an automorphism, then while considering the family of idempotents $\{e^g_\alpha\}$, we have $r^g_\alpha e^g_\alpha = (r_\alpha e_\alpha)^g = (re_\alpha)^g = r^g e^g_\alpha$, i.e., $r^g = \lim r^g_\alpha$.

And, finally, let $r=\lim_A r_\alpha$, $s=\lim_A s_\alpha$. In this case, if $\{e_\alpha\}, \{f_\alpha\}$ are the corresponding families of idempotents, then for the sets $\{r_\alpha \pm s_\alpha\}, \{r_\alpha s_\alpha\}$ we shall consider the family of idempotents $\{e_\alpha f_\alpha | \alpha \in A\}$. It is evident that this family is directed and, if $ee_\alpha f_\alpha = 0$ for all α, then, for an arbitrary $\beta \in A$ choosing an element $\gamma \geq \alpha, \beta$, we get $(e e_\alpha) f_\beta = e(e_\alpha e_\gamma)\cdot(f_\beta e_\gamma) = (e e_\gamma f_\gamma)\cdot e_\alpha f_\beta = 0$. Therefore, $ee_\alpha = 0$ and, hence, $e=0$, i.e., $\sup\{e_\alpha f_\alpha\} = 1$. Further on, $(r \pm s) e_\alpha f_\beta = (r\, e_\alpha) f_\alpha \pm (s\, f_\alpha) e_\alpha = (r_\alpha \pm s_\alpha) e_\alpha f_\alpha$, and, accordingly, $(rs) e_\alpha f_\alpha = (r_\alpha s_\alpha) e_\alpha f_\alpha$. The proposition is proved.

1.6.4. It should be remarked that the latter proof of proposition **1.6.3** does not imply that R_F is a topological ring, as, possibly, there exist families convergent in the above determined topology but having no limit (in the above determined sense). Therefore, there arises a question (inessential for the further material but of its own interest): if R_F is a topological ring with the topology determined above ?

1.6.5. Let $F(X)$ be a polynomial of a set of variables X with coefficients from R_F which do not commute with the variables, in the presentation of which not only the operations of addition and multiplication, but possibly the unary operations of derivation as well as automorphisms take part.

1.6.6.Lemma. *The mapping* $F(X): R_F \oplus ... \oplus R_F \to R_F$ *is quite* *continuous. In particular, if* $F(X)$ *turns to zero on the set* $T \subseteq R_F$, *then* $F(X)$ *turns to zero on the closure* $\hat{T} \subseteq R_F$.

Proof. Let us first remark that the difference, product and composition of quite continuous mappings $\varphi_1, \varphi_2 : M \to M_1$, $\psi : M_1 \to M_2$ are quite continuous. If $m = \lim_A m_\alpha$, then $\varphi_1(m) = \lim \varphi_1(m_\alpha)$, $\varphi_2(m) = \lim \varphi_2(m_\alpha)$ and, therefore, according to proposition **1.6.3**, we have

$$\varphi_1(m) \pm \varphi_2(m) = \lim_A ((\varphi_1(m_\alpha) \pm \varphi_2(m_\alpha))$$

$$\varphi_1(m)\varphi_2(m) = \lim_A \varphi_1(m_\alpha)\varphi_2(m_\alpha)$$

In the same way, by the definition of a continuous transformation, we get

$$\psi(\varphi(m)) = \psi(\lim_A \varphi(m_\alpha)) = \lim_A \psi(\varphi(m_\alpha))$$

We should now remark that $F(X)$ can be presented as a sum of products of the compositions of the projections $\pi_i : R_F \oplus \ldots \oplus R_F \to R_F$ with automorphisms and derivations $R_F \to R_F$, as well as with the operators of multiplications by the elements from R_F.

And, finally, if $F(X)$ turns to zero on T, then the mapping $F(X)$ is equal to zero on $T \oplus \ldots \oplus T$, the closure of which is certain to contain $\hat{T} \oplus \ldots \oplus \hat{T}$. The lemma is proved.

The above definition does not imply that the directed family of elements $\{m_\alpha\}$ has not more than one limit.

1.6.7. Lemma. *The following conditions on the C-module M are equivalent.*

(a) *Any directed family of elements of the module M has not more than one limit.*

(b) *Any one-element set is closed.*

(c) *The zero submodule is closed.*

(d) *M is a nonsingular module.*

It should be recalled that a *singular submodule* of the module M is a union of all the elements having essential annihilators in the ring of operators (an ideal of a ring is called *essential* if it has a nonzero intersection with any other nonzero ideal). And, accordingly, a module is called *nonsingular* if its singular submodule equals zero.

Proof. The implications of (a)→(b)→(c) are obvious. Let us prove that the closure of a zero submodule coincides with the singular ideal.

It should be remarked the essentiality of the ideal I for the ring C is equivalent to the fact that its support is equal to the unit, $e(I)=1$. Indeed, if I is essential, $f=1-e(I)$, then $I\cap fC=0$, i.e., $f=0$ and, hence, $e(I)=1$. Inversely, if $e(I)=1$ and $I\cap J=0$, then, choosing an arbitrary element f in J, we get $If\subseteq I\cap J=0$, i.e., $f=f\cdot e(I)=0$ and, hence, $J=0$.

Let $m\in Z(M)$. In this case $ann_C(m)$ is an essential ideal. Let A be a set of all idempotents of this ideal. It is a directed set, in which case $m\cdot\alpha=0\cdot\alpha$ for all $\alpha\in A$, so that $m=\lim\limits_{\alpha\in A} m_\alpha$, where $m_\alpha=0$. Consequently, $Z(M)\subseteq\overline{(0)}$. Let us show $Z(M)$ to be a closed submodule. Let $m=\lim\limits_{\alpha\in A} m_\alpha$, where $m_\alpha\in Z(M)$, and let $m\notin Z(M)$, i.e., $f=1-e(ann_C\, m)\neq 0$. As for the family of idempotents $\{e_\alpha,\ \alpha\in A\}$, which defines the limit, the condition $\sup\limits_A e_\alpha=1$ is valid, there is an element $\beta\in A$, such that $e_\beta f\neq 0$. Since $m_\beta\in Z(M)$, then $\sup(ann_C\, m_\beta)=1$, i.e., we can find an idempotent $u_\beta\in ann_C m_\beta$, such that $e_\beta fu_\beta\neq 0$. We have $me_\beta fu_\beta=m_\beta e_\beta fu_\beta=(m_\beta u_\beta)e_\beta f=0$. It means that $e_\beta fu_\beta\in ann_C m$. Since the idempotent f annihilates $ann_C m$, then $0=(e_\beta fu_\beta)f=e_\beta fu_\beta\neq 0$ which is a contradiction.

Let us, finally, show that (d)→(e). Let us assume that a certain directed family of idempotents m_α has two limits, $m^{(1)}\neq m^{(2)}$. Let $\{e_\alpha\},\{f_\alpha\}$ be the corresponding families of idempotents. In this case we have

$$(m^{(1)}-m^{(2)})e_\alpha f_\alpha=(m^{(1)}e_\alpha-m^{(2)}f_\alpha)e_\alpha f_\alpha=$$
$$=m_\alpha(e_\alpha-f_\alpha)e_\alpha f_\alpha=0.$$

If we now recall that $\sup\limits_A\{e_\alpha f_\alpha\}=1$ (see **1.6.3**), then the lemma is proved.

1.6.8. If A is a ring, then we as usual denote the ring of endomorphisms of the additive group of the ring A through $End(A,+)$. The ring $End(A,+)$ is the right module over the center of A, but it can be not an algebra over it. The action of the central elements is governed by the formula $x(\varphi c) = (x\varphi)c$.

1.6.9.Lemma. *The modules* R_F, $End(R_F,+)$ *are nonsingular.*

Proof. Let us make use of condition (a) of lemma **1.6.7**. Let r, s be the limits of the family $\{r_\alpha, \alpha \in A\}$, and let $\{e_\alpha\}, \{f_\alpha\}$ be the corresponding systems of idempotents. In this case $(re_\alpha)f_\alpha = r_\alpha e_\alpha f_\alpha = sf_\alpha e_\alpha$. Hence, $(r-s)e_\alpha f_\alpha = 0$. If $r, s \in R_F$, then $r - s = (r-s)\sup\{e_\alpha f_\alpha\} = 0$. If, finally, $r, s \in End(R_F,+)$, then for any $x \in R_F$ the equalities $(r-s)(x)e_\alpha f_\alpha$ yield $r(x) = s(x)$. The lemma is proved.

1.6.10.Lemma. *If* T *is a subring of* R_F, *then its closure* $\hat{T}$ *is also a subring.*

Proof. Let us denote through $k(T)$ a set of all the limiting points of T. Let

$$T_1 = T,\ T_{\alpha+1} = k(T_\alpha),\ T_\beta = \bigcup_{\alpha<\beta} T_\alpha,$$

where β is a limiting number. In this case the union of all T_α is a closed set and, hence, it coincides with $\hat{T}$. Therefore, according to the transfinite induction, it suffices to prove that $k(T)$ is a subring. Let $r = \lim_A r_\alpha, s = \lim_B s_\beta, r_\alpha, s_\beta \in T$, $\{e_\alpha\}, \{e_\beta\}$ be the corresponding systems of idempotents. Let us consider the family $\{e_\alpha e_\beta | (\alpha,\beta) \in A \times B\}$, assuming that $(\alpha,\beta) \le (\alpha_1,\beta_1) \Leftrightarrow \alpha \le \alpha_1, \beta \le \beta_1$. In this case we get $r \pm s = \lim_{A\times B}(r_\alpha \pm s_\beta)$, $rs = \lim_{A\times B}(r_\alpha s_\beta) \in k(T)$, which is the required proof.

1.6.11. Lemma. *Let* I *be an ideal of the subring* $T \subseteq R_F$. *Then the closure* $\hat{I}$ *is an ideal of* $\hat{T}$.

Proof. Like in the lemma above, it suffices to prove that $k(I)$ is an ideal of $k(T)$. Let $r = \lim_A r_\alpha \in k(T)$, $s = \lim_B s_\beta \in k(I)$, then $rs = \lim_{A\times B} r_\alpha s_\beta \in k(I)$ and $sr \in k(I)$.

1.6.12. Definition. The module M is called *complete* if any family of its elements $\{m_\alpha | \alpha \in A\}$, for which there exists a directed family of idempotents $\{e_\alpha\}$, such that $\sup\{e_\alpha\} = 1$ and at $\alpha \geq \beta$ the relations $m_\alpha e_\beta = m_\beta e_\beta$, $e_\alpha \geq e_\beta$ are valid, has a limit.

1.6.13. Lemma. *The modules R_F and $End(R_F, +)$ are complete.*

Proof. Let us first assume that $r_\alpha \in R_F$, and let $\{e_\alpha\}$ be a family of idempotents, such that $r_\alpha e_\alpha = r_\beta e_\beta$, $e_\alpha \geq e_\beta$ at $\alpha > \beta$ and $\sup\{e_\alpha\} = 1$. Let $N_\alpha = \{x \in R | xe_\alpha \in R, xr_\alpha e_\alpha \in R\}$. Then N_α is a left ideal of the ring R, in which case $N_\alpha \supseteq I_\alpha \in F(R)$. The union $N = \bigcup_\alpha N_\alpha e_\alpha$ is also a left ideal. Indeed, if $a_\alpha e_\alpha$, $a_\beta e_\beta \in N$, and let us find an element $\gamma \geq \alpha, \beta$, then $(a_\alpha e_\alpha + a_\beta e_\beta) r_\gamma e_\gamma = a_\alpha e_\alpha r_\alpha + a_\beta e_\beta r_\beta \in R$. Analogously, $(a_\alpha e_\alpha + a_\beta e_\beta) e_\gamma \in R$, and, hence, $a_\alpha e_\alpha + a_\beta e_\beta \in N_\gamma e_\gamma$.

As $N_\gamma e_\gamma \supseteq I_\alpha e_\alpha$, then $N \supseteq \sum I_\alpha e_\alpha = I$ is an ideal of R. Besides, the annihilator of I equals zero. Indeed, if $Ix = 0$, then $I_\alpha e_\alpha x = 0$, i.e., $e_\alpha x = 0$ and, hence, $x = 0$.

Let us determine the homomorphism of left R-modules $\xi: N \to R$ by the formula $a_\alpha e_\alpha \xi = a_\alpha e_\alpha r_\alpha$. If $a_\alpha e_\alpha = a_\beta e_\beta$ and $\gamma \geq \alpha, \beta$, then $a_\alpha e_\alpha r_\alpha = a_\alpha e_\alpha r_\gamma = a_\beta e_\beta r_\gamma = a_\beta e_\beta r_\beta$, i.e., ξ is defined correctly. Since the domain of the definition of ξ contains the ideal $I \in F$, then there is an element $r \in R_F$, such that $a_\alpha e_\alpha r = a_\alpha e_\alpha r_\alpha$, i.e., $I(e_\alpha r - e_\alpha r_\alpha) = 0$. This implies that $re_\alpha = r_\alpha e_\alpha$ and, hence, $r = \lim r_\alpha$.

If now $r_\alpha \in End(R_F, +)$, then for any $x \in R_F$ there exists a limit of $r_\alpha(x)$, and we can set $r(x) = \lim r_\alpha(x)$, in which case $r = \lim r_\alpha$. The lemma is proved.

1.6.14. Lemma. *The subrings* Q, C *are closed in* R_F *and, hence, they are complete modules over* C.

Proof. Let $r = \lim r_\alpha$ and $e_\alpha r_\alpha I_\alpha \subseteq R$, $e_\alpha I_\alpha \subseteq R$, where $I_\alpha \in F$. Then $J = \sum e_\alpha I_\alpha \in F$, in which case $re_\alpha I_\alpha = r_\alpha e_\alpha I_\alpha$, and, hence, $rJ \subseteq R$, i.e., $r \in Q$. The subring C is closed as an intersection of all the kernels of quite continuous mappings $ad\ a\colon x \to xa - ax$. The lemma is proved.

1.6.15. Let us recall that the module M is called *projective* if it has the following property. Let π be the epimorphism of a module B on a module A, then any homomorphism $\varphi\colon M \to A$ can be "raised" to the homomorphism $\psi\colon M \to B$, such that $\pi \cdot \psi = \varphi$

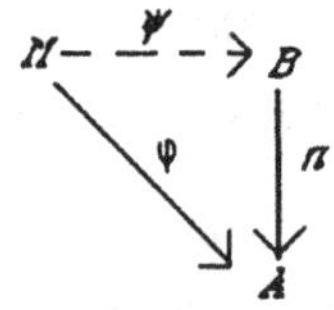

This property is equivalent to the fact that for any epimorphism $\theta\colon M' \to M$ there exists a decomposition $M' = \ker \theta \oplus M''$. It can be readily shown and it is well known that a direct sum of modules is projective iff all the summands are projective.

Accordingly, the module M is called *injective* if it has a dual property. Let π be a monomorphism of a module A into a module B, then any homomorphism $\varphi\colon A \to M$ allows an extension to the homomorphism $\psi\colon B \to M$, such that $\varphi \cdot \pi = \varphi$

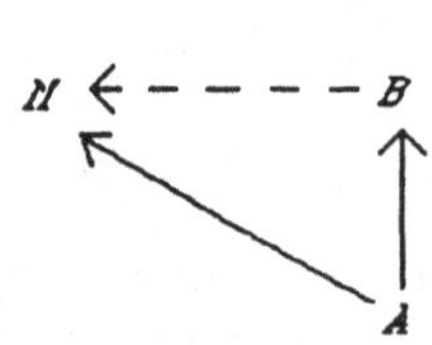

This property is equivalent to the fact that for any monomor-phism $\theta\colon M \to M'$ there exists a decomposition $M' = im\theta \oplus M''$, i.e., the module M is extracted as a direct summand of any module containing it. It can be easily shown, and this is also a well known fact, that a direct product of modules is injective iff all the cofactors are injective.

It should be recalled that a ring is called *(right) self-injective* if it is injective as a (right) module over itself.

1.6.16.Lemma. *Any complete nonsingular module over* C *is injective and, vice versa, any injective module over* C *is complete.*

Proof. Let T be a complete nonsingular submodule of the module M. Let us show that T is extracted from M by a direct summand.

In M let us consider a set of all submodules having zero intersection with T. This set is directed by inclusion, i.e., according to the Zorn lemma, this set has at least one maximal element M'.

Let us show that M' is a closed submodule in M. Let $m = \lim\limits_{A} m_\alpha$, $m_\alpha \in M'$ and let $\{e_\alpha\}$ be the corresponding families of idempotents. If $m \notin M'$, then $(M' + mC) \cap T \neq (0)$. Let $t = m' + mc \neq 0$ for a suitable $m' \in M'$, $c \in C$. Then for any $\alpha \in A$ we have $te_\alpha = m'e_\alpha + m_\alpha e_\alpha c \in M' \cap T = (0)$, i.e., due to nonsingularity of the module T the element t is equal to zero, which is a contradiction.

Let now m be an arbitrary element of M. Let us show that $m \in M' + T$. Let us assume that $m \notin M'$ and consider a set A of all idempotents $\alpha \in C$, such that $m\alpha \in M' + T$. This set is directed: if $\alpha_1, \alpha_2 \in A$, then $\alpha_1 \vee \alpha_2 = \alpha_1 + \alpha_2 - \alpha_1\alpha_2 \in A$, so that $m(\alpha_1 \vee \alpha_2) = m\alpha_1 + (m\alpha_2)(1 - \alpha_1)$. Let us prove that $\sup A = 1$.

Let, on the contrary, $f = 1 - \sup A \neq 0$. Then $f \notin A$ and, hence, $mf \notin M' + T$ and, in particular, $mf \notin M'$, i.e., $(M' + mfC) \cap T \neq (0)$. Let us find an element $c \in C$, such that $0 \neq mfc + m' \in T$ for a suitable $m' \in M$. In this case mfc is a nonzero element of $M' + T$. Since C is a regular ring, there is an element c', such that $e_1 = cc'$ is an idempotent, and $e_1 c = c$. The latter equality shows, in particular, that $fe_1 \neq 0$. On the other hand, $mfe_1 = (mfc)c'$ lies in $M' + T$, i.e., $fe_1 \in A$. However, $fA = (0)$ and, hence, $fe_1 = f(fe_1) = 0$, which is a contradiction.

Thus, $\sup A = 1$. Let $m\alpha = m_\alpha + t_\alpha$, where $m_\alpha \in M, t_\alpha \in T$. If $\beta \leq \alpha$, then $m\beta = (m\alpha)\beta = m_\alpha\beta + t_\alpha\beta$. As the sum $M' + T$, is direct, then $m_\beta = m_\alpha\beta, t_\beta = t_\alpha\beta$. Since the module T is complete, there exists a limit $t = \lim\limits_{\alpha \in A} t_\alpha$. We now have $(m - t)\alpha = m_\alpha \in M'$ and, hence,

$m - t = \lim_A \in M'$, as the submodule M' is closed. Therefore, $m \in M' + T$, which is the required proof.

Inversely, let M be an injective C-module. Let us consider a direct sum $M_1 = M \oplus aC$, where $aC \cong C$ is a free one-generated module. Let, then, $\{m_\alpha\}$ be a family of elements from M, and let $\{e_\alpha\}$ be a directed family of idempotents, such that $\sup_\alpha e_\alpha = 1$ and for $\beta \geq \alpha$ the equalities $m_\beta \cdot e_\alpha = m_\alpha e_\alpha$ hold. In M_1 let us consider a submodule N generated by the elements $m_\alpha e_\alpha \oplus ae_\alpha$. It should be remarked that $N \cap M = (0)$. Indeed, if

$$m = \sum_\alpha (m_\alpha e_\alpha c_\alpha \oplus ae_\alpha c_\alpha) \in M,$$

then $a\sum_\alpha e_\alpha c_\alpha = 0$, i.e., $\sum_\alpha e_\alpha c_\alpha = 0$.

Let β be the. upper boundary of elements α included in the latter sum. Then $0 = m_\beta \sum_\alpha e_\alpha c_\alpha = \sum_\alpha m_\beta e_\alpha c_\alpha = \sum_\alpha m_\alpha e_\alpha c_\alpha$, i.e., $m = 0$. Therefore, the natural homomorphism $\varphi\colon M \to M_1 / N$ is an embedding, and we can write the relations $\varphi(m_\alpha) e_\alpha + ae_\alpha = 0$ for all α. Let us apply the definition of injectivity to the following diagram:

$$\begin{array}{ccc} 0 \to M & \xrightarrow{\varphi} & M_1 / N \\ \downarrow & & \downarrow \psi \\ M & \equiv & M \end{array}$$

We shall find a homomorphism $\psi\colon M_1 / N \to M$, such that $\varphi\psi = 1$. Let $m = -\psi(a)$. Then $m_\alpha e_\alpha - me_\alpha = \psi(a) e_\alpha = \psi[\varphi(m_\alpha) e_\alpha + a e_\alpha] = 0$. Consequently, $m = \lim m_\alpha$, which is the required proof.

1.6.17.Corollary. *A generalized centroid of a semiprime ring is a regular self-injective ring.*

1.6.18.Note. It would be expedient to add here that any self-injective commutative ring A coincides with its generalized centroid (and is, according to theorem **1.5.3**, regular). Indeed, let $Q(A)$ be a Martindale ring of quotients. Let us consider it as a left A-module. Since $A \subseteq \mathcal{Q}(A)$, then the direct decomposition $\mathcal{Q}(A) = A \oplus M$ is valid. If $m \in A$, then, by the definition of a ring of quotients, there is an ideal $I \in F(A)$, such that $Im \subseteq A$. On the other hand, $Im = (0)$. Hence, $Im = (0)$ and, consequently, $M = (0)$. Thus, $A = \mathcal{Q}(A)$.

1.6.19.Note. We have seen the definition of topology to depend only on the structure of E-module (here E is the set of all central idempotents), and it would, therefore, be natural to expect that the statements formulated for C modules will also be valid for E-linear subsets. And this is really the case. Let us convert E into a ring by introducing an addition by the formula $e \oplus f = e + f - 2ef$. Then the E-linear sets will be converted into modules over the ring E. And, finally, it should be remarked that the ring E coincides with its generalized centroid.

1.6.20. Lemma. $\mathcal{Q}(E) = E$.

Proof. Let I be an essential ideal of E. In this case $\sup I = 1$ and, if $Iq \subseteq E$, $q \in \mathcal{Q}(E)$, let us assume that $e_i = iq$ and consider I as a directed set of idempotents. Since C is complete, we find $e = \lim_I e_i$. It is evident that e is an idempotent and $i(e - q) = e_i - iq = 0$, i.e., $I(e - q) = 0$ and, hence, $q = e \in E$. The lemma is proved.

1.6.21. Lemma. *Let M be a nonsingular C-module. In this case any finitely-generated C-submodule in M is projective and injective, and is a direct sum of a finite number of cyclic modules.*

Proof. Let us first prove that this is true for a one-generated submodule. We have $mC \cong C / \operatorname{ann} m$, where $\operatorname{ann} m = \{c \in C, \ mc = 0\}$. Since M is a nonsingular module, then $m \cdot \lim c_\alpha = \lim mc_\alpha$, where c_α is a converging family of elements from $\operatorname{ann} m$. This means that $\operatorname{ann} m$ is a closed ideal in C. Any closed ideal in C is principal (and vice versa). Indeed, a set of all idempotents from a closed ideal forms a directed family, the limit of which generates the given ideal. This limit exists because C is a closed submodule of the complete module Q. Therefore, $\operatorname{ann} m = eC$ and, hence, $mC \cong (1 - e)C$, which affords $C \cong mC \oplus (1 - e)C$. As C is both an injective and projective C-module, then mC is also both an injective and projective C-module.

Let us now assume that the lemma is also valid for the submodules with $n - 1$ generating elements. Let N be an n-generated submodule, and m be a generating element. In this case the chain $0 \to mC \to N \to N / mC \to 0$ splits, since, according to the above proved, mC is an injective module. We have $N \cong mC \oplus K$, where the submodule $K \cong N / mC$ is generated by the

$(n-1)$-th elements. Thus, the lemma is proved by induction.

In lemma **1.6.21** we, in particular, have shown that if m is an element of a nonsingular module, then $mC \cong eC$ for a certain idempotent $e \in C$. This idempotent will herefrom be denoted by $e(m)$ and called a *support of element* m, which fact does not contradict the denotation $e(s)$ for elements $s \in R_F$ assumed above.

1.6.22. Lemma. *Let* M *be a complete nonsingular* C*-module,* N *be a closed submodule of* M*. Then factor-module* M/N *is complete nonsingular module.*

Proof. By lemma **1.6.16** the module N is injective and, hence, it is singled out from M by a direct addent: $M = N \oplus K$. One, has, finally, to remark that $M/N \cong K$ is an injective nonsingular module.

1.6.23. Remark. *Let* V_1, V_2 *be nonsingular* C*-modules,* $v_1 \in V_1$, $v_2 \in V_2$. *Then* $v_1 \otimes v_2 = 0$ *in* $V_1 \otimes V_2$ *iff* $e(v_1)\, e(v_2) = \ = 0$.

Proof. We have $v_1 C \cong e(v_1)C$, $v_2 C \cong e(v_2)C$. Therefore, $v_1 C \otimes v_2 C \cong e(v_1)e(v_2)C$, which is the proof due to the embedding $v_1 C \otimes v_2 C \to V_1 \otimes V_2$.

1.6.24. Remark. *Let* V_1, V_2 *be nonsingular* C*-modules and let* $v_1, v_2, \ldots, v_n \in V_1$. *Then there exist elements* $c_2, c_3, \ldots, c_n \in C$, *such that for any* $a_1, \ldots, a_n \in V_2$ *the equality* $\sum v_i \otimes a_i = 0$ *implies* $a_1 e(v_1) = \sum_{i \geq 2} a_i c_i$.

The proof will be carried out by induction over n. At $n = 1$ remark **1.6.23** affords $a_1 e(v_1) = a_1 e(a_1) e(v_1) = 0$.

Let now $n > 1$. By lemma **1.6.21**, the submodule $v_1 C$ is extracted from the module $\sum v_i C$ by a direct summand. Let $\sum v_i C = v_1 C \oplus W_2 C \oplus \ldots$ and $v_i = v_1 c_i + \ldots, i \geq 2$. Then

$$0 = \sum v_i \otimes a_i = v_1 \otimes (\sum a_i c_i + a_1) + W_2 \otimes a'_2 + \ldots.$$

This affords $v_1 \otimes (a_1 + \sum a_i c_i) = 0$, and, by remark **1.6.23**, $a_1 e(v_1) = -\sum a_i c_i e(v_1)$, which is the required proof.

1.6.25. Lemma. *Let D_{inn} be a submodule of the module $End(R_F, +)$, which consists of all inner for Q derivations. Then D_{inn} is a closed submodule and, hence, the factor module $End(R_F, +)\,/\,D_{inn}$ is nonsingular.*

Proof. We have $D_{inn} \cong Q\,/\,C$. Since Q is a complete module and C is a closed submodule, then, according to lemma **1.6.22**, Q/C is a complete nonsingular module. We have already remarked that any module isomorphism is quite continuous. Hence, D_{inn} is also a complete module and, consequently, it is closed in $End(R_F, +)$. The lemma is proved.

Let us call the sum of elements $a+b$ of a nonsingular module M *orthogonal* if the supports of these elements are orthogonal, $e(a)\cdot e(b)=0$. The subset S of the module M will be referred to as an *E-subset* (here E is a set of all central idempotents), if it is closed relative to multiplication by the idempotents and, together with any two elements, contains their sum, provided it is orthogonal. Accordingly, the mapping $f\colon S \to N$ is called an *E-mapping* if for any idempotents $e_1, e_2 \in E$ and the orthogonal sum $a_1e_1 + a_2e_2$ the equality $f(a_1e_1 + a_2e_2) = f(a_1)e_1 + f(a_2)e_2$ holds, i.e., if f preserves the multiplication by the central idempotents and the orthogonal sums.

1.6.26. Lemma. *Let M be a complete nonsingular module, S be a closed E-subset in M. Then, if $f\colon S \to R_F$ is an E-mapping, then there is an element $s \in S$, such that the support of the f image coincides with that of $f(s)$.*

Proof. It should be first recalled that f either reduces or preserves the supports: $e(f(a)) \le e(a)$. Indeed, $f(a)\cdot(1-e(a)) = f(a-a) = 0$ and, hence, $e(f(a))(1-e(a)) = 0$.

Let us consider a set N of all pairs (s, e), such that $se = s \in S$, $e = e(f(s))$, and in N let us, as usual, determine the order relation $(s, e) \le (s_1, e_1) \Leftrightarrow s_1e = s$, $e \le e_1$. Since f either preserves or reduces the supports, then for any $s \in S$ the pair $(s\,e(f(s))), e(f(s)))$ lies in N, so that N is nonempty.

If $\{(s_\alpha, e_\alpha) \mid \alpha \in A\}$ is a linearly ordered subset in N, then, according to lemma **15** and since S is closed, there exists a limit

$s = \lim_A s_\alpha \in S$. In this case $e(f(s))e_\alpha = e(f(s)e_\alpha) = e(f(s_\alpha)) = e_\alpha$, and, therefore, $e \stackrel{df}{=} e(f(s)) \geq e_\alpha$. Hence, the pair (s, e) is the upper boundary of the set considered.

According to Zorn lemma, the set M has maximal elements, so let $(s, \hat{e})$ be one of them.

If now $f(S)(1 - \hat{e}) \neq 0$, then there is an element $s_1 \in S$, such that $f(s_1)(1 - \hat{e}) \neq 0$. Let $s_2 = s_1(1 - \hat{e})$. Then $s_2\hat{e} = 0$, $f(s_2) = f(s_1)(1 - \hat{e}) \neq 0$. The first of these equalities shows the sum $s_2 + s$ to be orthogonal. The sum $s_2 e(f(s_2)) + s$ is also orthogonal. Therefore, we have

$$(s, \hat{e}) \leq (s_2 e(f(s_2)) + s, e(f(s_2)) + \hat{e}) \in N,$$

since $e(f(s_2 e(f(s_2)) + s)) = e(f(s_2) + f(s)) = e(f(s_2)) + e(f(s)) = e(f(s_2)) + \hat{e}$. Hence, the support e_0 of the image f annihilates $(1 - \hat{e})$, i.e., $e_0 \leq \hat{e}$; on the other hand, $\hat{e}$ is the support of $f(s)$ and, hence, $\hat{e} \leq e_0$, i.e., $e_0 = \hat{e}$, which is the required proof.

1.7. Extension of Automorphisms to a Ring of Quotients. Conjugation Modules

Let us fix notations R, R_F, Q and C for a semiprime ring R, its rings of quotients and a generalized centroid, respectively. As above, $F = F(R)$ will denote a union of all essential ideals of the ring R.

1.7.1. Lemma. *Let g be an isomorphism of the ideal $I \in F$ into the ring R, such that $I^g \supseteq I_1 \in F$. Then g is uniquely extended to an isomorphism of the ring Q and to that of the ring R_F*

Proof. Let a be an arbitrary element from R_F. It should be remarked that $I_2 = I_1(II_a)^g I_1 \in F$, where I_a is an ideal from F, such that $I_a a \subseteq R$. Indeed, if $I_2 r = 0$, then $0 = I_1(II_a A)^g$, where A is such a set that $A^g = I_1 r$. This set exists since I_1 is an ideal and $I_1 \subseteq I^g$. As g is an isomorphism, then $II_a A = 0$, which affords $A = 0$ and, hence, $I_1 r = 0$, but $I_1 \in F$, i..e., $r = 0$, which fact proves that $I_2 \in F$.

Let us now define the mapping $a^g: I_2 \to R$ in the following way. If $i \in I_2$, then $i = j^g$ for a certain $j \in I^2 I_a I$, and we set $ia^g = (ja)^g$, in which the right part of the equality is determined because $ja \subseteq I$. It is this mapping that determines the element a^g due to property (3) (see 1.4.9). We can now easily check if $a \to a^g$ is the sought extension of g. In this case the formula $j^g a^g = (ja)^g$ shows its uniqueness, since for the other extension, g', we would have the equality $j^g a^g - j^g a^{g'} = (ja)^g - (ja)^g = 0$, i..e., $I_2(a^g - a^{g'}) = 0$, and, hence, $g = g'$. And, finally, if $a \in Q$ and $aJ \in R$, then $a^g I_1 (IJ)^g I_1 \subseteq a^g (I^2 JI)^g \subseteq (aJI)^g \subseteq I^g \subseteq R$, i..e., g is an automorphism of ring Q. The lemma is proved.

1.7.2. It follows from the lemma that any automorphism of the ring R has the only extension on Q and on R_F. Therefore, herefrom we shall always consider the automorphisms to be determined on R_F. Let us introduce the following notation:

$$A(R) = \{g \in Aut Q \mid \exists\ I, I_1 \in F, I_1 \subseteq I^g \subseteq R\}.$$

It is evident that the group of automorphisms of the ring R is contained in $A(R)$. All the further considerations on automorphisms are valid not only for the automorphisms of ring R, but also for those from $A(R)$ and, hence, if the word "automorphism" is not followed by a special indication of the ring, the automorphism will be assumed to belong to $A(R)$.

1.7.3. Lemma. *The set* $A(R)$ *forms a group which contains all the inner automorphisms of the ring* Q.

Proof. Let a be an inverse element from Q. Let us denote through $\tilde{a}$ an inner automorphism corresponding to this element $\tilde{a} : x \to a^{-1} xa$.

According to the definition, we can find an ideal $I \in F(R)$, such that $Ia^{-1} \subseteq R$, $a^{-1} I \subseteq R$, $aI \subseteq R$, $Ia \subseteq R$. In this case we have

$\tilde{a}(I^2) = (a^{-1}I)\cdot(Ia) \subseteq R$. Besides, $aI^4a^{-1} = (aI)\,I^2(Ia^{-1}) \subseteq I^2$, and, hence, $\tilde{a}(I^2) \supseteq I^4$. Therefore, $\tilde{a} \in A(R)$.

Let $g, h \in A(R)$ and we assume $R \supseteq I^g \supseteq I_1$, $R \supseteq J^h \supseteq J_1$, where $I, I_1, J, J_1 \in F$. Then $J_1 I_1 \subseteq I^g$ and, therefore, $J_1 I_1 = A^g$, where $A \subseteq I$. Let us consider the ideal $I_2 = IAI$. If $I_2 v = 0$, then $0 = (I_2 \cdot v)^g = I^g J_1 I_1 I^g v^g \supseteq I_1 J_1 I_1^2 v^g$, i..e., $v^g = 0$, which implies $I_2 \in F$. Then we have

$$I_2^{gh^{-1}} = (I^g J_1 I_1 I^g)^{h^{-1}} \subseteq (RJ_1 I_1 R)^{h^{-1}} \subseteq J_1^{h^{-1}} \subseteq J \subseteq R.$$

Besides,

$$I_2^{gh^{-1}} \supseteq (I_1 J_1 I_1^2)^{h^{-1}} \supseteq (RI_1 J_1 I^2 R)^{h^{-1}} \supseteq$$
$$\supseteq J(I_1 J_1 I_1^2)^{h^{-1}} J \in F(R).$$

The latter inclusion results from the fact that if $(I_1 J_1 I_1^2)^{h^{-1}} v = 0$, then $I_1 J_1 I_1^2 v^h = 0$ and, hence, $v^h = 0 = v$. Thus, $gh^{-1} \in A(R)$. The lemma is proved.

1.7.4. Lemma. *Let* $I \in F$. *Then* $A(I) = A(R)$.

Proof. Let $g \in A(R)$ and $R \supseteq J^g \supseteq J_1$ for suitable ideals. In this case $IJ_1 \subseteq J_1 \subseteq J^g$, and, therefore, $IJ_1 = A^g$ for a certain subset $A \subseteq J$. The ideal $JAIJ$ belongs to $F(I)$, since the equality $Av = 0$ implies $IJ_1 v^g = 0$. Then we have $(JAIJ)^g = J^g IJ_1 (IJ)^g \subseteq I$. Finally,

$$(JAIJ)^g \supseteq J_1 IJ_1 (IJ^2)^g \supseteq J_1 IJ_1 (IJ)^g I_1 \in F(R),$$

since $(IJ)^g \subseteq J^g \subseteq R$ and the equality $(IJ)^g v = 0$ implies $IJv^{g^{-1}} = 0$, i..e., $v = 0$. Therefore, $g \in A(I)$.

Inversely, if $g \in A(I)$ and $I \supseteq J^g \supseteq J_1$ for suitable ideals from $F(I)$, then

$$R \supseteq I \supseteq (IJI)^g \supseteq (J^3)^g \supseteq J_1^3 \supseteq IJ_1^3 I < R,$$

and $IJI,\ IJ_1^3 I \in F(R)$. The lemma is proved.

1.7.5. Definition. Let g be an automorphism from $A(R)$. Let us denote by Φ_g a set of all the elements of ring R_F, such that for any $x \in R$ the equality $\varphi x^g = x\varphi$ holds

$$\Phi_g = \{\varphi \in R_F \mid \forall\ x \in R\ \ \varphi x^g = x\varphi\}.$$

If h is another automorphism, then one can easily prove that

$$\Phi_g \Phi_h \subseteq \Phi_{gh}, \tag{11}$$

and, besides, if h is an identical automorphism, then $\Phi_h = C$ and, hence, Φ_g is a module over C. The module Φ_g will be called a *conjugation module* for the automorphism g (see lemma **1.7.7** below).

1.7.6. Lemma. *$\Phi_g \subseteq Q$ and the equality $\varphi x^g = x\varphi$ holds for all $x \in R_F$, $\varphi \in \Phi_g$.*

Proof. Let $x \in R_F$ and I be an ideal from F, such that $Ix \subseteq R$. Then for all $\varphi \in \Phi_g$, $i \in I$ we have $(ix)\varphi = \varphi(ix)^g = (\varphi i^g)\, x^g = i\varphi\, x^g$, herefrom we get $I(x\varphi - \varphi x^g) = 0$, i.e., $x\varphi = \varphi x^g$.

Let φ be a nonzero element from Φ_g. Then one can choose an ideal $I \in F$, such that $I\varphi \subseteq R$. By lemma **1.7.4**, the automorphism g lies in the group $A(I)$, i..e., one can find an ideal $J \in F(I)$, such that $J^g \supseteq J_1 \in F(I)$. We get

$$\varphi I J_1 I \subseteq \varphi J_1 \subseteq \varphi J^g = J\varphi \subseteq I\varphi \subseteq R,$$

and, hence, $\varphi \in Q$, which is the required proof.

1.7.7. Lemma. *Let g be an automorphism from group $A(R)$. Then Φ_g is a cyclic C-module $\Phi_g = \varphi_g C$ in which case the element φ_g*

is invertible in the ring $e(\Phi_g)C$ *and the action of* g *on* $e(\Phi_g)R_F$ *coincides with conjugation by the element* φ_g.

Proof..Let us show that any element $\varphi \in \Phi_g$ is invertible in the ring $e(\varphi)Q$ and the action of g on $e(\varphi)R_F$ coincides with conjugation by φ.

Let us consider a mapping $\zeta: x\varphi \to xe(\varphi)$, where x runs through Q. This mapping is well defined: if $x\varphi = 0$, then $xR\varphi = x\varphi R^g = 0$, and, hence, $xe(\varphi) = 0$ (see **1.5.6**). The mapping obeys all the requirements of point (4) of **1.5.9**. Indeed, by lemma **1.7.3** we can find an ideal $I \in F$, such that $I^{g^{-1}} \subseteq R$ and, by the definition of R_F an ideal $J \in F$, such that $J\varphi \subseteq R$. Then $JI^{g^{-1}}\varphi = J\varphi I$ is an ideal of the ring R, lying in the domain of the ξ definition, in which case its annihilator in R coincides with that of φ.

Therefore, one can find an element $b \in R_F$, such that $x\varphi b = xe(\varphi)$. Substituting into it $x = 1$, we get $\varphi b = b^{g^{-1}}\varphi = e(\varphi)$. Thus, φ is right and left invertible, i..e., $b^{g^{-1}} = b = \varphi^{-1}$, where we have, by construction, $e(b) = e(\varphi)$. The latter equality affords $e(\varphi)^g = e(b)^g = e(b^g) = e(b) = e(\varphi)$. If now $x \in e(\varphi)R_F$, then we have

$$x^g = e(\varphi)^g x^g = e(\varphi)x^g = b\varphi x^g = bx\varphi = \varphi^{-1}x\varphi.$$

In order to find the generator of the module Φ_g, in Φ_g let us introduce a relation of a partial order by the formula $\varphi_1 \le \varphi_2 \Leftrightarrow e(\varphi_1)\varphi_2 = \varphi_1$. If $\{\varphi_\alpha, \alpha \in A\}$ is a linearly ordered subset, then we put $e_\alpha = e(\varphi_\alpha) + 1 - \sup\{e(\varphi_\alpha)\}$. In this case $\sup e_\alpha = 1$ and $\varphi = \lim\limits_A \varphi_\alpha \in \Phi_g$, as Φ_g is a closed set (it is equal to the intersection of all the kernels of quite continuous mappings $\beta_x: a \to ax^g - xa$), in which case $\varphi \ge \varphi_\alpha$ for all α.

Therefore, according to the Zorn lemma, one can find a maximal element $\varphi \in \Phi_g$. Let $\psi \in \Phi_g$. Then $\varphi + (1 - e(\varphi))\psi \ge \varphi$, and, therefore, $\psi = e(\varphi)\psi$, i..e., $\psi = \varphi(\varphi^{-1}\psi)$. It remains to remark that $\varphi^{-1}\psi \in C$, as

$\varphi^{-1} \in \Phi_{g^{-1}}$. The lemma is proved.

1.7.8. Definition. Hereafter the idempotent $e(\Phi_g) = e(\varphi)$, where $\Phi_g = \varphi C$ will be denoted by $i(g)$. This idempotent singles out the part of the ring R_F on which the automorphism g acts in an inner way, i..e., there exists a decomposition

$$R_F = R_1 \oplus R_2, \quad R_1 = i(g) R_F, \quad R_2 = (1 - i(g)) R_F,$$

such that $g|_{R_1} \in IutR_1$, $\Phi_{g|R_2} = 0$. In particular, we have the equality $i(g) = i(g^{-1})$.

1.7.9. Corollary. *If R is a prime ring, then $\Phi_g \neq 0$ iff g is an inner automorphism for Q.. In this case the space Φ_g is one-dimensional over C.*

The importance of modules Φ_g and of the above mentioned decomposition is determined by the following theorem.

1.7.10.Theorem. *Let the elements $a_1, \dots, a_n$ of the ring R_F be such that $a_1 \notin a_2\Phi_{g_2^{-1}g_1} + \dots + a_n\Phi_{g_n^{-1}g}$ where $g_1, g_2, \dots, g_n$ are automorphisms from $A(R)$. Then there exists a natural k and elements $v_1, \dots, v_k, t_1, \dots, t_k \in R$, such that*

$$a = \sum_{j=1}^{k} v_j a_1 t_j^{g_1} \neq 0, \quad \sum_{j=1}^{k} v_j a_i t_j^{g_j} = 0, \text{ where } 2 \leq i \leq n.$$

The proof of this theorem will be deduced from lemma **1.7.12.** But let us first make some definitions.

1.7.11. Definition. Let Z be a ring of integer numbers. Let us denote through $L = L(R)$ a subring in the tensor product $Q \otimes_Z Q^{op}$ generated by elements of the type $1 \otimes r^{op}$, $r \otimes 1$ where the rings Q and Q^{op} are mutually antiisomorphic but have the same additive groups, while the elements r and r^{op} run through the rings R and R^{op}, respectively. We also assume that $r \to r^{op}$ is an identical antiisomorphism. If $g = A(R)$,

$r \in R_F$ $\beta = \sum_i r_i \otimes a_i^{op} \in \mathbb{L}$ then let us set $r \cdot \beta = \sum_i a_i r r_i$, and $\beta^g = \sum_i r_i^g \otimes a_i$.

If Λ is a subset of $\mathbb{L}$, then let

$$\Lambda^{\perp g} = \{ r \in R_F \mid \forall_\beta \in \Lambda,\ r \cdot \beta^g = 0 \}.$$

If $r \in R_F$, then let us denote by $r^{\perp g}$ a totality of all elements $\beta \in \mathbb{L}$, such that $r \cdot \beta^g = 0$

1.7.12. Lemma. *Let Λ be a right ideal of the ring $\mathbb{L}$; $g, g_1, \ldots, g_n$ be a sequence of (not necessarily different) automor-phisms. Then for elements $r_1, \ldots, r_n$ of the ring R_F the following equality holds:*

$$\Lambda^{\perp g} + \sum_{i=1}^{n} r_i \Phi_{g_i^{-1}} = (\bigcap_i r_i^{\perp g_i} \cap \Lambda)^{\perp g}. \tag{12}$$

Proof. Let us first show that the left part is contained in the right one. To prove it, it is sufficient to show that if $\beta = \sum s_k \otimes f_k \in r_i^{\perp g_i}$, then $r_i \varphi \cdot \beta^g = 0$ for all $\varphi \in \Phi_{g_i^{-1} g}$. We have

$$r_i \varphi \cdot \beta^g = \sum_k f_k r_i \varphi s_k^g = (\sum_k f_k r_i s_k^{g_i}) \varphi = 0.$$

Proof of an inverse inclusion will be carried out by induction over the number n. If $v \in (r_1^{\perp g_1} \cap \Lambda)^{\perp g}$ and $\beta \in \Lambda \cap r_1^{\perp g_1}$ then $v \cdot \beta^g = 0$ and, therefore, $\varphi\colon r_1 \cdot \beta^{g_1} \to v \cdot \beta^g$, $\beta \in \Lambda$ is a correct mapping from the subset $r_1 \cdot \Lambda^{g_1}$ to R_F. Moreover, φ is a homomorphism of left R-modules. Indeed,

$$[r(r_1 \cdot \beta^{g_1})]\varphi = [r_1 \cdot (\beta(1 \otimes r))^{g_1}]\varphi =$$
$$= v(\beta(1 \otimes r))^g = r(v \cdot \beta^g) = r[(r_1 \cdot \beta^{g_1})\varphi].$$

Let, then, I_1 be an ideal from $\boldsymbol{F}$, such that $R \supseteq I_1^{g^{-1}}$. Then for all $r \in I_1$ we have the equalities

$$\begin{aligned} &[(r_1 \cdot \beta^{g_1}) r]\varphi = [r_1(\beta(r^{g_1^{-1}} \otimes 1))g_1]\varphi = \\ &= v \cdot (\beta(r^{g_1^{-1}} \otimes 1))^g = (v \cdot \beta^g) r^{g_1^{-1}} = \\ &= [(r_1 \cdot \beta^{g_1})\varphi] r^{g_1^{-1}g}. \end{aligned} \tag{13}$$

Let us now choose an ideal $I \in \boldsymbol{F}$ in such a way that the inclusions $R \supseteq J^g \supseteq I$, $R \supseteq J_1^{g_1} \supseteq I$, $Ir_1 \subseteq R$, $Iv \subseteq R$ be fulfilled, where J, J_1 are suitable ideals from $\boldsymbol{F}$. In this case $I^{g^{-1}} \subseteq J$ and $I^{g_1^{-1}} \subseteq J_1$, and, therefore, $IR^g \subseteq (I^{g^{-1}}R)^g \subseteq (JR)^g \subseteq J^g \subseteq R$, and, in the same way, $IR^{g_1} \subseteq R$. Let us restrict the domain of the definition of the mapping φ, i..e., let us assume that β runs through $\Lambda(1 \otimes I^2)$. In this case the domain of the definition V forms an ideal of the ring R, while the domain of the values lies in R. Let us extend φ on $V + ann\, V \in \boldsymbol{F}$, assuming $\varphi(ann\, V) = 0$. Then, due to property (3) of the ring $R_{\boldsymbol{F}}$ (see 1.4.9), one can find an element $\xi \in R$, such that $(ann\, V)\xi = 0$, $a\xi = a\varphi$ for all $a \in V$. Now equality (10) shows that for all $\beta \in \Lambda(1 \otimes I^2)$ the following equality holds:

$$(r_1 \cdot \beta^{g_1})(r\xi - \xi r^{g_1^{-1}g}) = 0.$$

Besides, $(ann\, V)(r\xi - \xi r^{g_1^{-1}g}) = 0$ and, hence, $r\xi = \xi r^{g_1^{-1}g}$, i..e., $\xi \in \Phi_{g_1^{-1}g}$.

Then, if $\beta \in \Lambda(1 \otimes I^2)$ we have $(v - r_1\xi) \cdot \beta^g = v \cdot \beta^g - (r_1 \cdot \beta^{g_1})\xi = v \cdot \beta^g - (r_1 \cdot \beta^{g_1})\varphi = 0$, and, therefore, $v \in r_1\Phi_{g_1^{-1}g} + (\Lambda(1 \otimes I^2))^{\perp g}$. It remains now to remark that $\Lambda^{\perp g} = (\Lambda(1 \otimes I^2)^{\perp g}$. This is valid because $r \cdot [\Lambda(1 \otimes I^2)]^g = I^2(r \cdot \Lambda^g)$.

Therefore, the lemma is proved for $n=1$.

Let us now assume that

$$\Lambda^{\perp g} + \sum_{i=1}^{n-1} r_i \Phi_{g_i^{-1} g} = \Big(\bigcap_1^{n-1} r_i^{\perp g_i} \cap \Lambda \Big)^{\perp g} \overset{df}{=} \Lambda_1^{\perp g} \tag{14}$$

Then, due to the case $n=1$

$$\Lambda_1^{\perp g} + r_n \Phi_{g_n^{-1} g} = (r_n^{\perp g_n} \cap \Lambda_1)^{\perp g} .$$

However, $\Lambda_1 \cap r_n^{\perp g_n} = \bigcap_1^n r_i^{\perp g_i} \cap \Lambda$ and, by (11), we get

$$\Lambda^{\perp g} + \sum_{i=1}^{n-1} r_i \Phi_{g_1^{-1} g} + r_n \Phi_{g_n^{-1} g} = \Big(\bigcap_1^n r_i^{\perp g_i} \cap \Lambda \Big)^{\perp g} ,$$

which is the required proof.

If we now set $g = g_1$, $\Lambda = \mathbb{L}(R)$, then the theorem results from the lemma proved above. Indeed, in this case the condition of the theorem implies that a_1 does not belong to the left part of (9) at $i \geq 2$, while the conclusion implies that a_1 does not belong to the right part of equality (9) at $i \geq 2$.

1.7.13. Theorem. *If elements* $a_1, \dots, a_n$ *of the ring* R_F *are such that* $a_1 \notin \sum_{i \geq 2} \Phi_{g_1^{-1} g_i} a_i$ *then there exist elements* $v_j, t_j \in R$, $1 \leq j \leq m$, *such that*

$$a = \sum_j v_j^{g_1} a_1 t_j \neq 0, \quad \sum_j v_j^{g_i} a_i t_j = 0, \quad 2 \leq i \leq n .$$

This theorem is obtained from the preceding one by transition to an antiisomorphic ring R^{op}, since $\Phi_h(R^{op}) = \Phi_{h^{-1}}(R)$, where $r \to r^{op}$ is an identical antiisomorphism.

The above theorems are also valid in the case when all the automorphisms are identical, in which case all the modules Φ_g coincide with the generalized centroid, and we get the characterization of a linear independence.

1.7.14. Corollary. *If elements* $a_1, \dots, a_n$ *belong to* R_F *and* $a_1 \notin a_2 C + \dots + a_n C$ *then there can be found elements* $v_j, t_j \in R$, $1 \le j \le k$, *such that*

$$\sum v_j a_1 t_j \neq 0; \ \sum v_j a_i t_j = 0, \ \ 2 \le i \le n.$$

Herefrom one can easily deduce the proof of lemma **1.5.8** promised earlier. Indeed, let $e(a) \ge e(b)$ and $axb = bxa$ for all $x \in R$. Then, if $b \notin aC$, there can be found elements $v_j, t_j \in R$, $1 \le j \le k$, such that $b_1 = \sum v_j b t_j \neq 0$; $\sum v_j a t_j = 0$. We have

$$axb_1 = \sum_j axv_j bt_j = \sum bxv_j at_j = bx \sum v_j at_j = 0.$$

It means that $e(a) \cdot e(b_1) = 0$, which contradicts the fact that $e(b_1) \le e(b) \le e(a)$.

Inversely, if $b = ac$, then $axb = axac = acxa = bxa$ and $e(b) = e(a)e(c) \le e(a)$. The lemma is proved.

1.7.15. Definition. An element $a_1 \in R_F$ is said to be *right independent of elements* $a_2, \dots, a_n \in R_F$ *with respect to a sequence of automorphisms* $g_1, \dots, g_n \in A(R)$, if

$$a_1 \notin \sum_{i \ge 2} a_i \Phi_{g_i^{-1} g_1}.$$

Theorem **1.7.10** can now be reformulated in the following way: *if* a_1 *is right independent of* $a_2, \dots, a_n$ *with respect to* $g_1, \dots, g_n \in A(R)$, *then there exists an element* $\beta \in \mathbb{L}(R)$, *such that*

$a_1 \cdot \beta^{g_1} \neq 0,\ a_2 \cdot \beta^{g_2} = 0, \ldots, a_n \cdot \beta^{g_n} = 0.$

1.8. Extension of Derivations to a Ring of Quotients

1.8.1. Proposition. *Let μ be a derivation determined on the essential ideal I of a semiprime ring R, with its values in R. In this case μ can be uniquely extended to the derivation of ring Q and that of ring R_F.*

Proof. Let $r \in R_F$. Let us consider an ideal $I_r \in F$, such that $I_r r \subseteq R$ and denote the ideal $(II_r)^2$ through J. Let us consider the homomorphism of the left I-modules $\xi: J \to R$, acting by the formula $\xi(j) = (jr)^{\mu} - j^{\mu} r$. It should be remarked that $Jr \subseteq II_r \cdot r \subseteq I$, and, hence, $(jr)^{\mu}$ is determined. Besides, $j^{\mu} \in [(II_r)^2]^{\mu} \subseteq II_r I^{\mu} + I^{\mu} II_r \subseteq I_r$, and, therefore, $j^{\mu} r \in R$. Hence, due to property (3) of the ring R_F (see **1.4.9**), we find an element $r_1 \in R_F$, such that $jr_1 = (jr)^{\mu} - j^{\mu} r$.

Let us set $r^{\mu} = r_1$ and show that μ is a derivation of R_F. Let $r, s \in R_F$ and let us choose the ideal $J_1 = (II_r I_s)^2$. Then for any $j \in J_1$ we have the relations

$$(js)r^{\mu} = (jsr)^{\mu} - (js)^{\mu} r,$$
$$js^{\mu} r = (js)^{\mu} r - j^{\mu} sr,$$
$$j(sr)^{\mu} = (jsr)^{\mu} - j^{\mu} sr.$$

Adding the first two equalities and subtracting the third one, we get

$$J_1(sr^{\mu} + s^{\mu} r - (sr)^{\mu}) = 0,$$

from which, by property (2) of the ring R_F (see **1.4.9**), we get

$$(sr)^{\mu} = s^{\mu} r + sr^{\mu},$$

i.e., μ is a derivation of R_{F}.

Then, if $r \in Q$ and $rI' \subseteq R$, we have

$$r^{\mu} \cdot i' = (r \cdot i')^{\mu} - r(i')^{\mu} \in R,$$

where $i' \in (II')^2$. Therefore, $r^{\mu} \in Q$ and μ is a derivation of the ring Q.

And, finally, the uniqueness of the extension follows from the equalities

$$j(r^{\mu} - r^{\mu_1}) = jr^{\mu} - jr^{\mu_1} =$$
$$= (jr)^{\mu} - (jr)^{\mu_1} - j^{\mu} r + j^{\mu_1} r = 0,$$

where $j \in (II_r)^2$. The lemma is proved.

1.8.2. It follows from the lemma proved above that any derivation of the ring R has a unique extension on Q and R_{F}. From now on we shall therefore consider that the derivations are determined on Q. Let us introduce the following notation:

$$D(R) = \{\mu \in Der\, Q \mid \exists\ I \in F,\ I^{\mu} \subseteq R\}. \tag{15}$$

It is clear that the derivations of the ring R are contained in $D(R)$. All the further statements are valid not only for the derivations of the ring R, but also for those from $D(R)$ and, hence, if the word 'derivation' is not followed by the notation of a ring, the derivation will be assumed to belong to $D(R)$.

1.8.3. Lemma. *Let a ring R have a simple or zero characteristic. Then the set $D(R)$ forms a differential Lie C-algebra which contains all the inner derivations of ring Q.*

Proof. Let $q \in Q$ and I be an ideal from F, such that $Iq \cup qI \subseteq R$, in which case

$$I^{ad\, q} \subseteq qI + Iq \subseteq R$$

and, hence, the inner derivation conforming to the element q lies in $D(R)$.

A set of all derivations of the ring Q forms a differential Lie algebra over its center C. It, therefore, remains to check if $D(R)$ is a module over C closed with respect to the Lie operations.

If $c \in C$ and $I_c c \subseteq R$, then

$$((II_c)^2)^{\mu c} = ((II_c)^2)^{\mu} c \subseteq I_c c \subseteq R.$$

Moreover, if $\mu, \nu \in D(R)$ and I, J are the corresponding ideals, then

$$((IJ)^2)^{\mu\nu} \subseteq ((IJ)^{\mu} IJ + IJ(IJ)^{\mu})^{\nu} \subseteq J^{\nu} \subseteq R.$$

Hence, $D(R)$ is closed relative to the Lie operations $[\mu, \nu] = \mu\nu - \nu\mu$, $\mu^{[p]} = \mu \cdot \ldots \cdot \mu$. The lemma is proved.

1.8.4. Lemma. *Let* $I \in F$. *Then* $D(I) = D(R)$.

Proof. If $\mu \in D(R)$ and $J^{\mu} \subseteq R$, then

$$(IJ^2 I)^{\mu} \subseteq (IJ)^{\mu} JI + IJ(JI)^{\mu} \subseteq I$$

and, hence, $\mu \in D(I)$. The inverse inclusion is obvious to the same extent.

1.8.5. Lemma. *The automorphisms from* $A(R)$ *act on* $D(R)$ *by the formula*

$$\mu^g = g^{-1} \mu g.$$

Proof. The fact that μ^g is a derivation of ring Q is proved by direct checking.

Let us show that $\mu^g \in D(R)$. Let us choose an ideal $I \in F$, such that $I^{\mu} \subseteq R$. Reducing, if necessary, I one can assume that at the same time $I^g \subseteq R$. Using the equality $A(I) = A(R)$ we can find an ideal $J \in F$, such that $J^{g^{-1}} \subseteq I$. Now we have

$$(J^2)^{\mu^g} \subseteq (I^2)^{\mu g} \subseteq I^g \subseteq R.$$

The lemma is proved.

1.9. The Canonical Sheaf of a Semiprime Ring

In this paragraph we shall present a closure $\hat{RE}$ in ring R_F as a ring of global sections of a certain sheaf over the structure space of a generalized centroid. Such a presentation is useful since it enables one to see clearly the theoretic-functional intuition employed when studying a semiprime ring. It is worth while adding that in the case of a prime ring all these constructions degenerate and their essence and meaning is to reduce the process of studying semiprime rings to that of studying prime rings (i.e., stalks of a canonical sheaf).

Roughly speaking, a ring of global sections of a sheaf is a set of all continuous functions on a topological space, the only difference being that at every point these functions assume the values of their own rings. Let us give the exact definitions.

1.9.1. Let there be a topological space X, with its any open set U put in correspondence to a ring (group or, more generally, an object of a certain fixed category) $\mathfrak{R}(U)$, and let for any two open sets $U \subset V$ be given a homomorphism

$$\rho_U^V : \mathfrak{R}(V) \to \mathfrak{R}(U).$$

This system is called a *presheaf of rings* (groups or, more generally, objects of a fixed category), provided the following conditions are met:

(1) if U is empty, then $\mathfrak{R}(U)$ is a zero ring (a unit group, a trivial object);

(2) ρ_U^U is an identical mapping;

(3) for any open sets $U \subseteq V \subseteq W$ we have $\rho_U^W = \rho_U^V \rho_V^W$.

This presheaf will be denoted by one letter, $\mathfrak{R}$. The simplest example of a presheaf is a presheaf of all functions on X with the values in a ring A. In this case $\mathfrak{R}(U)$ consists of all functions on U with the values in A, and for $U \subset V$ the homomorphism ρ_U^V is a restriction of the function determined on V to the subset U.

In order to extend the intuition of this example to the case of any presheaf, the homomorphisms ρ_U^V are called *restriction* homomorphisms.

The elements (of a ring) $\mathfrak{R}(U)$ are called *sections* of the presheaf

$\mathfrak{R}$ over U. Sections of $\mathfrak{R}$ over X are called *global*. Therefore, $\mathfrak{R}(U)$ is a ring of sections of the presheaf R over U, $R(X)$ is a ring of global sections.

Returning to the example of the presheaf of all functions on X, let us assume the topological space X to be a union of open sets U_α. Then any function on X is uniquely determined by its restrictions on sets U_α and, besides, if on every U_α a function f_α is given, and the restrictions of f_α and f_β coincide on $U_\alpha \cap U_\beta$, then there exists a function on X, such that every f_α is its restriction on U_α. These properties may be formulated for any presheaf and they single out an extremely important class of presheaves.

1.9.2. Definition. A presheaf (of rings) $\mathfrak{R}$ on a topological space is called a *sheaf* (of rings) if for any open set $U \subset X$ and its any open cover $U = \cup U_\alpha$ the following conditions are met:

(1) if $\rho^U_{U_\alpha} s_1 = \rho^U_{U_\alpha} s_2$ for $s_1, s_2 \in \mathfrak{R}(U)$ and all α, then $s_1 = s_2$;

(2) if $s_\alpha \in \mathfrak{R}(U_\alpha)$ are such that for any α, β the restrictions of s_α and s_β on $U_\alpha \cap U_\beta$ coincide, then there exists an element $s \in \mathfrak{R}(U)$, whose restriction on U_α is equal to s_α for all α.

Now we shall try to consider an arbitrary sheaf as a sheaf of functions on space X. To this end it is necessary to determine the value of $s(x)$ for the section $s \in \mathfrak{R}(U)$ at any point $x \in U$. We can determine elements $s(V) = s\rho^U_V$ for open neighborhoods V of point x which are contained in U and, it would hence be natural to consider their 'limit', and so we have to introduce a direct limit

$$\mathfrak{R}_x = \varinjlim_{x \in V} \mathfrak{R}(V)$$

relative to the system of homomorphisms

$$\rho^U_V : \mathfrak{R}(U) \to \mathfrak{R}(V).$$

This limit is called a *stalk* or the sheaf (or a presheaf) $\mathfrak{R}$ at the point x. Therefore, a stalk element at the point x is set by any section over the open neighborhood of x, but in this case the two sections, $u \in \mathfrak{R}(U)$, $v \in \mathfrak{R}(V)$ are identified if their restrictions on an open neighborhood of x coincide.

For any open set $U \ni x$ a natural homomorphism

$$\rho^U_x : \mathfrak{R}(U) \to \mathfrak{R}_x$$

can be determined, which maps a section to the stalk element determined by it. Now we can determine the *value of section* s *at the point* x as $\rho^U_x(s)$.

If $\mathfrak{R}$ is a sheaf and if two sections, $u_1, u_2 \in \mathfrak{R}(U)$ are such that $\rho^U_x(u_1) = \rho^U_x(u_2)$ for all points $x \in U$, then $U_1 = U_2$. Therefore, for sheaf $\mathfrak{R}$ the elements from $\mathfrak{R}(U)$ can be given by the function $s : U \to \bigcup_{x \in U} \mathfrak{R}_x$, $s(x) \in \mathfrak{R}_x$ (or, equivalently, by the elements of a direct product $\prod_U \mathfrak{R}_x$). It is evident that not every such function s determines the section. The following condition is necessary and sufficient:

> *for any point* $x \in U$ *there exists such an open neighborhood* W *of the point* x *contained in* U, *and an element* $w \in \mathfrak{R}(W)$, *such that* $s(y) = \rho^W_y(w)$ *for all points* $y \in W$. (*)

In fact, a topology can be introduced on the set $\bigcup_{x \in X} \mathfrak{R}_x$ in such a way that condition (*) becomes equivalent to the continuity of function s.

1.9.3. Let us now go over to constructing a *canonical sheaf.* Let C be a generalized centroid of a semiprime ring R. Let us denote through X a set of all its prime ideals other than C. This set is called a *spectrum* of ring C and is often denoted by $Spec\ C$. The elements of the spectrum of ring C will be called *points of the spectrum* or simply *points.* In the theory of commutative rings there is a difference between a simple spectrum and a maximal one, but, however, as is shown by the following lemma, there is no such a difference in our case.

1.9.4. Lemma. *If* $P \in Spec\ C$, *then the factor-ring* $C_p = C / p$ *is a field, i.e., any prime ideal of ring* C *proves to be maximal.*

Proof. Let a be a preimage of a nonzero element at a natural homomorphism $C \to C_p$. In this case, due to regularity of ring C we can find an idempotent e, such that $ea = a$ and $e = aa'$ for some $a' \in C$. In terms of images it means $\bar{a} \cdot \bar{a}' = \bar{e}$. However, $\bar{e}C_p(1 - \bar{e}) = 0$, and, hence, either $\bar{e} = 0$ or $\bar{e} = 1$. The first case is impossible, as $\bar{a} = \bar{e} \cdot \bar{a} \neq 0$ and, hence, $\bar{a} \cdot \bar{a}' = 1$. Thus, C_p is a field, which is the required proof.

1.9.5. In order to set a topology on X, it is necessary to define the operation of closure. For the set $A \subset X$ let us define the closure $\bar{A}$ as a union of all the points containing the intersection $\bigcap_{p \in A} p$. The topology obtained in this way is called *spectral topology*. With the view of constructing a sheaf over X, it is useful to know the structure of open sets in X (they are also called *domains*).

1.9.6. Definition. If e is a central idempotent, then through $U(e)$ we denote a set of all points p, such that $e \notin P$.

1.9.7. Allowing for the fact that the product $e(1-e)=0$ belongs to any prime ideal, we see any point of the spectrum contain either e or $1-e$, but not both simultaneously. Therefore,

$$U(e) \cup U(1-e) = X, \quad U(e) \cap U(1-e) = \varnothing \tag{16}$$

On the other hand, the closure of $U(e)$ contains only the point q containing the intersection $\bigcap_{e \notin p} p = \bigcap_{1-e \in p} p$. The latter intersection contains the element $1-e$ and, hence, $1-e \in q$, which affords $e \notin q$, i.e., by the definition, $q \in U(e)$. Now relations (13) show $U(e)$ to be an open and closed set simultaneously.

1.9.8. Lemma. *The sets* $U(e)$, $e \in E$ *form a fundamental system of open neighborhoods of* X, *i.e., any open set is presented as a union of sets of type* $U(e)$.

Proof. Let W be an open set of points, $A = \bigcap_{p \notin W} p$. By $E(A)$ let us denote a set of all idempotents from A, and let us consider a family of neighborhoods $U(e)$, $e \in E(A)$. If $p \in U(e)$, $e \in E(A)$, then $e \notin p$, and, hence, $p \not\subset A$, i.e., p does not belong to the closure of the complement of W. Since the complement is closed, the condition $p \in U(e)$ implies $p \in W$. Therefore, $\bigcup_{e \in E(A)} U(e) \subseteq W$.

Let us show that the inverse inclusion is also valid. Let, on the contrary, $p \in W$, $p \notin \cup U(e)$. In this case all the idempotents from A lie in p. Since A is an ideal, then, since the ring C is regular, it contains, together with the element a, its support $e(a) = a \cdot a'$. This support lies in p and, hence, $a = e(a)\, a \in p$. Therefore, p contains a and, hence, it belongs to the closure of the complement of W. As the complement is closed, $p \notin W$, which is a contradiction.

The lemma proved shows, in particular, that the space X is *quite disconnected,* and its any open set being a union of closed domains.

1.9.9. Lemma. *Any closed domain in* X *has the form* $U(e)$ *for an idempotent* e.

Proof. Let U be an open and closed set. Let us consider its complement U' up to X. Let $A = \bigcap_{p \in U} p$, $B = \bigcap_{p \in U'} p$. The intersection of these ideals is zero and they, hence, form a direct sum. If a proper prime ideal q of ring C contains the sum $A + B$, then it belongs to the closure U and the closure U' simultaneously, which is impossible. It implies $A + B = C$. In particular, the unit of the ring C is presented as a sum $1 = e_1 + e$, where $e_1 \in A$, $e \in B$. Since the sum of the ideals $A + B$ is direct, then e_1, e are idempotents, and, hence, $B = eC$, $A = (1 - e)C$. Since U is closed, then $p \in U$ iff $p \supset A$, i.e., $p \ni 1 - e$, which is equivalent to the condition $e \notin p$. Thus, $U = U(e)$. The lemma is proved.

1.9.10. The latter lemma shows a set of central idempotents to be in a one-to-one correspondence with a set of closed domains of X. One can easily prove that this correspondence preserves the lattice operations and order relation:

$$U(e_1 e_2) = U(e_1) \cap U(e_2);\ U(e_1 + e_2 - e_1 e_2) = U(e_1) \cup U(e_2),\ U(e_1) \subseteq U(e_2) \Leftrightarrow e_1 \leq e_2. \tag{17}$$

This peculiarity allows one in a number of cases to identify idempotents with closed domains and consider idempotents as the ones consisting of points, which makes many considerations extremely vivid. The correspondence under discussion also preserves the exact upper boundaries

$$U(\sup_\alpha \{e_\alpha\}) = \overline{\bigcup_\alpha U(e_\alpha)}, \tag{18}$$

where on the right we have a closure of the union of the domains $U(e_\alpha)$.

Let us prove this equality. Let $B_\alpha = \bigcap_{p \in U(e_\alpha)} p$. Then point q belongs to the right part of equality (15) iff $q \supset \cap B_\alpha$. It should be remarked that $B_\alpha = (1 - e_\alpha)C$, since $p \in U(e_\alpha) \Leftrightarrow 1 - e_\alpha \in p$, and the factor-ring $C/(1 - e)C \cong e_\alpha C$ is semiprime.

Let us consider now in C the topology determined by the module structure. The ideals B_α are closed in this topology and, hence, their intersection B is also closed. The idempotents from B form a directed family whose limit e lies in B. As for any element $b \in B$ its support $e(b) = bb'$ lies in B, then $e(b) \le e$, and, hence, $B = eC$.

It should be remarked that $e = 1 - \sup_\alpha e_\alpha$. Indeed, on the one hand, $1 - \sup_\alpha e_\alpha \le 1 - e_\alpha$ for all α and, hence, $1 - \sup_\alpha e_\alpha \in B$, i.e., $1 - \sup_\alpha e_\alpha \le e$. On the other hand, $e \in B$ and, hence, $e \in (1 - e_\alpha)C$, i.e., $ee_\alpha = 0$ for all α. Therefore, $e \sup_\alpha e_\alpha = 0$ and, hence, $e \le 1 - \sup_\alpha e_\alpha$. Thus, $e = 1 - \sup_\alpha e_\alpha$ and, hence, the ideal q belongs to the right part of equality (15) iff $1 - \sup_\alpha e_\alpha \in q$. The latter inclusion is equivalent to $q \in U(\sup_\alpha e_\alpha)$ and thus formula (15) is proved.

1.9.11. Lemma. *The closure of an open set of space* $X = Spec\ C$ *is open (and, naturally, closed).*

The proof results immediately from lemma **1.9.8** and formula (15).

1.9.12. It should be recalled that a topological space is called *extremely disconnected* if a closure of any domain is open. We thus showed $Spec\ C$ to be extremely disconnected.

1.9.13. It should be also recalled that the space X is *compact and Hausdorff*. If $X = \bigcup_\alpha U(e_\alpha)$, then any point does not contain at least one of the idempotents e_α. It means that the sum of ideals $e_\alpha C$ is not contained in any proper prime ideal and, hence, $\sum_\alpha e_\alpha C = C$. In particular, $1 = e_{\alpha_1} c_1 + \dots + e_{\alpha_n} c_n$. Therefore, any ideal containing all e_{α_i}, also contains the unit. Thus, $X = \bigcup_i U(e_{\alpha_i})$, and the compactness is established.

If now $\mathfrak{p}_1, \mathfrak{p}_2$ are different points, then we can find elements $a_1 \in p_1 \setminus p_2$, $a_2 \in p_2 \setminus p_1$ (it should be recalled that all prime ideals are maximal). For the supports we also have inclusions $e(a_1) \in p_1 \setminus p_2$, $e(a_2) \in p_2 \setminus p_1$. Let us consider the idempotents $e_1 = e(a_1)(1 - e(a_2))$; $e_2 = e(a_2)(1 - e(a_1))$. These are orthogonal idempotents and $e_1 \notin p_2$, $e_2 \notin p_1$, i.e., $\mathfrak{p}_2 \in U(e_1)$, $\mathfrak{p}_1 \in U(e_2)$. It now remains to remark that the intersection of domains $U(e_1)$ and $U(e_2)$ is equal to $U(e_1 e_2) = U(0) = \emptyset$.

1.9.14. Now we are completely ready to determine the *canonical sheaf* $\Gamma = \Gamma(R)$.

Let U be an open set. By formula (15), its closure $\bar{U}$ is open and has the form $U(e)$ for some $e \in E$. Let us set

$$\Gamma(U) \overset{df}{=} \Gamma(\bar{U}) \overset{df}{=} e\,\hat{RE}\,. \tag{19}$$

Since the inclusion $W \subseteq U$ implies $\bar{W} \subseteq \bar{U}$, and this inclusion, by formula (14) gives the inequality $f \leq e$ for the corresponding idempotents $U(f) = \bar{W}$, $U(e) = \bar{U}$, then defined is the homomorphism $x \to xf$ acting from $e\,\hat{RE}$ onto $f\,\hat{RE}$, which will be viewed as the restriction homomorphism ρ^U_W.

The validity of axioms (1)-(3) of a presheaf for Γ is absolutely obvious.

1.9.15. Theorem. *The determined above presheaf* Γ *is a sheaf.*

Proof. Let $U = \bigcup_\alpha U_\alpha$ and $\rho^U_{U_\alpha}(s) = 0$ for all α and for a given $s \in \Gamma(U) = e\,\hat{RE}$. Let also $\bar{U}_\alpha = U(e_\alpha)$. The equalities $\rho^U_{U_\alpha}(s) = 0$ will be written as $se_\alpha = 0$. Since by formula (15) we have $e = \sup e_\alpha$, then $s = se = s \cdot \sup_\alpha e_\alpha = 0$, which is the required proof, of the first condition.

Let now $s_\alpha \in \Gamma(U_\alpha)$ be such that for any α, β the restrictions s_α and s_β coincide on $U_\alpha \cap U_\beta$. It means that $s_\alpha e_\alpha = s_\alpha$ and

$s_\alpha e_\beta = s_\beta e_\alpha$ for all α, β. Let us consider the set M of all pairs (s, e'), such that $se' = s$, $e' \le e$ and $se_\alpha = s_\alpha e'$ for all α. In this set let us introduce the order relation $(s, e') \le (s_1, e'_1) \Leftrightarrow s_1 e' = s$, $e' \le e'_1$, and show that M transforms into an inductive partially-ordered subset.

If $\{(s_\gamma, e_\gamma)\}$ is a linearly-ordered subset, then, due to the completeness of $\hat{RE}$ the limit $\tilde{s} = \lim s_\gamma$ with respect to the system of idempotents $e_\gamma + 1 - \sup\limits_\gamma e_\gamma$ is also determined. In this case $(\tilde{s}, \tilde{e})$, where $\tilde{e} = \sup e_\gamma$, is the upper boundary for $\{(s_\gamma, e_\gamma)\}$ and $\tilde{s} e_\alpha = (\lim\limits_\gamma s_\gamma) e_\gamma = \lim\limits_\gamma s_\alpha e_\gamma = s_\alpha \lim\limits_\gamma e_\gamma = s_\alpha \cdot \tilde{e}$, i.e., $(\tilde{s}, \tilde{e}) \in M$.

Let $(\hat{s}, \hat{e})$ be a maximal element from M.

Let us show that $\hat{e} = e$. If this is not the case, then $\hat{e} e_\alpha \ne e_\alpha$ for some e_α (it should be recalled that $e = \sup\limits_\alpha e_\alpha$) and the pair $(\hat{s} + s_\alpha(e_\alpha - \hat{e} e_\alpha), \hat{e} + e_\alpha - \hat{e} e_\alpha)$ strictly exceeds $(\hat{s}, \hat{e})$ and lies in M. Therefore, $\hat{e} = e$, i.e., $\hat{s} \in \Gamma(U)$ and $\rho^U_{U_\alpha}(\hat{s}) = \hat{s} e_\alpha = s_\alpha \hat{e} = s_\alpha e = s_\alpha$. Thus, $\hat{s}$ is the sought element. The theorem is proved.

1.9.16. So, we have achieved the posed goal, i.e., the ring $\hat{RE}$ is presented as a ring of global section of the sheaf Γ. This sheaf will henceforth be called a *canonical sheaf* of the ring R and, if there is a necessity to underline its relation with ring R it will be denoted by $\Gamma(R)$. This sheaf obeys another important condition: its any section can be extended up to the global one (the sheaves obeying this condition are called *flabby*) and, moreover, the homomorphisms of the restrictions are retractions, so that the ring of global sections naturally contains all the rings of local sections $e\,\hat{RE} \subseteq \hat{RE}$.

1.9.17. We have seen the structure of a canonical sheaf to be completely defined by the structure of the E-module of the ring $\hat{RE}$. In an analogous way, to any complete (injective) module M over any self-injective regular ring E there can be associated a certain sheaf $\mathbb{M}$:

$$\mathbb{M}(U) = \mathbb{M}(\bar{U}) = Me, \; \rho^V_U(x) = xe,$$

where $\bar{U} = U(e)$.

In conclusion of this paragraph we shall make two remarks on stalks of a canonical sheaf.

1.9.18. Lemma. *Section s belongs to the kernel of the natural homomorphism ρ_p iff the support of the element s belongs to p or, which is equivalent, $s \in p\,\hat{RE}$. The stalk Γ_p can be associated with the factor-ring*

$$\hat{RE} \,/\, \hat{RE} \cap p\,\hat{RE}.$$

Proof. Let section s lie in kernel ρ_p. Then, by definition, there is a neighborhood $W = U(e)$ of point p, such that the restriction of s onto its neighborhood is equal to zero, i.e., $se = 0$. It affords $e(s)\,e = 0$ and, hence, $e(s) \le 1 - e \in p$, since $e \notin p$. Therefore, $e(s) \in p$ and $s = s\,e(s) \in p\,\hat{RE}$.

Inversely, if $s \in p\,\hat{RE}$, then element s is presented as a sum $\alpha_1 s_1 + \dots + \alpha_n s_n$, where $\alpha_i \in p$. Allowing for the fact that, together with the two idempotents e_1, e_2, the ideal p also contains their upper boundary $e_1 \vee e_2 = e_1 + e_2 - e_1 e_2$, we get

$$\sup e(\alpha_i) = \sup(\alpha_i \alpha'_i) \in p,$$

where the elements α'_i exist due to the regularity of ring C. Moreover, $s \cdot \sup e(\alpha_i) = s$, which implies $e(s) = e(s) \cdot \sup e(\alpha_i) \in p$. Thus, point p belongs to the neighborhood $U(1 - e(s))$, the restriction of s on which equals zero. The lemma is proved.

Using the metatheorem (see **1.11**) we shall show below that all the stalks of a canonical sheaf are prime rings. It can be proved with no methods of the theory of models used. We recommend the reader to try to do it, and the more efforts are made in order to achieve this purpose, the more pleasure will be derived from reading the paragraph on the metatheorem.

1.10 Invariant Sheaves

In this paragraph we shall adjust the canonical sheaf to studying groups of automorphisms. Let R be a semiprime ring, G be a group of automorphisms $G \subseteq A(R)$. Choosing an arbitrary closed domain U of the spectrum we see that the group G cannot be a group of automorphisms for a ring of sections if $U^g \not\subset U$ for a certain $g \in G$ and, hence, the canonical sheaf stops being a sheaf if the ring is considered together with the action of the group G.

This obstacle will be eliminated by going over to another similar sheaf, which will be called *G-invariant* or simply *invariant* provided the group G is fixed.

It should be remarked that the group G acts on the space $X = Spec\, C$ in such a way that the corresponding transformations are homeomorphisms of the topological space C.

1.10.1. Definition. *The orbit* $\overline{p}$ of a point $p \in X$ is a set of all its images under the action of the group G.

$$\overline{p} = \{ p^g,\ g \in G \}.$$

One can easily see that any two orbits either coincide or do not intersect, i.e., the whole space X is presented as a disjunctive union of orbits. Let us denote by X / G a set of all orbits and consider the mapping $\pi\colon X \to X / G$, which puts the point p into correspondence to its orbit $\overline{p}$.

1.10.2. Definition. A *space of orbits* X / G is a set of orbits having the weakest topology, for which the mapping π is continuous and open (i.e., X / G is a factor-space of X in the topological sense).

1.10.3. Lemma. *A set is open in* X / G *iff it is an image of a domain under the mapping* π.

Proof. The images of domains must be open since π is an open mapping. These images define a topology and it suffices to remark that π is continuous in this topology; if U is a domain from X, then $\pi^{-1}(\pi(U)) = \bigcup_{g \in G} U^g$ is also a domain. The lemma is proved.

1.10.4. Definition. A *G-invariant sheaf* is a sheaf over the space of orbits X/G, the ring of sections of which over a domain W is equal to $\Gamma(\pi^{-1}(W))$, while the homomorphisms ρ^V_W are equal to $\rho^{\pi^{-1}(V)}_{\pi^{-1}(W)}$.

Since $\pi^{-1}(W)$ is an invariant set of points, and the automorphisms from G are the homomorphisms of the space X, then the closure $\overline{\pi^{-1}(W)}$ is an invariant domain. Consequently, the idempotent e which corresponds to this domain is fixed relative the action of the group G. We see the structure of an invariant sheaf to be determined by that of the module on the ring $\hat{RE}$ over the ring E^G of invariant idempotents.

1.10.5. Lemma. *The ring* E^G *is self-injective.*

Proof. As E^G is an intersection of kernels of quite continuous mappings $x \to x - x^g$, $g \in G$ then this set is closed in the topology determined by the idempotents from E, and, moreover, it is closed in the topology determined by its own idempotents. By lemma **1.6.16** and lemma **1.6.20**, E^G is a self-injective Boolean ring. The lemma is proved.

According to remark **1.9.17**, one can define a sheaf over the space $Spec\, E^G$, the section rings and restriction homomorphisms for which coincide with those of an invariant sheaf. In essence, they are the same sheaves, though the spaces X/G and $Spec\, E^G$ may be not homeomorphic. The natural mapping $\overline{p} \to p^G$ is open and continuous but it can prove to be no one-to-one correspondence.

1.10.6. Notations. Since rings of section of an invariant sheaf are exactly those of a canonical sheaf over invariant domains, it would be reasonable to denote an invariant sheaf with the same letter, Γ, as a canonical one. It does not result in any ambiguity, since the stalk of an invariant sheaf over the orbit $\overline{p}$ will be designated by $\Gamma_{\overline{p}}$, while the stalk of a canonical sheaf at the point p by Γ_p.

1.10.7. Lemma. *A section* s *belongs to the kernel of a natural homomorphism* $\rho_{\overline{p}}$ *iff the support of the element* s *belongs to the ideal* $p^G C$ *or, which is equivalent,* $s \in p^G \hat{RE}$. *The stalk* Γ_p *can be identified with a factor-ring*

$$\hat{RE} / p^G \hat{RE} \cap \hat{RE}.$$

Proof. Let section s lie in the kernel $\rho_{\bar{p}}$. Then, by definition, there can be found a neighborhood W of the orbit $\bar{p}$, such that the restriction of s on $\overline{\pi^{-1}(W)}$ equals zero, i.e., $se = 0$, where $U(e) = \overline{\pi^{-1}(W)}$. It affords $e(s)\,e = 0$ and, hence, $e(s) \leq 1 - e \in p^G$, since on the one hand, $\overline{\pi^{-1}(W)}$ is the neighborhood of the point p, while on the other hand, $1 - e$ is a fixed idempotent. Consequently, $e(s) \in p^G E$ and $s = se(s) \in p^G \hat{RE}$.

Inversely, if $s \in p^G \hat{RE}$, then the element s can be presented as a sum $\alpha_1 s_1 + \ldots + \alpha_n s_n$ where $\alpha_i \in p^G$. It should be remarked that the supports e_i of an element α_i also belong to p^G: if $\alpha^g = \alpha$, then n $e(\alpha)^g = e(\alpha^g) = e(\alpha)$ and, besides, $e(\alpha) = \alpha\alpha'$, where the element α' is determined by the condition of regularity of the ring C.

Then, together with its two idempotents, e_1, e_2, the set p^G also contains its exact upper boundary, $e_1 \vee e_2 = e_1 + e_2 - e_1 e_2$, and, therefore, $e = \sup e_i \in p^G$.

Besides, $s = s \cdot \sup e_i$, i.e., the restriction of s on the invariant neighborhood $U(1 - e)$ equals zero. This neighborhood contains the point p, as $e(s) \leq \sup e_i \in p$. Hence, the restriction of the section s on the neighborhood $W = \pi(U(1 - e))$ of the orbit $\bar{p}$ equals zero. The lemma is proved.

1.10.8. Lemma. *The space X / G is extremely disconnected.*

Proof. Let W be a domain of the space of an orbit. In this case $U = \pi^{-1}(W)$ is an invariant domain of the spectrum. Since the elements of G act on X as homeomorphisms, the closure $\bar{U}$ will be an invariant domain, i.e., $\pi(\bar{U}) \cap \pi(X \setminus \bar{U}) = \varnothing$ and, hence, $\pi(\bar{U})$ is a closed domain of the space of orbits.

It is evident that the domain contains a closure $\bar{W}$. Moreover, since π is continuous, then $\pi^{-1}(\bar{W})$ is a closed set containing $U = \pi^{-1}(W)$ and, hence, $\bar{U}$ as well, i.e., $\bar{W} = \pi(\bar{U})$ is a closed domain, which is the required proof.

1.11 The Metatheorem

Our task now is to describe a possibly widest class of properties, theorems, statements, which can be transferred from prime rings onto semiprime ones using either canonical or invariant sheaves. Let us limit ourselves by the statements, which can be presented in the elementary language, i.e., those pertaining to elements. If a statement on n elements has already been presented and we know a way of elucidating whether this statement is true for a given n-tuple, then this situation is equivalent to setting an n-ary predicate.

1.11.1. It should be recalled that an *n-ary predicate* on a set A is a mapping P of the Cartesian degree A^n onto a two-element set $\{T, F\}$ ($T \leftrightarrow$ 'truth', $F \leftrightarrow$ 'falsity') and, an *n-ary operation* is a mapping f from A^n to A.

Predicates of the same name (i.e., those with one name, P, and the same arity, n) or operations of the same name can be defined on different sets. In this case we speak about the values of the same predicate, P (operation f) on different sets.

A *signature* is a set Ω of the names of predicates and operations put in correspondence to arities. An *algebraic system* of the signature Ω is a set with the given values of the predicates and operations from Ω of the corresponding arity.

Sometimes both zero-ary operations and zero-ary predicates are considered. A zero-ary operation on set A is a fixed element of this set, while a zero-ary predicate is either truth or falsity.

The predicates and operations from Ω defined on the algebraic system of this signature are called *principal* or *basic*. With their help one can construct new operations, the so-called *termal functions*, and new predicates, i.e., *formula* ones, termal functions and terms being defined through induction by construction.

To begin with, projections of an arbitrary arity $(a_1,..., a_n) \to a_i$ which have standard designations $x_i, y_i, z_i,..$ pertain to termal functions. Besides, if F is an n-ary basic operation and $\varphi_1,..., \varphi_n$ are termal functions of the same arity m, then $F(\varphi_1,..., \varphi_n)$ is an m-ary termal function, which has a standard designation, $F(\phi_1,..., \phi_n)$, where $\phi_1,..., \phi_n$ are standard designations for $\varphi_1,..., \varphi_n$. A standard designation of a termal function is called a *term*. It is common belief that there are no other termal functions.

For instance, *if* $\Omega = \{+, -, \cdot\}$ is a ring signature, then terms are the arbitrary non-associative polynomials of variables x_i, while termal functions are the mappings $(a_1,\ldots, a_n) \rightarrow f(a_1,\ldots, a_n)$, where an element $f(a_1,\ldots, a_n)$ is obtained from the polynomial f through the substitution $x_i = a_i \in R$.

Generalized polynomials can be also viewed as terms. For this purpose it is necessary to extend the signature, putting in correspondence to every coefficient $r \in R$ a zero-ary operation a_r, whose value on R is set equal to r.

The simplest formula predicates are those set by atomic formulas

$$P(x_1,\ldots, x_n), \quad x_1 = x_2, \quad F(x_1,\ldots, x_n) = x_{n+1},$$

where P, F are basic predicates. The formula predicate is given by a formula of the elementary language, i.e., it is 'obtained' from simplest formula predicates by employing logical connectives, $\&, \vee, \neg, \rightarrow$ and by quanti-fication by $\exists, \forall$ of subject variables x_i (which are, in our case, identical with the names of certain termal functions).

1.11.2. Let $f_1,\ldots, f_n$ be terms of the arity m, $P \in \Omega$. Then the formulas

$$P(\hat{f}_1,\ldots, \hat{f}_n) \quad , \quad \hat{f}_1 = \hat{f}_2 \tag{20}$$

give certain predicates: $P(\hat{f}_1,\ldots, \hat{f}_n)(a_1,\ldots, a_m) = T$ iff $P(f_1(a_1,\ldots, a_m), \ldots, f_n(a_1,\ldots, a_m)) = T$, and, analogously, $(\hat{f}_1 = \hat{f}_2)\ (a_1,\ldots, a_m) = T$, iff $f_1(a_1, \ldots, a_m) = f_2(a_1,\ldots, a_m)$.

By induction through the construction of terms one can easily prove that these predicates can be obtained from simplest formula predicates only through quantification by $\exists$ and the use of the connective $\&$. For instance, if $f_1 = F_1(\varphi_1,\ldots, \varphi_k)$, $f_2 = F_2(\psi_1,\ldots, \psi_2)$, then the formula $\hat{f}_1 = \hat{f}_2$ is equivalent to the following formula

$$\exists\, y_1 \ldots y_{k+1} \exists\, z_1 \ldots z_{l+1} (y_{k+1} = F_1(y_1,\ldots, y_{l+1}) \;\&\; z_{l+1} = F_2(z_1,\ldots, z_l) \;\&\; z_1 = \phi_1 \& \ldots \& z_l = \phi_l \;\&\; y_1 = \phi_1 \& \ldots \& y_k = \phi_k).$$

The subformulas $z_i = \Phi_i$ and $y_i = \Phi_i$ incorporated in the above formula, set the formula predicates by the inductive supposition.

1.11.3. Let us describe the most important for us class, that of Horn formulas. The simplest Horn formulas are the following ones:

$$A_1 \& A_2 \& \ldots \& A_p \to A_{p+1}; \quad A_i; \quad \neg A_1 \vee \neg A_2 \vee \ldots \vee \neg A_p,$$

where A_i are formulas of type (17). An arbitrary Horn formula is a quantorization of the conjunction of simplest Horn formulas. The predicate set by a Horn formula is called a *Horn predicate*.

1.11.4. Definition. A flabby sheaf $\mathfrak{R}$ over an extremely disconnected space X is called *correct* if $\mathfrak{R}(U) = \mathfrak{R}(\bar{U})$ for any domain U, i.e., the mappings $\rho_U^{\bar{U}}$ are isomorphisms.

It should be recalled that a sheaf is called *flabby* if for any domain $U \subset V$ the mappings ρ_U^V (or, equivalently, ρ_U^X) are epimorphisms.

According to definitions **1.9.14** and **1.10.4**, canonical and invariant sheaves are correct (see also **1.9.1** and **1.10.8**).

1.11.5. Let us now fix a correct sheaf $\mathfrak{R}$ of algebraic systems of a signature Ω (the reader can view $\mathfrak{R}$ as a sheaf of rings in an extended signature).

1.11.6. Definition. An n-ary predicate $P(x_1, \ldots, x_n)$ will be called *strictly sheaf* if its values are given on all the systems (rings) of sections and stalks of the sheaf $\mathfrak{R}$, in which case $\mathfrak{R}$ remains to be a sheaf in the category of algebraic systems with the additional predicate P.

Analogously, an n-ary operation f will be referred to as *strictly sheaf*, if the $(n+1)$-ary predicate $P(x_1, \ldots, x_{n+1})$ given by the formula $f(x_1, \ldots, x_n) = x_{n+1}$ is strictly sheaf.

These conditions mean that both restrictions ρ_U^V and natural homomorphisms ρ_t remain homomorphisms in the category of algebraic systems obtained by adding the predicate symbol P (or the functional one, f) into the signature Ω, in which case ρ_t remains a direct limit of ρ_U^V, $t \in U$, and for any covering $U \subseteq \cup U_\alpha$ the system of homomorphisms $\rho_{U_\alpha}^U$ remains approximating.

1.11.7. Let us write the definition of a strictly sheaf predicate in more detail.

(a) If $P(s_1,..., s_n) = T$ for $s_i \in \mathfrak{R}(V)$, then $P(s_1\rho^V_U,..., s_n\rho^V_U) = T$ for all $U \subseteq V$.

(b) $P(\bar{s}_1,..., \bar{s}_n) = T$ for $\bar{s}_i \in \mathfrak{R}_t$ iff there exists a neighborhood W of a point t and preimages $s_i \in \mathfrak{R}(W)$, such that $P(s_1,..., s_n) = T$.

(c) If $U = \cup U_\alpha$ and $s_1,..., s_n \in \mathfrak{R}(U)$, in which case $P(s_1\rho_\alpha,..., s_n\rho_\alpha) = T$ for all $\rho_\alpha = \rho^U_{U_\alpha}$, then $P(s_1,..., s_n) = T$.

The importance of strictly sheaf predicates and operations is governed by the fact that they can be included into a signature, in which case all the theorems proved for sheaves (the metatheorem given below, in particular) remain valid, thus expanding significantly the possibilities of their application.

1.11.8. Zero-ary operations. Let s be a zero-ary strictly sheaf operation; $s(U)$, s_t be its values on $\mathfrak{R}(U), \mathfrak{R}_t$, respectively. If $s_0 = s(X)$, then, by conditions (a), (b), we have $s(U) = s_0\,\rho^X_U$, $s_t = s_0\,\rho_t$. Inversely, if s is a global section, then the zero-ary operation $s(U) = s\,\rho^X_U$, $s_t = s\,\rho_t$ will be strictly sheaf.

Therefore, setting a zero-ary strictly sheaf operation is equivalent to singling out a global section.

1.11.9. Unary predicates and operations. A unary predicate is determined by a set of the elements on which it is true. Inversely, with any subset S in an algebraic system there can be associated a predicate P_S: $P_S(x) = T$ iff $x \in S$.

Let us now elucidate for what sets $S \subseteq \widehat{RE}$ the predicate P_S will be strictly sheaf for an invariant sheaf. Condition (a) implies $SE^G \subseteq S$. Condition (b) implies that the value of the predicate P_S on a stalk is determined by the set $\rho_{\bar{p}}(S)$. Finally, in the terms of idempotents condition (c) looks as follows: if $\sup_\alpha \{e_\alpha\} = e$ and $e_\alpha x \in S$ for all $\alpha \in A$, then $ex \in S$, where $e, e_\alpha \in E^G$.

In particular, if S is a closed submodule of the E^G-module $\widehat{RE}$, then the predicate P_S will be strictly sheaf provided its values on the stalks are determined by the formula $P_S(\bar{x}) = T \leftrightarrow \bar{x} \in \rho_{\bar{p}}(s)$. Here only condition (c) can be doubtful, since A is not a directed set in it. Let B be a set of all finite subsets A, ordered by inclusion. Let us set $e_\beta = \sup_{\alpha \in \beta} \{e_\alpha\} + (1 - e)$. Then $xe_\beta \in S$ since $x \sup\{e_1, e_2\} = xe_1 + xe_2 - xe_1 e_2$. Hence, $xe = \lim_B xe_\beta \in S$, which is the required proof.

Going now over to the module viewpoint, we come to the conclusion that any *injective* E^G*-submodule* S *in* $\hat{RE}$ *determines a strictly sheaf predicate* P_S.

We know that an injective submodule is singled out by a direct addent $\hat{RE} = S \oplus W$ and the E^G-module projection $\pi_S\colon \hat{RE} \to S$ is, therefore, defined. Viewing π_S as a unary operation, one can easily see that it is a strictly sheaf operation.

Thus, *any closed* E^G*- submodule in* $\hat{RE}$ *determines a strictly sheaf predicate* P_S *and a strictly sheaf projection* $\pi_S\colon \hat{RE} \to S$ *for an invariant sheaf.*

Setting $G = \{1\}$, we get an analogous statement for a canonical sheaf.

1.11.10. Composition, support. According to the definition, strictly sheaf are the operations of addition, subtraction and multiplication, i.e., the predicates given by the formulas

$$x + y = z,\ x - y = z,\ xy = z,\ x = y.$$

Propositions **1.11.15** and **1.11.16** to be proved below and remark **1.11.2** show that any termal operation, as well as any predicate of type $\hat{f}_1 = \hat{f}_2$ or $P(\hat{f}_1, \ldots, \hat{f}_n)$, where P is a strictly sheaf predicate, will be strictly sheaf, i.e., one can say that the composition of strictly sheaf operations and predicates will be strictly sheaf.

Strictly sheaf will also be the function of support, $\boldsymbol{e}\colon s \to \boldsymbol{e}(s)$. Incidentally, if the ring $\hat{RE}$ has no unit, then this function will not be an operation and, strictly speaking, it cannot be included into the signature. Nevertheless, in this case defined is the binary operation $(x, y) \to \boldsymbol{e}(x)\, y$, which will be strictly sheaf.

The values of operation $\boldsymbol{e}$ on the stalks must be defined by condition (b). As the element x belongs to the kernel ρ_p iff its support belongs to

p, then for a canonical sheaf we have $e(x) = 1$ for all nonzero elements of the stalk.

1.11.11. Definition. Let us call an n-ary predicate P *sheaf* provided its values are given on all the rings of sections and on all the stalks of sheaf $\mathfrak{R}$ and the following conditions are met.

(b') If $P(\bar{s}_1,..., \bar{s}_n) = T$ for $\bar{s}_i \in R_t$, then there exists a neighborhood W of the point t and preimages $s_i \in \mathfrak{R}(W)$, such that $P(s_1\rho^W_U,..., s_n\rho^W_U) = T$ for any domain U included in W.

(c') If $U = \cup U_\alpha$ and $s_1,... s_n \in \mathfrak{R}(U)$, in which case $P(s_1\rho^U_{V_\alpha},..., s_n\rho^U_{V_\alpha}) = T$ for any domains $V_\alpha \subseteq U_\alpha$, then $P(s_1,..., s_n) = T$.

The following statement shows that it is sheaf predicates that are of primary interest for us.

1.11.12. Proposition. *Let P be a sheaf predicate, $s_1,... s_n$ be sections over a domain U. Then, if a set of these points $t \in U$, for which $P(s_1\rho_t,..., s_n\rho_t) = T$ is dense in U, then $P(s_1,..., s_n) = T$. In particular, if a certain theorem is set by a zero-ary sheaf predicate and is valid on all (or nearly all) stalks, then it is also valid on the ring of global sections.*

The proof can be directly obtained by first applying condition (b') to the stalks on which predicate P is true, followed by employing condition (c').

1.11.13. Metatheorem. *Any Horn predicate is a sheaf one. In particular, if the formulation of a theorem can be presented as a Horn formula, then the truthfulness of this theorem on nearly all the stalks of a correct sheaf implies its truthfulness on a system (ring) of global sections of this sheaf.*

The importance of this theorem is strengthened by the fact that in the class of prime rings any elementary formula is equivalent to a Horn one. Indeed, when constructing Horn formulas, it is forbidden to use only disjunction, but in the class of prime rings $f = 0 \vee g = 0$ is equivalent to $\forall\, x\; fxg = 0$. Or, in more exact terms, it is necessary to reduce the quantorless part to a conjunctive normal form, and then the subformulas of the type

$$f \neq 0 \vee f_2 \neq 0 \vee ... \vee f_k \neq 0 \vee g_1 = 0 \vee ... \vee g_n = 0$$

should be replaced with Horn formulas

$$\forall\, x_1 x_2 .. x_{n-1}(f_1 = 0 \& .. \& f_n = 0 \to g_1 x_1 g_2 .. x_{n-1} g_n = 0).$$

The proof of the metatheorem will follow from the lemmas and propositions presented below.

1.11.14. Lemma. *Let* $\{U_\alpha, \alpha \in A\}$ *be a nonempty family of closed domains of an extremely disconnected space* X. *Then there exists a set of mutually disjoint closed domains* $\{U_\beta, \beta \in B\}$, *such that the union* $\bigcup_B U_\beta$ *is dense in* $\bigcup_A U_\alpha$ *in which case for every* $\beta \in B$ *one can find* $\alpha(\beta) \in A$, *such that* $U_\beta \subseteq U_{\alpha(\beta)}$.

Proof. Let us consider a set M of all the sets satisfying the lemma conditions with a possible exception of the density in the union. This set is nonempty, since any one-element set $\{U_{\alpha_0}\}$ belongs to M, where $\alpha_0 \in A$. Let us consider M as a partially ordered set by inclusion. In this case M is inductive and, according to the Zorn lemma, there exists a maximal element $\mathcal{M} = \{U_\beta, \beta \in B\}$. If now a point t belongs to the union $\bigcup_{\alpha \in A} U_\alpha$, but does not belong to the closure of $\bigcup_{\beta \in B} U_\beta$, then there is a domain $W \ni t$, which does not intersect with the domain $\overline{\cup U_\beta}$. Let $t \in U_\alpha$ and $W_\alpha = W \cap U_\alpha$. Then $\overline{W_\alpha} \cap \overline{\cup W_\beta} = \varnothing$ and the set $\{\bar{W}_\alpha, U_\beta | \beta \in B\}$ belongs to M and is strictly greater than $\mathcal{M}$. The lemma is proved.

1.11.15. Proposition. *If* P, Q *are (strictly) sheaf predicates, then* $P \& Q$ *is a (strictly) sheaf predicate.*

Proof. By definition, predicate $P \& Q$ is true on the n-tuple $s_1,.., s_n$ iff both predicates, P and Q, are true on it. Herefrom follows the validity of statements (a) and (b) for predicate $P \& Q$ when P and Q are strictly sheaf, as well as the validity of (c') for sheaf predicates P and Q.

Let us check (b') for sheaf predicates. Let $(P \& Q)(\bar{s}_1,.., \bar{s}_n) = T$ for $\bar{s}_i \in \mathfrak{R}_t$. Then we have $P(\bar{s}_1,.. \bar{s}_n) = T$ and $Q(\bar{s}_1,.., \bar{s}_n) = T$ at the same time. Consequently, there can be found neighborhoods U, U' of point t and preimages $s'_1,.., s'_n \in \mathfrak{R}(U)$, $s''_1,.., s''_n \in \mathfrak{R}(U')$, such that $P(s'_1 \rho^U_W, .., s'_n \rho^U_W) = T$ and $Q(s''_1 \rho^{U'}_{W'},.., s''_n \rho^{U'}_{W'}) = T$ for all domains $W \subset U$,

$W' \subset U'$. Since elements s'_i and s''_i determine the same element on the stalk, they coincide on a certain neighborhood. Consequently, there is a neighborhood $V \subseteq U \cap U'$ of the point t, such that $s_i \stackrel{df}{=} s'_i\, \rho^U_V = s''_i\, \rho^{U'}_V$ for all i. Therefore, for any domain $W \subseteq V$ both predicates are true on $s_1\rho^V_W,\ldots, s_n\rho^V_W$, i.e., $P\&\, Q(s_1\rho^V_W,\ldots, s_n\rho^V_W) = T$.

Let us check the inverse statement (b) for strictly sheaf predicates. If $P\&\, Q(s_1,\ldots, s_n) = T$, then both predicates, P and Q are true on $s_1,\ldots, s_n$, and, hence, they are true on $\bar{s}_1,\ldots, \bar{s}_n$. Hence, $P\&\, Q(\bar{s}_1,\ldots, \bar{s}_n) = T$. The proposition is proved.

1.11.16. Proposition. *Let* $P(x, x_1,\ldots, x_n)$ *be a (strictly) sheaf* $(n+1)$*-ary predicate. Then the* n*-ary predicates* $\exists\, xP(x, x_1,\ldots, x_n)$ *and* $\forall\, xP(x, x_1,\ldots, x_n)$ *are also (strictly) sheaf.*

Proof. By definition, the predicate $\exists\, xP$ is true on the n-tuple $s_1,\ldots, s_n$ iff there exists an element s from a corresponding ring of sections or a stalk, such that predicate P is true on $s, s_1,\ldots s_n$. The validity of statement (a) for predicate $\exists\, xP$ follows obviously from the same statement for predicate P.

The validity of $\exists\, xP(x, \bar{s}_1,\ldots, \bar{s}_n)$ is equivalent to the existence of an element $\bar{s} \in \mathfrak{R}_t$, such that $P(\bar{s}, \bar{s}_1,\ldots, \bar{s}_n) = T$. Since predicate P is strictly sheaf, there exists a neighborhood $W \ni t$ and preimages $s_1,\ldots s_n$, such that $P(s\rho^W_V,\ldots, s_n\rho^W_V) = T$ for all domains $V \subseteq W$, i.e., it implies the validity of $\exists\, xP(x, s_1,\ldots s_n)$ on $\mathfrak{R}(V)$.

Let us, finally, check condition (c'). Let $U \subset \bigcup_{\alpha \in A} U_\alpha$ and $s_1,\ldots s_n \in \mathfrak{R}(U)$, in which case $\exists\, xP(x, s_1\rho^U_{V_\alpha},\ldots, s_n\rho^U_{V_\alpha}) = T$ for any domains $V_\alpha \subset U_\alpha$. Then for any α there can be found a section $s_\alpha \in \mathfrak{R}(U_\alpha)$, such that $P(s_\alpha\rho^{U_\alpha}_{V_\alpha}, s_1\rho^{U_\alpha}_{V_\alpha},\ldots) = T$.

By lemma **1.11.14**, U contains a dense disjunctive union of domains U_β, in which case $U_\beta \subseteq U_{\alpha(\beta)}$ for suitable $\alpha(\beta) \in A$. This inclusion and the fact that the predicate P is sheaf show that $P(s_{\alpha(\beta)}\rho^{U_{\alpha(\beta)}}_{V_\beta},...) = T$ for any domains $V_\beta \subset U_\beta$. Since the domains U_β are disjoint, then by the definition of a sheaf, there is an element $s \in \mathfrak{R}(\bigcup_\beta U_\beta) = \mathfrak{R}(\overline{\bigcup U_\beta}) = \mathfrak{R}(U)$, such that $s\rho^U_{V_\beta} = s_{\alpha(\beta)}\rho^{U_{\alpha(\beta)}}_{V_\beta}$ and, therefore,

$$P(s\rho^U_{V_\beta}, s_1\rho^U_{V_\beta}, ..., s_n\rho^U_{V_\beta}) = T$$

Condition (c') applied to predicate P shows the latter to be true on the $(n+1)$-tuple $s, s_1,..., s_n$, i.e., $\exists xP(x, s_1,..., s_n) = T$, which is the required proof.

Let us consider a predicate $\forall xP$. By definition, it is true on the n-tuple $s_1,..., s_n$ iff for any s from the corresponding ring of sections or a stalk, the predicate P is true on $s, s_1,..., s_n$. Since the mapping ρ^U_V, because the sheaf is flabby, are epimorphisms, then the validity of statement (a) for predicate $\forall xP$ is obvious when P is strictly sheaf. Statements (c) and (c') for $\forall xP$ follows from the same statement for P.

Let us check condition (b'). Let us assume $\forall xP(x, \bar{s}_1... \bar{s}_n) = T$ on stalk $\mathfrak{R}(t)$. Let us fix some global sections $s_1,..., s_n$, such that $s_i\rho_t = \bar{s}_i$. Then for any global section s the predicate P is true on $\bar{s} = s\rho_t, \bar{s}_1,..., \bar{s}_n$, i.e., there exists a neighborhood V_s of the point t, such that for any domain $V \subseteq V_s$ we have

$$P(s\rho^T_V, s_1\rho^T_V, ..., s_n\rho^T_V) = T \tag{21}$$

According to condition (c') for the predicate P, one can find the biggest open set V_s with such a property. Since, by the definition of a correct sheaf, a ring of sections over the domain coincides with that over its closure, then V_s is a closed domain. Let W_s be a complement of V_s in the space X. The domain W_s satisfies the following property.

If W is a nonempty subset of W_s then it contains a nonempty open *subset* W', *such that*

$$P(s\rho^{X}_{W'}, \ldots, s_n\rho^{X}_{W'}) = F. \tag{22}$$

Indeed, in the opposite case the predicate P, by property (c'), will be true on $s\rho^{X}_{V}, \ldots, s_n\rho^{X}_{V}$ for any domain $V \subseteq W \cup V_s$.

Let us show that the closure of a union of W_s, where s runs through all the global sections, does not contain a point t. Let, on the contrary, $t \in \overline{\cup W_s}$. By lemma **1.11.14**, the domain $\cup W_s$ contains a dense disjunctive union of the domains W_β, $\beta \in B$, such that $W_\beta \subseteq W_{s(\beta)}$ for any β and suitable $s(\beta)$. Since the union is disjunctive, there exists a section $\sigma \in \Re(\cup W_\beta) = \Re(\overline{\cup W_\beta})$, such that its restriction on W_β coincides with $s(\beta)\rho^{X}_{W_\beta}$ at any β. As the sheaf is flabby, we can assume σ to be a global section with the same properties.

Moreover, the intersection $V_\sigma \cap \overline{\cup W_s}$ is a neighborhood of the point t and, hence, it has a nonempty intersection W with one of the sets W_β. It results in a contradiction, since, on the one hand, condi-tion (19), combined with the equalities $\sigma\beta^{X}_{W} = \sigma\rho^{X}_{W_\beta}\rho W^{W_\beta}_{W} = s(\beta)\rho^{X}_{W}$ implies the existence of a nonempty domain $V \subseteq W$, such that

$$P(s(\beta)\rho^{X}_{V}, s_1\rho^{X}_{V}, \ldots, s_n\rho^{X}_{V}) = F$$

but, on the other hand, the inclusion $V \subseteq W_\beta \subseteq W_{s(\beta)}$, combined with condition (18) implies that the same predicate is true.

Thus, t does not belong to the closure $\overline{\cup W_s}$, and, hence, the complement U of its closure is a neighborhood of t, contained in the intersection of all V_s. Therefore, for any s and any domain $V \subset U$ the predicate P is true on $s\rho^{X}_{V}, \ldots, s_n\rho^{X}_{V}$, i.e., by definition, $\forall x P(x, s_1\rho^{X}_{V}, \ldots, s_n\rho^{X}_{V}) = T$, which is the required proof.

Finally, let us check the remaining part of statement (b) for a strictly sheaf predicate. Let there exist a neighborhood W of a point t and elements $s_1, \ldots, s_n \in \Re(W)$, such that $\forall x P(x, s_1, \ldots, s_n) = T$. Then for any $s \in \Re(W)$

the predicate P is true on $s, s_1,\dots, s_n$, i.e., by condition (b) this predicate is true on $\bar{s}, \bar{s}_1,\dots, \bar{s}_n$. As ρ_t is an epimorphism, then the predicate $\forall\, xP$ is true on the stalk $\mathfrak{R}_t$. The proposition is completely proved.

1.11.17. Proposition. *If P is a strictly sheaf n-ary predicate and Q is a sheaf one, then $P \to Q$ is a sheaf predicate.*

Proof. By definition, the predicate $P \to Q$ is true on $s_1,\dots, s_n$, if the fact that the predicate P is true implies that the predicate Q is also true. In other words, this predicate is true in two cases: either if P is false, or if Q is true.

So, let $P \to Q(\bar{s}_1,\dots, \bar{s}_n) = T$ for $\bar{s}_i \in \mathfrak{R}_t$. If $P(\bar{s}_1,\dots, \bar{s}_n) = T$, then the predicate Q on $\bar{s}_1,\dots, \bar{s}_n$ should also be true. In this case, since Q is a sheaf predicate, one can find a neighborhood W and preimages $s_i \in \mathfrak{R}(W)$, such that $Q(s_1\rho^W_V,\dots, s_n\rho^W_V) = T$ for all domains $V \subseteq W$ and, moreover,

$$P \to Q(s_1\rho^W_V,\dots, s_n\rho^W_V) = T,$$

which is the required proof.

When $P(\bar{s}_1,\dots, \bar{s}_n) = F$, then, since P is strictly sheaf, for any neighborhood $W \ni t$ and any preimages $s_1,\dots, s_n$ we have

$$P(s_1, \dots, s_n) = F.$$

According to properties (a) and (b) of a strictly sheaf predicate, one can find, by fixing global preimages $s_1,\dots, s_n$, the biggest domain V on which P is true. This domain is closed and contains no point t, and, hence, its complement W is a neighborhood of the point t. Since V is maximal, the predicate P is false on any domain $U \subseteq W$, and, hence,

$$P \to Q(s_1\rho^X_V, \dots, s_n\rho^X_V) = T,$$

which is the required proof.

Let us check up statement (c'). Let $U = \cup\, U_\alpha$ and let the predicate $P \to Q$ be true on $s_1\rho^U_{V_\alpha},\dots$ for any domains $V_\alpha \subset U_\alpha$, where $s_i \in \mathfrak{R}(U)$. Let us assume that $(P \to Q)(s_1,\dots) = F$. Then $P(s_1,\dots) = T$ and $Q(s_1,\dots) =$

F. As the predicate P is strictly sheaf, we have $P(s_1\rho^U_{V_\alpha},..) = T$ and, hence, $Q(s_1\rho^U_{V_\alpha},..) = T$. Since Q is a sheaf predicate, then $Q(s_1,..) = T$, which is a contradiction. The proposition is proved.

1.11.18. Proposition. *Let P be a sheaf, and Q be a strictly sheaf predicate, and let us assume that the predicate $P \to Q$ is true on all systems (rings) of sections. Then this predicate is true on all the stalks as well.*

Proof. If the predicate P is false on the stalk $\mathfrak{R}_t$, then the implication $P \to Q$ is true. Let the predicate P be true on the stalk $\mathfrak{R}_t$. As P is sheaf, we can find a neighborhood W of the point t, where P is true. By the condition, the predicate Q must be also true on W. As Q is a strictly sheaf predicate, it is also true in the stalk over t. The proposition is proved.

1.11.19. We can now easily complete the proof of the metatheorem. It should be remarked that the negation $\neg P$ of a strictly sheaf predicate P is equivalent to the implication $P \to x \neq x$ and, hence, $\neg P$ is a sheaf predicate. Propositions **1.11.16** and **1.11.17** prove the predicates given by the simplest Horn formulas to be sheaf. The same propositions imply that conjunctions and quantorizations do not violate sheafness.

1.11.20. Corollary. *Let $P_1, P_2,.., P_n,...$ be a sequence of m-ary Horn predicates, such that the implications $P_n \to P_{n+1}$ are true on certain global sections $s_1,... s_m$ of an invariant sheaf. Then, if for any point $\overline{p}$ one of predicates P_i is true on $\bar{s}_1,..., \bar{s}_m$, then one of predicates P_i is true on $s_1,..., s_m$..*

Proof. By the metatheorem, each point $\overline{p}$ has a neighborhood $W_{\overline{p}}$, such that a predicate $P_{i(\overline{p})}$ is true on $s_1\rho^{W_{\overline{p}}}_U,...$ for any domain $U \subset W_{\overline{p}}$. The domains $W_{\overline{p}}$ cover the whole compact space of orbits. Consequently, it is possible to find a finite number of orbits $\{\overline{p}_1,..\overline{p}_n\}$, such that $X / G = W_{\overline{p}_1} \cup .. \cup W_{\overline{p}_n}$. The predicate P_s is now proved to be fulfilled on $\widehat{RE}$, where $s = \max\{i(\overline{p}_k)\}$.

1.11.21. Setting $G = \{1\}$, we come to the conclusion that an analogous statement is also true for a canonical sheaf.

1.12. Stalks of Canonical and Invariant Sheaves

1.12.1. Lemma. *All stalks of a canonical sheaf are prime rings.*

Proof. Let us fix a point $\boldsymbol{p}$ of the spectrum and assume $\bar{s}_1 \Gamma_{\boldsymbol{p}} \bar{s}_2 = 0$. Then the predicate $P(\bar{s}_1, \bar{s}_2)$ defined by the formula $\forall\, x\ \bar{s}_1 x \bar{s}_2 = 0$ is true on the stalk $\Gamma_{\boldsymbol{p}}$. According to proposition **1.11.16**, it is a strictly sheaf predicate, i.e., there is a neighborhood $U(e)$, $e \notin \boldsymbol{p}$, and preimages $s_1, s_2 \in e\hat{RE}$, such that $s_1(e\hat{RE})s_2 = 0$ and, hence, $Rs_1Rs_2Re = 0$, i.e., $\bullet(s_1) \cdot \bullet(s_2) \cdot \cdot e = 0 \in \boldsymbol{p}$. Since $\boldsymbol{p}$ is a simple ideal and $e \notin \boldsymbol{p}$, then one of the supports, $e(s_1)$ or $e(s_2)$ lies in $\boldsymbol{p}$, i.e., by lemma **1.9.18**, we have $\bar{s}_1 = 0$ or $\bar{s}_2 = 0$, which is the required proof.

1.12.2. Lemma. *All stalks of an invariant sheaf are semiprime rings.*

Proof. Let us fix an orbit $\overline{\boldsymbol{p}}$ and assume $\bar{s}\Gamma_{\overline{\boldsymbol{p}}}\bar{s} = 0$. Then the predicate $P(\bar{s})$ defined by the formula $\forall\, x\ \bar{s}x\bar{s} = 0$ is true on the stalk $\Gamma_{\overline{\boldsymbol{p}}}$. By proposition **1.11.6**, this is a strictly sheaf predicate, i.e., there is a neighborhood W and a preimage $s \in \Gamma(\pi^{-1}(W))$, such that $s(e\hat{RE})s = 0$, where the fixed idempotent e is determined by the domain $\pi^{-1}(W)$. Therefore, $sReRs = 0$, i.e., $e(s)e = 0$, which affords $\bullet(s) \le 1 - e \in \boldsymbol{p}^G$, since the point $\overline{\boldsymbol{p}}$ belongs to the neighborhood $U(e) = \overline{\pi^{-1}(W)}$. By lemma **1.10.7**, the section s belongs to the kernel $\rho_{\overline{\boldsymbol{p}}}$ which is the required proof.

1.12.3. Proposition. *Let $\boldsymbol{p}$ be an arbitrary point of the spectrum. Then the following inclusions are valid:*

$$\Gamma_{\boldsymbol{p}}(R) \subseteq \Gamma_{\boldsymbol{p}}(Q) \subseteq \Gamma_{\boldsymbol{p}}(R_F) \subseteq (\Gamma_{\boldsymbol{p}}(R))_F, \quad \Gamma_{\boldsymbol{p}}(Q) \subseteq Q(\Gamma_{\boldsymbol{p}}(R)).$$

This proposition will result from an analogous statement on an invariant sheaf, since for a trivial group an invariant sheaf coincides with a canonical one.

1.12.4. Proposition. *Let* $\overline{p}$ *be an arbitrary orbit. Then the following inclusions occur:*

$$\Gamma_{\overline{p}}(R) \subseteq \Gamma_{\overline{p}}(Q) \subseteq \Gamma_{\overline{p}}(R_F) \subseteq (\Gamma_{\overline{p}}(R))_F,$$

$$\Gamma_{\overline{p}}(Q) \subseteq Q(\Gamma_{\overline{p}}(R)).$$

1.12.5. Lemma. *Let* I *be an essential ideal of the ring* R, $\overline{p}$ *be an arbitrary orbit. Then* $\rho_{\overline{p}}(\hat{IE})$ *is an essential ideal of the stalk* $\Gamma_{\overline{p}}$.

Proof. Since $\hat{IR}$ is an ideal of the ring $\hat{RE}$, then it is sufficient to show that the annihilator of $\rho_{\overline{p}}(\hat{IE})$ in the stalk is equal to zero. If $\rho_{\overline{p}}(\hat{IE}) \cdot \bar{s} = 0$, then on the stalk the following predicate is true

$$\forall x(P_{\hat{IE}}(x) \rightarrow x\bar{s} = 0),$$

where $P_{\hat{IE}}$ is the predicate determined by the set $\hat{IE}$ (see **1.11.9**). According to the metatheorem, this is a sheaf predicate and, in particular, it is true on a neighborhood W of the orbit $\overline{p}$. Let $\overline{\pi^{-1}(W)} = U(e)$, where e is a fixed idempotent. In this case the fact that the predicate is true implies that for a preimage s of the element $\bar{s}$ the equality $e\hat{IE}s = 0$ holds, which affords $es = 0$, i.e., the restriction of the element s on the neighborhood W equals zero and, hence, $\bar{s} = 0$. The lemma is proved.

Proof of proposition 1.12.4. By lemma **1.10.7**, the kernel of the homomorphism $\rho_{\overline{p}}$ consists of the elements, the supports of which lie in $p^G C$. This condition does not depend on the fact in which of the rings $\hat{RE}$, Q, R_F the element itself lies, and, hence, for the sheaf $\Gamma(R)$ the homomorphism $\rho_{\overline{p}}$ is a restriction of the same homomorphism for the sheaf $\Gamma(Q)$ which is, in its turn, a restriction of $\rho_{\overline{p}}$ for the sheaf $\Gamma(R_F)$. This is a proof of the validity of the first two inclusions of the upper chain.

Let us now check the third inclusion of the upper chain. It would be natural to put into correspondence to an element of the stalk $\Gamma(R_{\mathbf{F}})$, determined by the global section $v \in R_{\mathbf{F}}$, an element of the ring $(\Gamma_{\bar{\mathbf{p}}}(R))_{\mathbf{F}}$ determined by the homomorphism $\tilde{v}\colon \rho_{\bar{\mathbf{p}}}(\widehat{IE}) \to \Gamma_{\bar{\mathbf{p}}}(R)$, which acts by the formula $\tilde{v}(x) = xv + \ker \rho_{\bar{\mathbf{p}}}$ (here I is an ideal from $\mathbf{F}(R)$, such that $Iv \subseteq R$).

It should be remarked that $\rho_{\bar{p}}(\widehat{IE})$ is an essential ideal of the ring $\Gamma_{\bar{\mathbf{p}}}(R)$, and it is necessary to check the correctness of the definition of the action of v and calculate the kernel of the mapping $v + \ker \rho_{\bar{p}} \to \tilde{v}$.

Let us check the correctness. If $a \in \ker \rho_{\bar{p}}$, then $\boldsymbol{e}(a) \le e \in \mathbf{p}^G$ and $\boldsymbol{e}(av) \le \boldsymbol{e}(a)\,\boldsymbol{e}(v) \le e \in \mathbf{p}^G$, and, hence, $(\ker \rho_{\bar{\mathbf{p}}})v \subseteq \ker \rho_{\bar{\mathbf{p}}}$, which proves the correctness.

Let us, finally, assume that $\tilde{v} = 0$. Then $\widehat{IE}\, v \subseteq \ker \rho_{\bar{p}}$, i.e., on the stalk $\Gamma_{\bar{\mathbf{p}}}(R_{\mathbf{F}})$ the predicate

$$\forall\, x(P_{\widehat{IE}}(x) \to x\bar{v} = 0),$$

is true, where $\bar{v} = \rho_{\bar{p}}(v)$. By the metatheorem, this predicate is true on a neighborhood W of the orbit $\bar{p}$. Let $\overline{\pi^{-1}(W)} = U(e)$ and $e \notin \mathbf{p}^G$. Then we have $e\,\widehat{IE} = 0$, i.e., $ev = 0$ and, hence, the restriction of v on the domain W equals zero, i.e., $v \in \ker \rho_{\bar{p}}$, which is the required proof.

The above mentioned embedding $\bar{v} \to \tilde{v}$ determines the inclusion in the lower chain as well: if $v \in Q$ and $vI \subseteq R$, then $v\,\widehat{IE} \subseteq \widehat{RE}$, i.e., $\tilde{v} \in Q(\Gamma_{\bar{\mathbf{p}}}(R))$.

1.12.6. Lemma. *A generalized centroid of the stalk $\Gamma_{\mathbf{p}}(R)$ of a canonical sheaf equals $\rho_p(C) = C/p$, where C is a generalized centroid of the ring R.*

Proof. Let λ be a nonzero element of a generalized centroid of a stalk. By the definition of a Martindale ring of quotients, one can find nonzero elements of the stalk, $\bar{a}, \bar{b}$, such that $\bar{a} = \lambda\bar{b}$. For these two elements let us consider the predicate

$$\forall\, x(\bar{a}x\bar{b} = \bar{b}x\bar{a}).$$

According to proposition **1.11.6**, this is a strictly sheaf predicate and, hence, there can be found preimages a, b of the elements $\bar{a}, \bar{b}$, respectively, such that the identity $axb = bxa$ is fulfilled on ring $\hat{RE}$. Lemma **1.5.8**, applied to the elements $e(b)a$ and $e(a)b$ shows them to be proportional: $e(b)a = cb$ for an $c \in C$. As $\bar{b}$ is a nonzero element, then the image of $e(b)$ equals the unit, i.e., $\bar{a} = \bar{c}\bar{b}$, where $\bar{c} = \rho_p($. In particular, the difference $(\lambda - \bar{c})$ annihilates the two-sided ideal of the stalk, generated by the element $\bar{b}$. For the prime ring $\Gamma_{\bar{\mathbf{p}}}(R)$ it implies $\lambda = \bar{c}$. The lemma is proved.

1.12.7. Lemma and notation. *The group G is induced on the stalk $\Gamma_{\bar{\mathbf{p}}}$ of an invariant sheaf. Let us designate the induced group by $G_{\bar{p}}$. Then $G_{\bar{\mathbf{p}}} \subseteq \mathbf{A}(\Gamma_{\bar{\mathbf{p}}})$.*

Proof. Since the kernel $\rho_{\bar{\mathbf{p}}}: Q \to \Gamma_{\bar{\mathbf{p}}}(Q)$ is equal to $p^G Q$, then it is G-invariant and, hence, G is induced on the stalk $\Gamma_{\bar{\mathbf{p}}}(Q)$. If $g \in G$ and I, J are essential ideals of the ring R, such that $J \subseteq I^g \subseteq R$, then $\hat{JE} \subseteq (\hat{IE})^g \subseteq \hat{RE}$, since the automorphisms are continuous in the topology determined by the idempotents of C and a set of all idempotents E is invariant. Applying ρ_p to the ideals of the latter chain and using lemma **1.12.5**, we come to the conclusion that the induced automorphism g lies in $\mathbf{A}(\Gamma_{\bar{\mathbf{p}}}(R))$. The lemma is proved.

We can now consider an invariant sheaf of rings $\Gamma(Q)$ together with the automorphisms from G, assuming them to be unary operations.

1.12.8. Proposition. *The invariant sheaf $\Gamma(Q)$ with the unary operations from G remains to be a correct sheaf, i.e., the automorphisms from G prove to be strictly sheaf operations.*

Proof. Let us check conditions (a), (b) and (c) on a strictly sheaf predicate.

(a) If $s_1^g = s_2$ and $s_1, s_2 \in eQ$ for a fixed idempotent e, then $(fs_1)^g = fs_2$ for any fixed idempotent $f \leq e$.

(b) Let $\bar{s}_1^g = \bar{s}_2$ and $\bar{s}_1, \bar{s}_2 \in \Gamma_{\mathbf{p}}$. Then, by the definition, for some preimages the inclusion $s_1^g - s_2 \in \ker \rho_{\mathbf{p}}$, i.e., $\mathbf{e}(s_1^g - s_2) \in \mathbf{p}^G C$. In more detail, $\mathbf{e}(s_1^g - s_2) = e_1 c_1 + \ldots + e_n c_n$ for some fixed idempotents $e_1, \ldots, e_n$ from $\mathbf{p}$.

Let $f = 1 - \sup(e_1, \ldots, e_n)$. Then f is a fixed idempotent, and $f \notin \mathbf{p}$, i.e., $U(f)$ is an invariant neighborhood of the point $\mathbf{p}$. By restricting s_1, s_2 on this neighborhood, we get $(s_1^g - s_2) f = (s_1^g - s_2) \cdot \mathbf{e}(s_1^g - s_2) \cdot f = 0$, i.e., $(s_1 f)^g = s_2 f$ and, hence, $U(f)$, $s_1 f$, $s_2 f$ are the sought neighbor-hoods and preimages.

The inverse statement results from the definition of the G action on a stalk: if $s_1^g = s_2$ in a ring of sections, then $\bar{s}_1^g = \bar{s}_2$ on a stalk.

(c) The checking is obvious.

The proposition is proved.

1.13 Martindale's Theorem

In this paragraph we shall study the construction of a central closure of a prime ring which satisfies a non-trivial generalized identity. A *generalized identity* is a polynomial of noncommutative variables with the coefficients from R (or from R_F), turning to zero when substituted instead of the variable elements of the ring R. Trivial are the identities incorporated into the definition of a ring (distributivity, associativity, etc.), identities of type $xc = cx$, where c is an element of a generalized centroid, and all their corollaries. Here we have an opportunity to give exact definitions in terms of free products.

1.13.1. It should be recalled that a *free product* of algebras with a unit, A, B, over a field C is the algebra $A * B$, containing A and B as subalgebras and generated by the latter, and such that for any homomorphisms $p\colon A \to D$, $q\colon B \to D$ there is a homomorphism $p * q\colon A * B \to D$, expanding p and q.

Such a definition does not yet guarantee the existence of a free product, but it can be, however, easily constructed as a sum of tensor products of spaces:

$$A * B = \sum_n D_1 \otimes D_2 \otimes ... \otimes D_n , \ D_i \in \{A, B\}, \ D_i \neq D_{i+1}, \tag{23}$$

the multiplication on which is determined by the natural formula

$$\begin{aligned}(D_1 \otimes ... \otimes D_n) \cdot (D'_{n+1} \otimes ... \otimes D'_m) = \\ D_1 \otimes ... \otimes D_n \hat{\otimes} D'_{n+1} \otimes D'_m,\end{aligned} \tag{24}$$

where $D_n \hat{\otimes} D'_{n+1} = D_n \otimes D'_{n+1}$ when $D_n \neq D'_{n+1}$, while if $D_n = D'_{n+1}$, then $\hat{\otimes}$ is a product in D_n.

1.13.2. A *generalized polynomial* over R is an element of the free product $R_F * C\langle X \rangle$ of algebras over C, where $C\langle X \rangle$ is a free algebra with free generators $\{x_1, x_2,..., x_n,..\} = X$.

By this definition, any generated polynomial f can be presented as a sum of monomials

$$f = \sum_i a_1^{(i)} x_{i1} a_2^{(i)} x_{i2} ... a_n^{(i)} x_{in} a_{n+1}^{(i)} , \quad a_k^{(i)} \in R_F , \ x_{ij} \in X \tag{25}$$

and, in particular, it depends only on the finite number of variables $f = f(x_1,..., x_m)$.

Let $a_1,..., a_n$ be some elements from R_F. Then the value of the polynomial $f(x_1,..., x_n)$ at $x_1 = a_1,... x_n = a_n$ is an element of ring R_F obtained when replacing x_i with a_i in formula (22). Strictly speaking, the value of f is the image of f at the homomorphism $1 * \varphi$, where $\varphi: C\langle X \rangle \to R_F$ is a homomorphism, such that $\varphi(x_i) = a_i$, while $1: R_F \to R_F$ is an identical homomorphism.

1.13.3. Definition. A *generalized identity* of the ring R is a generalized polynomial which turns to zero at all values of the variables from

R. The identity of f is called *non-trivial* if $f \neq 0$ as an element of $R_F * C\langle X \rangle$.

1.13.4. Theorem (Martindale). *If in a prime ring R the non-trivial generalized identity holds, then the central closure RC has an idempotent e, such that $eRCe$ is a finite-dimensional sfield over C.*

Proof. It should be first remarked that a non-trivial multilinear generalized identity holds in ring R, i.e., the identity f, in presentation (22) of which all the monomials depend on the same set of variables, $x_1,\dots, x_n$, each of these variables entering each of the monomials only once.

Let us consider the following operator:

$$\begin{aligned} \Delta^{y}_{x_i}: f(x_1,\dots, x_i,\dots, x_n) \to f(x_1,\dots, x_i + y,\dots, x_n) - \\ - f(x_1,\dots, x_n) - f(x_1,\dots, y,\dots x_n) \end{aligned} \tag{26}$$

where y is a variable not included into the set $\{x_1,\dots, x_n\}$. This operator is certain to transform any identity into another one. When it transforms f into a trivial identity, f is linear over the variable x_i. If a non-trivial identity results, then its degrees over x_i and over y are less than those of f over x_i. As a result, multiple action of operators Δ^t_z will finally give the required non-trivial polylinear identity.

For an arbitrary generalized polynomial f let us denote by V_f a linear space over C generated by all f values at $x_i \in R$. By induction over the degree of f we shall prove that if for a non-trivial multilinear f the dimension of V_f is finite, then the conclusion of the theorem is valid.

Let the power of f equal one. Then $f = \sum_{i=1}^{k} a_i x b_i$, and we can assume that $a_1,\dots, a_k$ are linearly independent, while $b_1,\dots, b_k$ are nonzero elements, since in the opposite case one could reduce the number k in the presentation of f without changing f proper. Corollary **1.7.14** states that there exist elements $r_1,\dots, r_m; t_1,\dots, t_m$ from R, such that $a_0 = \sum_{j=1}^{m} r_j a_1 t_j \neq 0$, $\sum_{j=1}^{m} r_j a_i t_j = 0$ at $i = 2, 3,\dots, k$. We have $a_0 x b_1 = \sum_{j=1}^{m} r_j f(t_j x)$, and, hence, $a_0 RC b_1 \subseteq \sum_j r_j V_f$. In particular, $a_0 RC b_1$ is a finite-dimensional space. Then, there is an element $r, s \in R$, such that $0 \neq ra_0 \in R \ni sb_1 \neq 0$, and, hence, $(ra_0)R\,C(sb_1)$ is finite-dimensional.

Let us choose nonzero elements $a, b \in RC$ in such a way that the space $aRCb$ has a minimal possible dimension l. As the ring R is prime, we can find an element u, such that $bua \neq 0$. In this case the dimension of the space $S = (aRCb)u$ is not greater and, hence, equal to l, and $S^2 = aRCbuaRCbu \neq 0$. Besides, for any nonzero element $s \in S$ we have $sRCs \subseteq S$ and again, since the dimension is minimal, $sRCs = S$. It, in particular, implies that $s^2 \neq 0$ (in the opposite case $S^2 = sRCs^2RCs = 0$) and, hence, $s^2RCs^2 = S$. Finally, $s \cdot S = s^2RCs \supseteq s^2RCs^2 = S$, i.e., S is a finite-dimensional sfield.

If e is the unit of the sfield S, then $eRCe \subseteq S$ and, since the dimensionality is minimal, $eRCe = S$, which is the required proof.

Let now the power of f be greater than one. An element f can be presented as

$$f = \sum_{i=1}^{m} a_i x_1 w_i + \sum_{i=1} p_i x_1 q_i + \sum_{i=1}^{k} v_i x_1 b_i,$$

where $a_i, b_i \in R_F$; w_i, p_i, q_i, v_i are non-trivial generalized polynomials. As before, one can assume that the elements $a_1, \ldots, a_n$ are linearly independent. Repeating the starting considerations for the case $n = 1$, we can find a polynomial f_1, such that $\dim V_{f_1} < \infty$ and f_1 has the form

$$a_1' x_1 w_1 + \sum_i p'_i x_1 q_i + \sum_i v'_i x_1 b_i.$$

This polynomial is not equal to zero in $R_F * C\langle X\rangle$, since in their presentation p_i and v_i have the first letter on the left other than x_1 and $a_1' \neq 0$.

Then, if for any $r_2, \ldots, r_n \in R$ the elements $w_1(r_2, \ldots, r_n)$ are linearly expressed through $b_1, \ldots, b_k$, then the dimensionality of V_{w_1} over C is finite and we can use the inductive supposition.

If this is not the case, then, by corollary **1.7.14**, one can find elements $s_1, \ldots, s_m, t_1, \ldots, t_m$, such that $\sum_i s_i b_j t_i = 0$ and the polynomial

$w'_1 = \sum_i s_i w_i t_i$ is not an identity of R.

Now the polynomial $f_2 = \sum_{i=1}^{m} f_1(xs_i)t_i$ has the form

$$a_1 x_1 w'_i + \sum p'_i x_1 q'_i,$$

in which case $V_{f_2} \subseteq \sum V_{f_1} t_i$, and f_2 is nonzero in $R_F * C\langle X\rangle$.

Extending these considerations to the letters $x_2, x_3, \ldots, x_r$, which are the first in the presentation of the monomials from f_2, one can find an element f_3, which is nonzero in $R_F * C\langle X\rangle$ and such that $\dim V_{f_3} < \infty$ and f_3 has the form

$$f_3 = \sum_i a_i x_1 s_i + \sum_i b_i x_2 u_i + \ldots + \sum_i d_i x_r h_i, \tag{27}$$

where any of the letters $x_1, \ldots, x_r$ is not the last for any monomial encountered in presentation (24) of the element f_3, and the sets of elements $\{a_i\}, \{b_i\}, \ldots, \{d_i\}$ are linearly independent.

Since $n \geq 2$ and at least one variable should be the last in the presentation of a monomial, then $r < n$. According to the inductive supposition, one can assume that for an $a_2, \ldots, a_n \in R$ the inequality $s_1(a_2, \ldots, a_n) \neq 0$ is valid. Let us set

$$s'_i = s_i(x_2, \ldots, x_{n-1}, a_n),\ u'_i = u_i(x_1, x_3, \ldots, x_{n-1}, a_n), \ldots$$
$$\ldots, h'_i = h_i(x_1, \ldots, x_{r-1}, x_{r+1}, \ldots, a_n),$$

where the element $f_4 = \sum a_i x_1 s'_i + \ldots + \sum d_i x_r h'_i$ has the power $n-1$. Besides, $V_{f_4} \subseteq V_{f_3}$, and, finally, f_4 is nonzero in $R_F * C\langle X\rangle$, since otherwise we have $\sum_i a_i x_1 s'_i = 0$ and, applying corollary **1.7.14**, we can find a nonzero element a, such that $ax_1 s'_1 = 0$. Allowing for the fact that

$s_1(a_2,\dots, a_n) \neq 0$, we come to a contradiction with the primarity of the ring R. According to the inductive supposition, the theorem is proved.

By way of concluding this paragraph, let us prove another three useful statements on prime rings satisfying a generalized identity.

1.13.5. Lemma. *A center of the sfield* $T = eRCe$ *is equal to* Ce.

Proof. Let t be a central element of T. Then for any $x \in R$ we have $f(x) \equiv txe - exet = 0$. If the elements e and t are linearly independent over C in ring Q, then, by corollary **1.7.14**, there can be found elements $v_i, r_i \in R$, such that $0 = \sum v_i e r_i \neq \sum v_i t r_i = a$. Then $0 = \sum v_i f(r_i x) = axe$ for all x, which is impossible. Therefore, $t \in Ce$, which is the required proof.

1.13.6. Lemma. *For any linear over C transformation* $l\colon T \to T$ *there exist elements* $a_i, b_i \in T$, *such that* $l(x) = \sum a_i x b_i$.

Proof. If n is the dimension of the sfield T over the center Ce, then the dimension of the space of all linear transformations is n^2. On the other hand, n^2 of the linear transformations $l_{ij}\colon x \to a_i x a_j$, where $a_1,\dots, a_n$ is a basis of T over the center, are linearly independent. Indeed, if

$$f(x) \equiv \sum c_{ij} a_i x a_j = 0 \quad , \quad x \in T,$$

then, applying corollary **1.7.14** to ring T, we get $c_{ij} = 0$. The lemma is proved.

1.13.7. Corollary. *If a ring with no zero divisors obeys non-trivial generalized identity, then its center is nonzero and the central closure is a finite-dimensional sfield.*

Proof. It should be remarked that if a ring R has no zero divisors, then the two-sided ring of quotients $Q(R)$ has them neither. Let, on the contrary, $q, v \in Q$, $qv = 0$. Then for a suitable ideal $I \in F$ we have $Iq \subseteq R$, $vI \subseteq R$. It yields $(Iq) \cdot (vI) = 0$ and, hence, $Iq = 0$ or $vI = 0$, i.e., $q = 0$ or $v = 0$.

Now, according to the Martindale theorem, RC has a primitive idempotent, but a ring with no zero divisors can have only one idempotent other than zero, i.e., the unit. Therefore, $RC = 1 \cdot RC \cdot 1 = T$ is a finitedimen-sional sfield over C.

Let us consider a linear over C projection $l\colon T \overset{on}{\to} C$ and assume that $l(T) \cap R = 0$. Then, by lemma **1.13.6**, there are elements $a_i, b_i \in T$, such that $l(x) = \sum a_i x b_i$ for all $x \in T$. For the elements a_i, b_i one can find a nonzero element $q \in R$, such that $a_i q$, $qb_i \in R$, since $T \subseteq Q$. We have $l(qRq) \subseteq R$ and, therefore, $C = l(T) = l(qTq) = l(qRCq) = l(qRq)C \subseteq (l(T) \cap R)C = 0$, this contradiction proving the corollary.

1.14. Quite Primitive Rings

In this paragraph we discuss in detail the structure of rings arising in the Martindale theorem. As a *quite primitive ring* we shall call a prime ring R with a nonzero idempotent e, such that eRe is a sfield. In literature these rings are called primitive with a nonzero one-sided ideal or primitive with a nonzero socle. The idempotent $e \neq 0$ of a prime ring R, such that eRe is a sfield, is called *primitive*.

A *socle* of a quite primitive ring R is a two-sided ideal generated by all the primitive idempotents. A socle is designated by $H(R)$ or simply by H.

1.14.1. Lemma. *The socle H of a quite primitive ring R is the least nonzero ideal of this ring. In particular, a socle is generated by any primitive idempotent and is a simple ring.*

Proof. If I is a nonzero ideal and e is an arbitrary primitive idempotent, then $0 \neq eIe \subseteq I \cap eRe$. As a sfield has no proper ideals, then $I \cap eRe \ni e$ and, hence, $I \supseteq H$.

If now A is a nonzero ideal of the ring H, then HAH is an ideal of the ring R contained in A. According to the above proved, this ideal contains a socle, i.e., $H = A$, which is the required proof.

It should be recalled that the module M over the ring R is called *irreducible* if it contains no proper submodules (i.e., submodules other than (0) or M).

1.14.2. Lemma. *Let e, f be primitive idempotents of a prime ring R. Then the sfields eRe and fRf are isomorphic. The right ideals, fR and eR, as well as the left ideals, Re and Rf, are mutually isomorphic as modules over R and are irreducible.*

Proof. Since the ring R is prime, there can be found an element u, such that $fue \neq 0$. By the same reason, the set $fueRf$ forms a nonzero right ideal of the sfield fRf, i.e., there exists an element u', such that

$$fue \cdot eu'f = f.$$

Squaring both parts of the latter equality, we see that $\xi = eu'f\ fue \neq 0$. Therefore, ξ is a nonzero idempotent $(\xi^2 = eu'3f(fue \cdot eu'f)\ fue = \xi)$ lying in the sfield eRe. Thus,

$$eu'f \cdot fue = e.$$

It is now absolutely clear that the mappings

$$ere \to fuereu'f, \qquad frf \to eu'frfue$$

give the sought isomorphism of sfields.

Let N be a nonzero right ideal contained in eR. Then $eN = N$ and, hence, since the ring R is prime, $eNe = Ne \neq 0$. However, eNe is the right ideal of the sfield eRe and, therefore, $e \in eNe \subseteq N$ and $eR = N$, so that eR (and, analogously, Re) is an irreducible module.

Let us now consider a mapping $\varphi\colon eR \to fR$ which acts by the rule $er \to fuer$, where u is an element determined from the onset. As $fue \neq 0$, then φ is a nonzero homomorphism of right R-modules. Therefore, the kernel of φ is a submodule of eR other than eR. Due to irreducibility, φ is an embedding. The image of φ is a nonzero submodule in fR, coinciding, due to its irreducibility, with fR. Thus, φ is an isomorphism of the modules. The lemma is proved.

1.14.3. The lemma proved above makes it possible to determine *the sfields of a quite primitive ring* R as a sfield T, which is isomorphic to eRe for a certain primitive idempotent e. Moreover, the lemma states that the module $V = eR$ is also independent of the choice of a primitive idempotent. Since $eRe \cdot eR \subseteq eR$, then this module can be viewed as a left vector space over the sfield $T = eRe$. In this case the elements of the ring R turn to transformations of the left vector space V over the sfield T.

Indeed, an element $vr \in V$ is unambiguously determined for every $v \in V$ and $r \in R$, in which case the mapping $v \to vr$ is a linear transformation of V over T, with which the element r is identified. This presentation is exact, since if $Vr = 0$ then $r = 0$, as R is prime.

There now naturally arises a ring $\mathbb{L}$ of all linear transformations of the space V over the sfield T. This ring has an explicit description with (infinite) matrices over T.

1.14.4. Let I be a set. An $I \times I$-matrix is an element of the Cartesian product $T^{I \times I}$ (i.e., the mapping $\varphi: I \times I \to T$), which will be designated by $\|t_{\alpha\beta}\|$, where $t_{\alpha\beta}$ is an element of the sfield occupying the (α, β)-th place (i.e., $t_{\alpha\beta} = \varphi(\alpha, \beta)$). The line numbered $\alpha_\nu \in I$ of the matrix $\|t_{\alpha\beta}\|$ is the line $(t_{\alpha_\nu, \beta})$, $\beta \in I$ (i.e., an element of $T^{\{\alpha_\nu\} \times I}$). Analogously, the column numbered β_ν is the column $(t_{\alpha\beta_\nu})$, $\alpha \in I$ (i.e., an element of $T^{I \times \{\beta_\nu\}}$).

A matrix is called *finite-rowed* if its every row contains only a finite number of nonzero components. On a set of finite rowed $I \times I$-matrices ring operations are naturally determined:

$$\|t'_{\alpha\beta}\| \pm \|t''_{\alpha\beta}\| = \|t'_{\alpha\beta} \pm t''_{\alpha\beta}\|, \quad \|t'_{\alpha\beta}\| \cdot \|t''_{\alpha\beta}\| = \Big\|\sum_{\gamma \in I} t'_{\alpha\gamma} t''_{\gamma\beta}\Big\|.$$

In the second formula the sum is determined, since it contains only a finite number of nonzero terms. It is also evident that these operations result again in finite-rowed matrices.

1.14.5. Lemma. *The ring* $\mathbb{L}$ *is isomorphic to the ring of finite-rowed* $I \times I$*-matrices over the sfield* T *for a set* I.

Proof. Let us choose a basis $\{v_\alpha \mid \alpha \in I\}$ of the space V over the sfield T. Then for any transformation $\varphi: V \to V$ the images $\varphi(v_\alpha)$ can be written as finite linear combinations of basic elements

$$\varphi(e_\alpha) = \sum_\beta t_{\alpha\beta} e_\beta$$

It is the correspondence $\varphi \to \|t_{\alpha\beta}\|$ that sets the sought isomorphism. The lemma is proved.

1.14.6. The embedding of a quite primitive ring R into the ring $\mathbb{L}$ is, by itself, not enough information. The most essential is the fact that at such an embedding R proves to be a *dense* subring in $\mathbb{L}$.

A ring S of linear transformations of the space V over the sfield T is called *dense* (in $\mathbb{L}$), if for any finite-dimensional subspace $W \subseteq V$ and any linear transformation $l \in \mathbb{L}$, there exists an element $s \in S$ which coincides with l on W.

In terms of matrices this is equivalent to the fact that for any finite subset $J \subset I$ and any $J \times J$-matrix $\|l_{\alpha\beta}\|$ there exists a matrix $\|s_{\alpha\beta}\| \in S$, extending l, i.e., such that $s_{\alpha\beta} = l_{\alpha\beta}$ at $\alpha, \beta \in J$.

It should also be added that on the ring $\mathbb{L}$ there ia natural topology, called *finite topology*, the density in which coincides with that given above. Namely, in finite topology the basis of zero neighborhoods are the sets $W^{\perp} = \{l \in \mathbb{L} \mid Wl = 0\}$ where W runs through a set of all finite-dimensional subspaces of V.

1.14.7. Among transformations from $\mathbb{L}$ of special importance are those of a finite rank, i.e., the ones having a finite-dimensional image. In terms of matrices such transformations are set by the matrices having only a finite number of nonzero columns.

1.14.8. Theorem. *A quite primitive ring R is dense in $\mathbb{L}$.. A set of all transformations of a finite rank lying in R coincides with the socle $\mathbb{H}$ of a ring R.*

Proof. In order to establish density, it is enough to show that for any linear independent over $T = eRe$ elements $w_1, \dots, w_n \in V = eR$ and any elements $v_1, \dots, v_n \in V$ there is an element $r \in R$, such that $w_1 r = v_1, \dots, w_n r = v_n$.

Let us carry out induction over n. At $n = 1$ an element w_1 is nonzero and, hence, $w_1 R$ is a nonzero submodule in V, i.e., due to irreducibility, there can be found an element $r \in R$, such that $w_1 r = v_1$.

Let the required statement be proved for a given n. Let us consider a linearly independent set $w_0, w_1, \dots, w_n$ and show, first, that there is an element $r_0 \in R$, such that $w_0 r_0 \neq 0$, $w_1 r_0 = 0, \dots, w_n r_0 = 0$.

Let us assume this is not the case. According to the inductive supposition, there are elements $r_1, \dots, r_n$, such that $w_i r_i = e$, $w_i r_j = 0$ at $i \neq j$. For an arbitrary $x \in R$ we have the following system of equalities:

$$w_i(x - \sum_{j=1}^{n} r_j e w_j x) = (w_i - \sum_{j=1}^{n} w_i r_j e w_j)x = 0, \ i = 1, 2, \ldots, n.$$

Therefore,

$$w_0(x - \sum_{j} r_j e w_j x) = (w_0 - \sum_{j=1}^{n} w_0 r_j e w_j) x = 0.$$

It affords $w_0 = \sum_{j=1}^{n} (w_0 r_j e) \cdot w_j$, but $w_0 r_j e \in eRe = T$ and this equality contradicts linear independence.

Thus, there exists an element r_0, such that $w_0 r_0 \neq 0, w_1 r_0 = 0, \ldots w_n r_n = 0$. Analogously, one can find elements $r_1, r_2, \ldots, r_n \in R$, which obey the conditions $w_i r_i \neq 0$, $w_i r_j = 0$ at $i \neq j$. Like it was the case for $n = 1$, there are elements $s_i \in R$, such that $w_i r_i s_i = v_i$. It is now evident that the element $r = \sum_{i=0}^{n} r_i s_i$ is the sought one.

Let us prove the second part. As $Ve = eRe$ is a one-dimensional subspace, then e is a transformation of a finite rank. The image of the sums of transformations is contained in the sum of images, and, therefore, the set H' of transformations of a finite rank is closed by addition. Right and left multiplications result in no increase in the dimension of the image. Consequently, H' is a nonzero ideal of the ring R and, by lemma **1.14.1**, we have $H' \supseteq H$.

Inversely, let r be a transformation of a finite rank from R. Since the image of r is finite-dimensional, its kernel K has a finite co-dimension and we can find a finite-dimensional subspace $W = Tw_1 + \ldots + Tw_n$, complementing K to $V = eR$. In this case the elements $w_1 r, \ldots, w_n r$ will be linearly independent over T and, due to density, for every $i, 1 \leq i \leq n$ there is an element $\xi_i \in R$, such that $w_1 r \xi_i = e$, $w_j r \xi_i = 0$ at $j \neq i$, as well as an element ψ_i with the condition $e\psi_i = w_i r$.

Let us consider an element $r' = r \sum \xi_i e \psi_i$. Its action on a linear transformation coincides with that of r on both K and all w_i, and, hence, $r = r' \in H$. The theorem is proved.

1.15. Rings of Quotients of Quite Primitive Rings

Here we shall characterize rings R_F and $Q(R)$ for a quite primitive ring R, preserving the denotations from the preceding paragraph, $\mathbb{L}, \mathbb{H}$.

1.15.1. Theorem. *If R is quite primitive, then R_F coincides with the ring $\mathbb{L}$ of all linear transformations of the space $V = eR$ over the sfield $T = eRe$.*

Proof. Since the ring R has the least nonzero ideal $\mathbb{H}$, then $R_F = Hom({}_R\mathbb{H}, R)$. If $h \in Hom({}_R\mathbb{H}, R)$, then the restriction of h on $eR \subseteq \mathbb{H}$ determines a linear transformation $\hat{h}: eR \to eR$. If $\hat{h} = 0$, then $(eR)h = 0$ and, hence, $(\mathbb{H})h = (\mathbb{H}\, eR)h = \mathbb{H}(eR)h = 0$, i.e., $h = 0$, so that $h \to \hat{h}$ is an embedding of R_F in $\mathbb{L}$.

Inversely, if $\hat{h} \in \mathbb{L}$ and $a \in \mathbb{H}$, then, by theorem **1.14.8**, eRa is a finite-dimensional over T subspace in V and, hence, due to density, there is an element $a_1 \in R$, the action of which on this subspace coincides with that of $\hat{h}$. Let us determine the action of h on an element a by the formula $ah = aa_1$. This definition is correct. Indeed, if $a_1 = a_2 = \hat{h}$ on eRa, then $eR(aa_1 - aa_2) = 0$ and, hence, $aa_1 = aa_2$. Therefore, $h \in Hom({}_R\mathbb{H}, R)$ and we get an inverse mapping, $\hat{h} \to h$. The theorem is proved.

1.15.2. Let us now make a small retreat into general topology. Let us consider an arbitrary set X and a set F of its transformations. One can set any topologies on the set X. As the set F has been singled out, of special interest are the topologies for which F proves to consist of continuous transformations. It is evident that if τ_α, $\alpha \in A$ is a class of such topologies (a topology is considered to be given by a set of open sets), then their inter-section $\bigcap_A \tau_\alpha$ will also be a topology where all the functions from F are continuous.

1.15.3. Under our conditions it would be natural to consider the weakest topology on V, which converts V into a topological linear space over the sfield T, supplied with a discrete topology and such that the set R consists of continuous transformations.

It should be recalled that the linear space V (supplied with a topology) over a topological sfield T is called a *topological linear space* , provided the linear operations (addition and multiplication, $T \times V \to V$), as well as the mappings $tv \to t$ for all $v \neq 0$ are continuous (here t runs through T).

1.15.4. Theorem.

(a) *There is a weakest topology on* V, *which turns it into a topological linear space over the sfield* T *supplied with a discrete topology, such that* R *consists of continuous transformations.*

(b) *A set of all linear continuous transformations in this topology equals* $\mathfrak{A}(R)$.

(c) *A set of all continuous transformations of a finite rank coincides with the socle of the ring* R.

Proof. Let τ' be a certain (possibly, not the weakest) topology obeying condition (a) of the theorem. As, by definition, the mappings $Tv \to T$, acting by the formula $tv \to T$, are continuous, then the topology induced on any one-dimensional subspace is discrete. If now r is an arbitrary element from R, e is a primitive idempotent, such that $V = eR$, $T = eRe$, then the continuous transformation re transforms V into a one-dimensional subspace. Therefore, the preimage of zero should be both open and closed.

Let us consider a topology τ, the subbase of zero neighborhoods for which consists of the kernels of the mappings $v \to vre$, $r \in R$, while the neighborhood of an arbitrary element v is determined as sums $v + W$, where W is the neighborhood of zero. As the co-dimensionality of the kernel of re equals one, there is an element $v \in V$, such that

$$V \setminus \ker re = \bigcup_{0 \neq t \in V} (tv + \ker re).$$

Hence, the complement to the kernel is open, while the kernel itself is closed in τ, i.e., the topology τ is not greater than τ'.

Let us show that the topology τ obeys all the requirements of the theorem. Since kernels are subspaces, then (V, τ) is a topological Abelian group and the multiplication $T \times V \to V$ is continuous. Moreover, if $0 \neq v = er \in V$, then, since R is prime, there is an element $r_1 \in R$, such that $er \cdot r_1 e \neq 0$. In this case the intersection of kernel of $r_1 e$ with the subspace Tv equals zero, i.e., the topology induced on Tv is discrete and the mapping $tv \to t$ is a homomorphism.

Let us check if $Q(R)$ consists of continuous transformations, assuming $Q(R) \subseteq R_F = L$ (see **1.15.1**). Let $q \in Q(R)$. Then q is a linear transformation and $qH \subseteq R$, since H is the least nonzero ideal of R. If U is a neighborhood of zero, then it contains a subset of type $\bigcap_{i=1}^{n} \ker r_i e$, $r_1 \in R$. Its complete preimage at the mapping q contains an intersection $\bigcap_i \ker qr_i e$. As $r_i e \in H$, then $qr_i e = (qr_i e)e \in Re$ and, therefore, the preimage is again a neighborhood of zero.

Inversely, let l be a continuous transformation. Then for any $r \in R$ the preimage of $\ker re$ is open. This preimage is equal to the kernel of transformations lre. Consequently, the kernel of lre contains an intersection $W = \bigcap_i \ker r_i e$ for suitable $r_i \in R$.

One can assume that this intersection is minimal relative to n, i.e., $\ker r_j e \not\supseteq \bigcap_{i \neq j} \ker r_i e$, $1 \leq j \leq n$. It enables one to find elements $v_j \in V$, such that $v_j r_j e \neq 0$, $v_j r_i e = 0$ at every $i, j, i \neq j$. Then the elements $v_j lre$ and $v_j r_j e$ belong to the sfield eRe and, therefore, there is an element $t_j \in R$, such that $v_j lre = v_j r_j e t_j e$. Let $r' = \sum_{j=1}^{n} r_j e t_j e$.

Let us show that $lre = r'$. Indeed, the action of these elements coincide on all v_j and on W. Moreover, for every $v \in V$ the images $vr_i e$ and $v_i r_i e$ lie in the same linear subspace eRe, i.e., $vr_i e = t'_i v_i r_i e$ for a certain $t'_i \in T$. Now the difference $v - \sum_i t'_i v_i$ lies in the intersection of the kernels of transformations $r_i e$. It implies that $V = W + \sum_j T v_j$ and, hence, $lre = r'$.

Hence, $lre \in R$, i.e., $lRe \subseteq R$. We have

$$lH = lReR = (lRe)R \subseteq R,$$

and, hence, $l \in Q(R)$.

Statement (c) results immediately from theorem **1.14.8.** The theorem is thus completely proved.

REFERENCES

S.Amitsur [4,6]
V,A.Andrunakievich, Yu.M.Ryabukhin [8]
K.I.Beidar [13,14]
K.I.Beidar, A.V.Mikhalev [15,16]
G.Bergman, I.Isaacs [21]
S.Burris, H.Verner [25]
N.Jacobson [65]
V.P.Elizarov [42]
J.Lambek [91]
G.Levitsky [98]
Lyubetsky, Gordon [100]
A.I.Mal'tsev [102,103]
W.Martindale [105]
V.K.Kharchenko [70,71,74,80]
G.Higman [59]

CHAPTER 2
ON ALGEBRAIC INDEPENDENCE OF AUTOMORPHISMS AND DERIVATIONS

2.0 Trivial Algebraic Dependences

Automorphisms of fields are well known to be algebraically independent, this fact playing an important role in the classical Galois theory. Therefore, the problem of algebraic dependence of automorphisms is of importance in a noncommutative situation as well. A simple example that follows shows that from a common point of view automorphisms can be algebraically (and even linearly) dependent.

Example. Let $R = C_2$ be an algebra of 2×2-matrices over a field of complex numbers and let φ be a conjunction by a diagonal matrix $\phi = diag(i, 1)$. Then

$$x^{\varphi^3} - x^{\varphi^2} + x^{\varphi} - x = 0 \tag{1}$$

for all matrices x, though the automorphisms φ^3, φ^2, φ, 1 are different in the group $Aut(R)$.

It is also evident that if g is an inner automorphism of an infinite order of any algebra R, while the element corresponding to it is algebraic, then the g powers will be linearly dependent.

We shall see below that algebraic dependences among the elements corresponding to inner automorphisms (derivations) prove to be the only reason which can result in an algebraic dependence of automorphisms (derivations).

Under noncommutative conditions it would be natural to consider the algebraic dependences set by generalized polynomials. If $g_1, \ldots, g_n$ is a set of automorphisms or derivations, then their dependence in such a sense means that for a nontrivial generalized polynomial $W(z_i)$ the equality $W(x^{g_i}) = 0$ is valid for all x from the ring. If we now consider the set $g_1, \ldots, g_n$ as a set of unary operations, then we come to the necessity to study identities with automorphisms (and differential identities as well).

We can now write some identities with automorphisms which evidently hold in any ring:

$$1.\ (xy)^g = x^g y^g,\quad (x+y)^g = x^g + y^g;$$

$$2.\ (x^g)^{g_1} = x^{(gg_1)};$$

$$3.\ x^g - \varphi^{-1} x\varphi = 0,$$

where g is an inner automorphism, and φ is an element corresponding to it.

These identities and their corollaries should naturally be considered trivial. From the viewpoint of studying algebraic dependences of automorphisms, generalized identities (without automorphisms) of a ring should also be referred to as trivial identities. In this sense equality (1) presented above is trivial, since after substituting $\varphi^{-i} x \varphi^{i}$ for x^{φ^i} we get a zero generalized polynomial (as an element of the ring of generalized polynomials $R * C\langle X\rangle$). The results presented in this chapter show, in particular, that a semiprime ring has no multilinear identities with automorphisms from $A(R)$, which is analogous to the theorem on algebraic independence of automorphisms of fields.

For the derivations of a semiprime ring R of the characteristic $p \geq 0$ we can also write identities determining the structure of the Lie ∂-algebra on the set of derivations $D(R)$ and its inner (for Q) part:

$$4.\ (x+y)^\mu = x^\mu + y^\mu;$$

$$5.\ (xy)^\mu = x^\mu y + xy^\mu;$$

$$6.\ ax^{[\mu,\delta]} = (x^\mu)^\delta - (x^\delta)^\mu,$$

where μ, δ are any derivations, and $[\mu, \delta]$ is a commutator in the algebra $D(R)$;

$$7.\ x^{\mu\alpha + \delta\beta} = \alpha x^\mu + \beta x^\delta,$$

where μ, δ are any derivations, $\alpha, \beta \in C$;

$$8.\ x^\mu = ax - xa,$$

where μ is an inner for Q derivation and a is a corresponding element;

$$9.\quad \overbrace{(\ldots(x^{\mu})^{\mu})^{\mu}\ldots)^{\mu}}^{p} = x^{\mu^{[p]}},$$

where μ is any derivation, $\mu^{[p]}$ is the value of a p-operation in the algebra $D(R)$ (when $p=0$ this identity assumes the form $x = x$).

If these identities and their corollaries , as well as the generalized identities of a ring are considered trivial, then the results obtained below for derivations imply that there are no nontrivial differential identities. This is an analog of the theorem on algebraic independence of derivations of sfields.

And, finally, there arises a problem on dependences between automorphisms and derivations. The remark on the fact that automorphisms from $A(R)$ operate on the Lie ∂-algebra $D(R)$.

$$10.\quad ((x^{g^{-1}})^{\mu})^{g} = x^{\mu^{g}},$$

which is also to be considered trivial. It is proved that there are also no nontrivial multilinear dependences between automorphisms and derivations.

No exact definition will be given on what a trivial identity is, or what a corollary of identities is (though, it is not, in fact, a problem), but let us instead show in what way an arbitrary polynomial $f(X)$, which includes in its presentation not only the variables, ring operations and coefficients from R_F but also the derivation operations and automorphisms, can be reduced using identities 1 - 10. Then we shall show that the reduced expression turns to zero at all the values of the variables from R only when it is obtained from generalized identities of a ring using substitutions $z_{ij} = x^{g_i \Delta_j}$.

As above, let us fix the following notations: R , Q , R_F and C for a semiprime ring, its two-sided and left ring of quotients and a generalized centroid, respectively.

2.1 Process of Reducing Polynomials

2.1.1. Let R be a semiprime ring. Let us designate by I_p an ideal of R, which consists of all the elements of an additive order p, where p is a simple number. Let $e_p = e(I_p)$ and $e_0 = 1 - \sup\limits_{p\geq 2} e_p$. Then all idempotents $e_0, e_2, e_3, e_5, \ldots, e_p, \ldots$ are mutually orthogonal and the characteristic of the ring $e_p R_F$ equals p. Moreover, $\sup_p e_p = 1$, i.e., the expression $f(X) = 0$ is an identity in R iff $e_p f(X) = 0$ is an identity of $e_p R$ for all $p \geq 0$. Since all idempotents e_p are invariants of all automorphisms and are constants of all derivations, then, when studying identities, one can limit oneself with the case when *a ring has a characteristic* $p \geq 0$ (the ring is assumed to have a zero characteristic if there is no additive torsion).

So, let R be a semiprime ring of a characteristic $p \geq 0$. Then, as we have seen above, $D(R)$ is a Lie ∂-algebra over a generalized centroid.

2.1.2. Definition. A set M of derivations from $D(R)$ will be called strongly independent if the fact that a linear combination $\sum_i \mu_i c_i$ of different derivations from M is an inner derivation for the ring Q, affords $\mu_i c_i = 0$ for all i.

Let us order a strongly independent set of derivations M and in a standard way expand this order on a set of words from M, i.e., we assume that a longer word is bigger than a shorter one, while the words of the same length are compared lexicographically.

2.1.3. Definitions. A word Δ from M is called correct if it does not contain subwords of the type $\mu_1 \mu_2$, where $\mu_1 > \mu_2$ and if $p > 0$, then it does not contain subwords of type μ^p. In other words, $\Delta = \delta_1 .. \delta_m$, where $\delta_1 \leq \delta_2 \leq .. \leq \delta_m$, and this chain should not contain p signs of equality in a row.

2.1.4. It should be recalled (lemma **1.6.21**) that for an element m of a nonsingular module we have determined $e(m)$ as an idempotent from C, such that $mC \cong e(m)C$. If $\mu \in D(R)$, then $e(\mu) = e(R^{\mu})$ is easily seen to be a support of the image of μ.

It should be also recalled (see 1.7.8) that by $i(g)$ we have denoted the support of Φ_g for automorphisms $g \in A(R)$, which singles out the part of the ring R_F on which g acts in an inner way.

2.1.5. Definition. The polynomial $f(x_1,..., x_n)$, in the presentation of which use is made of not only variables, addition operation, multiplication operation and coefficients from the ring of quotients, but also of derivation operations and automorphisms will be called, for short, a *DA-polynomial*.

2.1.6. A *DA*-polynomial $f(x_1,\dots, x_n)$ is called *reduced* if there exists a generalized polynomial $F(z_{ijk})$, $1 \le i \le n$, $1 \le k, j \le m$, such that

$$f(x_1,\dots, x_n) \equiv F(x_i^{g_{ik}\Delta_{ij}}\varepsilon_{ik}),$$

where ε_{ik} are central idempotents, such that $\varepsilon_{ik}\, i(g_{ik}^{-1}\cdot g_{it}) = 0$ at all $t > k$ and all i, in which case for every i the set $D_i = \{\Delta_{ij} | 1 \le j \le m\}$ consists of mutually different correct words of a strongly independent set of derivation M_i.

2.1.7. Theorem. *Any DA-polynomial can be transformed to a reduced form using identities* 1 - 10.

Proof. Using identities 1, 4 and 5, we can achieve a state when automorphisms and derivations are directly applied only to variables, i.e., a *DA*-polynomial $f(x_1,\dots, x_n)$ is transformed to $F'(x_i^{T_j})$, where $F'(z_{ij})$ is a generalized polynomial, and T_j are the words of automorphisms and derivations. Identity 10, in its equivalent form $x^{\mu g} = x^{g\mu^g}$ makes it possible to place all automorphisms in the beginning of the words, while identity 2 allows one to replace their product in every word with one automorphism. Therefore, $f(x_i)$ is transformed into a polynomial form

$$F''(x_i^{g_{ik}\Delta_{ij}}). \tag{2}$$

In order to better understand the essence of the phenomenon in question, a further reduction will be first carried out for a prime ring, and then for a general case.

Let R be prime. Then its generalized centroid C is a field, and $\Phi_g \ne 0$ for an automorphism $g \in A(R)$ iff g is an inner automor-phism for Q. Let G be a group generated by all automorphisms g_{ik}, G_{in} be a normal subgroup of inner for Q automorphisms. Let us choose a system $\{g_t\}$ of representatives of cosets of G_{in}. Then $g_{ik} = h_{ik}g_{t(i,k)}$, where $h_{ik} \in G_{in}$ and, hence, $x^{g_{ik}} + +(\varphi^{-1}x\varphi)^{g_{t(i,k)}}$. This enables one to reduce $f(x_i)$ using identities 3 and 1 to form 2, where different automorphisms differ by modulo G_{in}, i.e., $i(g^{-1}h) = 0$ at $g \ne h$. In particular, $i(g_{ik}^{-1}g_{it}) = 0$ at $k \ne t$, since g_{ik}, g_{it} are different automorphisms encountered with a variable x_i.

Let us make derivations. Let us fix a basis $\{1_\beta, \beta \in B\}$ of the subalgebra $adQ \subseteq D(R)$ and expand it to the basis $\{1_\alpha | \alpha \in A \supset B\}$ of the whole algebra $D(R)$. Then the derivations $\{1_\alpha | \alpha \in A \setminus B\} = M$ will be strongly independent. Now, writing the derivations in the presentation of words Δ_j with respect to this basis and replacing the occurrences u^{1_β} for $ud_\beta - d_\beta u$, where $ad\ d_\beta = 1_\beta$, one can reduce the DA-polynomial $f(x_1,...)$ to form 2, where Δ_{ij} are words of the strongly independent set M.

Let us fix a complete order on M. Using identities 5, 6, 8 and the table of multiplication in $D(R)$

$$[1_\alpha, 1_\gamma] = \sum_\delta \lambda_{\alpha\gamma}^{(\delta)} 1_\delta, \quad \lambda_{\alpha\gamma}^{(\delta)} \in C;$$

$$1_\alpha^p = \sum_\delta \rho_\alpha^{(\delta)} 1_\delta, \quad \rho_\alpha^{(\delta)} \in C.$$

we can transform the words Δ_j into a linear combination of correct words. Therefore, the initial polynomial is reduced to a reduced form (in which case all sets M_i will even coincide).

Let us now consider a semiprime ring. Let us first transform the DA-polynomial (2) to a form where Δ_{ij} are correct words of a strongly independent set of derivations.

For every monomial m from a polynomial presentation (2) of $f(x_1,...)$ let us denote by $l(m)$ a sum of the lengths of words of derivations occurring in the presentation of m. Let $D_k(f)$ be a set of all derivations occurring in the presentation of the monomial m, such that $l(m) = k$.

Let $D_l = D_l(f)C$ be a submodule in $End(R_F, +)$, generated by the derivations $D_l(f)$, where $l = l(f)$ is the maximum among the numbers $l(m)$ relative all monomials m from the presentation of f. According to lemmas **1,6,9**, **1.6.21** and **1.6.22**, the finitely-generated submodule $A = D_l / D_l \cap D_{in}$ of the module $End(R_F, +)/D_{in}$ is injective and, hence, the exact sequence

$$0 \longrightarrow D_l \cap D_{in} \longrightarrow D_l \longrightarrow A \longrightarrow 0$$

splits, i.e., $D_l = (D_l \cap D_{in}) \oplus A'$, where $A \cong A' \subseteq D_l$. Then, A' is a finitely-generated submodule and, hence, it splits into a direct sum of cyclic modules $A' = \sum\oplus\, \delta_i C$.

Therefore, using identities 6 and 7 one can reduce $f(X)$ to a polynomial form F', where derivations from $D_l(F') = \{\delta_i\}$ are strongly independent.

Let us order a strongly independent set $\{\delta_i\}$ and extend this order in a standard way onto a set of words from $\{\delta_i\}$. Then, using identities 5 and 8 one can reduce F' to a form F'', where all the words occurring in the presentation of monomials m, such that $l(m) = l$, are correct words from $\{\delta_i\}$ and $l(F'') \le l(F')$.

Let us consider a submodule A_1 in $End(R_{\mathbf{F}}, +)$ generated by derivations $D_l(F'') \cup D_{l-1}(F'')$. Let $A_{in} = A_1 \cap D_{in}$. Then, as has been shown above, A_{in} is singled out from A_1 by a direct addent and is, hence, a finitely-generated submodule. Therefore, $A_{in} + D_l(F')C$ is also a finitely-generated submodule and, hence, it is singled out from A_1 by a direct addent $A_1 = (A_{\text{int}} \oplus D(F')C) \oplus B$, where B is a finitely-generated submodule, i.e., by lemma **1.6.9**, $B = \sum\oplus\, \mu_j C$. Let us order a set $\{\delta_i\} \cup \{\mu_j\}$ in such a way that $\delta_i > \mu_j$, and, using identities 6 and 9, let us transform the expression F'' to a form F''', where all the words encountered in the presentation of the monomials m , such that $l(m) \ge l - 1$ are correct words from the strongly independent set $\{\delta_i\} \cup \{\mu_j\}$.

Continuing this process, we obtain an expression $\tilde{F}(X)$ of form (2), which is equivalent to the initial one by modulo of identities 1, 2, 4 - 10, and such that the words occurring in its presentation are correct words from the strongly independent set of derivations D .

Let us now transform the automorphisms. Let $g_{i1}, g_{i2}, \ldots, g_{in}$ be all the automorphisms occurring in the presentation $F(X)$ with a letter x_i. Let, then, $\Phi_{g_{ik}^{-1} g_{it}} = \varphi_{kt} C,\ e_{kt} = e(\varphi_{kt})$, in which case one can assume (see **1.7.8**) that $\varphi_{kt}\varphi_{tk} = e_{kt} = e_{tk}$. From the definition of Φ_g we get a system of equations:

$$e_{kj}\, x^{g_{ik}} = \varphi_{kt}\, x_i^{g_{it}}\, \varphi_{tk}, \quad k < t < n.$$

Let $\varepsilon_{ik} = 1 - e_{k,\,k+1} \circ \ldots \circ e_{kn}$, where, by the definition, $a \circ b = a + b - ab = \sup(a, b)$. Then

$$x_i^{g_{ik}} = \varepsilon_{ik} x_i^{g_{ik}} + \sum_{t > k} \lambda_k^{(t)} \varphi_{kt} x_i^{g_{it}} \varphi_{tk}, \tag{3}$$

where $\lambda_k^{(t)}$ are central idempotents. In this case the idempotents ε_{ik} are characterized by the following property: $\varepsilon_{ik} \cdot \Phi_{g_{ik}^{-1} g_{it}} = 0$ at all $t > k$.

Let us now substitute into $\tilde{F}(X)$ the right-hand part of 3 at $k = 1$ instead of $x^{g_{i1}}$. Then the automorphism g_{i1} will occur in the presentation only in the terms with the coefficients ε_{i1}, such that $i(g_{i1}^{-1} g_{it}) = 0$ at $t>1$. In the expression obtained all the occurrences $x_i^{g_{i2}}$ will be replaced with the right-hand part of 3 at $k=2$, etc.. As a result, we get a reduced DA-polynomial. The theorem is proved.

2.2. Linear Differential Identities with Automorphisms

2.2.1. Definition. Automorphisms g, h are called *mutually outer* if $i(gh^{-1}) = 0$ (see **1.7.8**).

2.2.2.Theorem. *Let $\Delta_1, \ldots \Delta_n$ be pairwise different correct words from a strongly independent set of derivations M, such that $e(\mu) = 1$ for all $\mu \in M$. Let, then, $g_1, \ldots, g_m$ be pairwise mutually outer automorphisms. In this case , if the following identity holds on R:*

$$f(x) = \sum_{i=1}^{m} \sum_{j=1}^{n(i)} \sum_{k=1}^{s(i,\,j)} a_{ij}^{(k)} x^{g_i \Delta_j} b_{ij}^{(k)} = 0, \tag{4}$$

then the following equalities are valid in the tensor product $R_{\mathbf{F}} \otimes_C R_{\mathbf{F}}$:

$$\sum_{k=1}^{s(i,\,j)} a_{ij}^{(k)} \otimes b_{ij}^{(k)} = 0 \tag{5}$$

for all $i, j, 1 \le i \le m, 1 \le j \le n(i)$.

The proof will be divided into several steps.

2.2.3. *One can assume the supports of all coefficients to coincide.*

In a Boolean ring of central idempotents E let us consider a subring T generated by the unit and the supports of the coefficients. This subring is finite (since any Boolean ring is locally-finite). Let $e_1,..., e_r$ be all minimal idempotents of the ring T . Since $e_q e_t \leq e_q, e_t$, then these idempotents are pairwise orthogonal and, as $1 \in T$, their sum equals the unit.

Therefore, it is enough to show in the theorem the validity of the following equalities:

$$e_t \sum_k a_{ij}^{(k)} \otimes b_{ij}^{(k)} = 0.$$

Let us multiply all coefficients of identity (4) by e_t. We have $e(a_{ij}^{(k)} e_t) = e(a_{ij}^{(k)}) \cdot e_t \leq e_t$. As a result, the supports of all nonzero coefficients of the obtained identity are equal to each other (and to e_t), and it is now sufficient to prove the theorem for the identity obtained.

2.2.4. *One can assume that none of equalities (5) takes place*, since in the opposite case one can eliminate a definite number of members without violating identity. Namely, if $\sum a^{(k)} \otimes b^{(k)} = 0$ then one can cross out the sum $\sum a^{(k)} x^{g\Delta} b^{(k)}$, since in this case $\sum a^{(k)} y b^{(k)} = 0$ at all $y \in R_F$.

The same considerations allow one to assume that for any fixed i, j the elements $\{a_{ij}^{(k)}\}$ are linearly independent over C , i.e., *none of them belongs to the module generated by others.*

Let us now carry out induction over a senior word Δ_1 which occurs in the presentation of identity (4).

2.2.5. Let Δ_1 be an empty word. Then derivations take no part in the presentation of the identity and it acquires the form:

$$f(x) = \sum_{i=1}^{m} \sum_{k=1}^{s(i)} a_i^{(k)} x^{g_i} b_i^{(k)} = 0.$$

As, by the condition $i(g_i g_t^{-1}) = 0$ at $i \neq t$, then this identity will be written as

$$f(x) = \sum_{i=1}^{k} a_i x^{g_i} b_i = 0, \tag{6}$$

where the element a_1 is right-independent of $a_2,\ldots a_n$ with respect to the sequence of automorphisms $g_1, g_2,\ldots, g_n$ (see **1.7.15**). By theorem **1.7.10**, one can find an element $\beta \in \mathbb{L}(R)$, such that $a \overset{df}{=} a_1 \cdot \beta^{g_1} \neq 0$, $a_i \cdot \beta^{g_i} = 0$ at $i \geq 2$. Let us recall the following definition.

2.2.6. Definition. If $f(x)$ is a DA-polynomial with a singled out variable x, $\beta = \sum r_i \otimes s_i \in \mathbb{L}(R)$, then we set $f(x) \cdot \beta = s_i f(r_i x)$.

2.2.7. If $\beta \in \mathbb{L}(R)$, g is an automorphism , and $\Delta = \delta_1 \ldots \delta_m$ is a correct word, then the following identity holds:

$$(ax^{g\Delta}b) \cdot \beta = (a \cdot \beta^g)\, x^{g\Delta} b + s \cdot (a \cdot \beta^{g\delta_1})\, x^{g\delta_2 \ldots \delta_m} + \ldots, \qquad (7)$$

where dots denote a sum of the terms $dx^{g\bar{\Delta}} b$, where $\bar{\Delta} < \delta_2 \ldots \delta_m$, and s is a natural number, such that $\delta_1 = \delta_2 = \ldots = \delta_s \neq \delta_{s+1}$ and, hence, $s \not\equiv 0 (\mathrm{mod}\ p)$.

This equality results from the Leibnitz formula

$$(xy)^{\delta_1 \ldots \delta_n} = \sum x^{\Delta_1} y^{\Delta_2},$$

where summation is carried out over all correct subwords Δ_1, obtained from $\delta_1 \ldots \delta_n$ by crossing out some letters, and where Δ_2 is a correct word extending Δ_1 to Δ. For instance,

$$(xy)^{\delta_1\delta_2\delta_3} = x^{\delta_1} y^{\delta_2\delta_3} + x^{\delta_2} y^{\delta_1\delta_3} + x^{\delta_3} y^{\delta_1\delta_2} +$$
$$+ x^{\delta_1\delta_2} y^{\delta_3} + x^{\delta_1\delta_3} y^{\delta_2} + x^{\delta_2\delta_3} y^{\delta_1}.$$

If Δ is empty, then formula (7) assumes the form

$$(ax^g b) \cdot \beta = (a \cdot \beta^g) x^g b. \qquad (8)$$

2.2.8. Applying the obtained in **2.2.5** element β to relation (6), we get

$$ax^{g_1} b_1 = f(x) \cdot \beta = 0.$$

Hence, $e(a)\cdot e(b_1)=0$. However, since the element a is expressed through a_1, then $e(a)\le e(a_1)=e(b_1)$ (see **2.2.3**). Thus, $e(a)=e(a)\cdot e(b_1)=0$, a contradiction which serves a basis for induction.

Let us carry out an induction step.

2.2.9. In addition to **2.2.4** one can assume that the *senior word* Δ_1 *can occur in the presentation of identity* (4) *but once.*

Indeed, an element $a_{11}^{(1)}$ is right-independent of

$$a_{11}^{(2)},\dots, a_{11}^{(n(1))}, a_{12}^{(1)},\dots, a_{12}^{(n(2))},\dots, a_{1i}^{(k)},\dots, a_{1\,m(1)}^{(n(m(1)))}$$

relative a sequence of automorphisms

$$g_1,\dots, g_1,\dots, g_2,\dots, g_2,\dots, g_i,\dots, g_{m(1)}$$

because $a_{11}^{(1)}$ is not expressed linearly over C through $a_{11}^{(2)},\dots, a_{11}^{n(1)}$ and $\Phi_{g_i g_j^{-1}}=0$ at $i\neq j$. According to theorem **1.7.10**, one can find an element $\beta\in \mathbb{L}(R)$, such that $a=a_{11}^{(1)}\cdot\beta^{g_1}\neq 0$, $a_{i1}^{(k)}\beta^{g_i}=0$, if $i\neq 1$ or $k\neq 1$. According to (7), $f\cdot\beta$ has the only term $ax^{g_1\Delta_1}b_{11}^{(1)}$, containing Δ_1, in which case $e(a)\le e(a_{11}^{(1)})=e(b_{11}^{(1)})$ and, hence, $e(a)=e(a)e(b_{11}^{(1)})\neq 0$, i.e., by lemma **1.6.23**, $a\otimes b_{11}^{(1)}\neq 0$. Therefore, substituting an element $x^{g_1^{-1}}$ instead of x one can assume $g_1=1$, $n(1)=m(1)=1$.

2.2.10. Let $\mu_1,\mu_2,\dots,\mu_n$ be all the derivations, such that $\mu_1\bar{\Delta},\dots,\mu_n\bar{\Delta}$ occur in the presentation of f, where $\bar{\Delta}=\delta_2\dots\delta_m$. If $\beta\in\mathbb{L}(R)$, then, according to (7), the sum of terms $f\cdot\beta$ containing $x^{\bar{\Delta}}$ will have the form

$$s\cdot(a_1^{(1)}\beta^{\delta_1})x^{\bar{\Delta}}b_1^{(1)}+\sum_{r,k}(a_{1,r}^{(k)}\cdot\beta^{\mu_r})x^{\bar{\Delta}}b_{1,r}^{(k)}+ \tag{9}$$
$$+\sum_k(a_{1,j}^{(k)}\cdot\beta)x^{\bar{\Delta}}b_{1,j}^{(k)},$$

where $\bar{\Delta} = \Delta_{j}$, $\mu_r\bar{\Delta} = \Delta_r$ and we have, for simplicity, omitted the second index 1 at all coefficients a, b (it should be recalled that $g_1 = 1$).

If $a_1^{(1)} \cdot \beta = 0$, then $f(x) \cdot \beta = 0$ is an identity to which one can apply the inductive supposition. It means that in the tensor product $R_F \otimes R_F$ valid is the equality obtained from (9) by substituting all occurrences of $x^{\bar{\Delta}}$ with the sign $\otimes$. According to lemma **1.6.24**, one can find elements $c_2,\ldots, c_v \in C$ which are β-independent and such that the first left coefficient a in (9) is linearly expressed through the remaining left coefficients using $c_2,\ldots, c_v$. Here account has been taken of the fact that $e(a) = \ e(sa_1^{(1)} \cdot \beta^{\delta_1}) \le e(a_1^{(1)}) = e(b_1^{(1)})$ and, hence, $a = a \cdot e(a) = a\, e(b_1^{(1)})$.

Using linearity of the operators β and β^{μ_1} and introducing new denotations, we get a relation of the following type

$$a_1^{(1)} \cdot \beta^{\delta_1} + \sum_r d_r \beta^{\mu_r} + h \cdot \beta = 0. \tag{10}$$

In this case the elements d_r, h are independent of the choice of the element β, since $c_2,\ldots, c_v$ are independent of it.

Equality (10) shows the mapping

$$\xi\colon a_1^{(1)} \cdot \beta \longrightarrow a_1^{(1)} \cdot \beta^{\delta_1} + \sum_r d_r \cdot \beta^{\mu_r} + h \cdot \beta\ ,$$

where β runs through the set $L(R)$ to be well-defined. Let us choose an ideal $I \in F$ in such a way that $Ia_1^{(1)} \subseteq R$, $Id_r \subseteq R$, $Ih \subseteq R$, $I^{\mu_r} \subseteq R$. If now β runs through $T = L(R)(1 \otimes I^2)$, then the domain of the definition of the corresponding mapping is an ideal in R , while the domain of values lies of R. Moreover,

$$\xi(v(a_1^{(1)} \cdot \beta)) = \xi(a_1^{(1)} \cdot \beta(1 \otimes v)) = a_1^{(1)} \cdot \beta^{\delta_1}(1 \otimes v) +$$
$$+ \sum_r d_r \cdot \beta^{\mu_r}(1 \otimes v) + h \cdot \beta(1 \otimes v) = v\xi(a_1^{(1)} \cdot \beta).$$

Therefore, property (3) of the ring R_F (see **1.4.9**) determines an element t, such that

$$(a_1^{(1)}\cdot\beta)t = a_1^{(1)}\cdot\beta^{\delta_1} + \sum d_r\beta^{\mu_r} + h\cdot\beta. \tag{11}$$

Let $x\in R$, then

$$\begin{aligned}(a_1^{(1)}\beta)\,xt &= (a_1^{(1)}\beta(x\otimes 1))t = a_1^{(1)}\cdot(\beta\otimes(x\otimes 1))^{\delta_1} + \\ &+\sum_r d_r\cdot(\beta(x\otimes 1))^{\mu_r} + h\cdot\beta(x\otimes 1) = (a_1^{(1)}\cdot\beta^{\delta_1})x + \\ &+(a_1^{(1)}\cdot\beta)x^{\delta_1} + \sum_r(d_r\cdot\beta^{\mu_r})x + \sum_r(d_r\cdot\beta)\,x^{\mu_r} + (h\cdot\beta)x\,.\end{aligned} \tag{12}$$

Let us multiply relation (11) from the right by x and subtract it from (12), getting

$$(a_1^{(1)}\cdot\beta)[x,t] = (a_1^{(1)}\cdot\beta)x^{\delta_1} + \sum(d_r\cdot\beta)\,x^{\mu_r}. \tag{13}$$

Let us now consider a module $M = a_1^{(1)}C + \sum_r d_rC$. By lemmas **1.6.9** and **1.6.21**, it can be presented as a direct sum of cyclic modules

$$M = a_1^{(1)}C \oplus a_2C \oplus \ldots \oplus a_nC.$$

In this case if $\beta\in T_1 = \bigcap_{i\geq 2} a_i^{\perp}\cap T$ (it should be recalled that $T = \mathcal{U}(R)\cdot(1\otimes I^2)$ and $d_r = \alpha_r a_r + \ldots$, $\alpha_r\in C$, then $d_r\cdot\beta = \alpha_r a_1^{(1)}\beta$, and equality (13) assumes the form

$$(a_1^{(1)}\cdot\beta)([x,t] - x^{\delta_1} - \sum_r\alpha_r x^{\mu_r}) = 0. \tag{14}$$

As $T^{\perp} = 0$ and the element $a_1^{(1)}$ is not expressed linearly over C through $a_2,\ldots,a_n$, then, by lemma **1.7.9**, at $g = g_i = 1$ we see that $a_1^{(1)}\cdot T_1$ is a nonzero two-sided ideal of R. Let e_0 be its support. Then equality (14) yields

$$e_0\delta_1 + \sum \alpha_r \mu_r \in D_{in}.$$

Since $\{\delta_1, \mu_r\}$ is a strongly independent set of derivations, then $e_0\delta_1 = 0$ which contradicts the supposition that $e(\delta_1) = 1$. Theorem **2.2.2** is proved.

This theorem immediately yields a corollary, which the reader can easily prove using corollaries **1.7.14.**

2.2.11. Corollary. *On a semiprime ring R the following linear generalized identity*

$$\sum_{i=1}^{n} a_i x b_i = 0$$

with the coefficients from R_F *holds,* iff $\sum a_i \otimes b_i = 0$ *in the tensor product* $R_F \otimes_C R_F$.

2.3. Multilinear Differential Identities with Automorphisms

2.3.1. Theorem. *If a multilinear reduced differential identity with automorphisms* $f(x_1,\ldots, x_n) \equiv F(x_i^{g_{ik}\Delta_{ij}} \varepsilon_{ik}) = 0$ *holds on a semiprime ring* R, *then on* R_F *the generalized identity* $F(z_{ijk}\varepsilon_{ik}e_{ij}) = 0$ *is valid, where* $e_{ij} = e(\mu_1)e(\mu_2)\ldots e(\mu_t)$, $\Delta_{ij} = \mu_1\ldots\mu_t$.

Proof. It is sufficient to show that for any nonzero central idempotent e there is a nonzero central idempotent $e_1 \leq e$, such that $e_1F(z_{ijk}\varepsilon_{ik}e_{ij}) = 0$ is an identity on R_F Indeed, in this case, if $a = F(a_{ijk}\varepsilon_{ik}e_{ij}) \neq 0$, we can find an idempotent $e_1 \leq e(a)$, such that $ae_1 = 0$, which is impossible (since $e_1 = e_1e(a) = 0$).

Moreover, it would be enough to consider the case when $f(x_1,\ldots, x_n)$ depends on a single variable. Indeed, giving concrete values to all the variables but the first one, we see that the identity $F(z_{1jk}\varepsilon_{1k}e_{ij}, x_i^{g_{ik}\Delta_{ij}}\varepsilon_{ik}) = 0$, $i \geq 2$. Extending these considera-tions to the letters $x_2,\ldots$, we get the required result. Therefore, herefrom the index i

can be omitted everywhere.

Let us fix a nonzero central idempotent e. Our nearest aim is to find a nonzero idempotent $u \leq e$ and divide $uf(x)$ into a sum of several addends $uf(x) = \sum_{\lambda} uf_{\lambda}(x)$, such that $uf_{\lambda}(x)$ is an identity of an ideal $I \in F$ and all the automorphisms participating in the presentation of $f_{\lambda}(x)$ act in the same manner on the set of all idempotents which are less than or equal to u.

If all the automorphisms act in the same way on $\{u_{\lambda} \leq e\}$, then one can set $u = e$. If this is not the case, then one can choose an idempotent $u_1 \leq e$, on which not all automorphisms act in the same way. Then $f(xu_1) = \sum_{\lambda} f_{\lambda}(x)u_1^{g_{\lambda}}$, where $u_1^{g_{\lambda}}$ are different (central) idempotents. On a finite Boolean ring generated by elements $u_1^{g_{\lambda}}$ let us choose a minimal idempotent u_2. Then for any λ we have either $u_1^{g_{\lambda}}u_2 = 0$ or $u_1^{g_{\lambda}}u_2 = u_2$, in which case u_2 can be chosen in such a way that neither the first nor the second equality hold for all λ. Therefore, we have the decomposition

$$u_2 f(x) = [f(x)u_2 - f(xu_1)u_2] + [f(xu_1)u_2],$$

where all the expressions in square brackets turn to zero on $I \in F$, and contain the least number of terms (here I is an ideal, such that $Iu_1 \subseteq R$). Thus, we have got the required decomposition by induction.

In different addends $uf_{\lambda}(x)$ no similar automorphisms can occur, and therefore, $uF(z_{kj}) = \sum_{\lambda} uF(z_{kj})$, where $f_{\lambda}(x) = uF_{\lambda}(x^{g_k \Delta_j}\varepsilon_k)$ is a reduced polynomial, and it is sufficient to establish the validity of the theorem for the addends $uf(x)$.

Let h be one of the automorphisms occurring in $uf_{\lambda}(x)$. Then all the automorphisms from $uf_{\lambda}(x^{h^{-1}})$ act on $\{e_{\lambda} \leq u\}$ in a trivial way. In this case $uf_{\lambda}(x)$ is a reduced DA-polynomial, as $(h^{-1}g_k)^{-1}h^{-1}g_j = g_k^{-1}g_j$.

Reducing, if necessary, the idempotent u, one can assume that for any k we have either $u\varepsilon_k = u$ or $e\varepsilon_k = 0$, and for any j we have either $ue_j = u$ or $ue_j = 0$. This fact, in particular, implies that the automorphisms

g_k for which $u\varepsilon_k \neq u$, as well as the words Δ_j, for which $ue_j \neq u$, take no part whatsoever in the presentation of $f''(x) \equiv uF_\lambda(x^{h^{-1}})$.

Therefore, going over to a ring uR, one can assume that all ε_k and e_j are equal to the unit; all the automorphisms occurring in $f''(x)$ act on E in an identical way, and $f''(x)$ turns to zero on $I \in F$.

Let us now apply theorem **2.2.2** to the identity $f''(x)$ of the ring I. We have every possibility to do so, since $\varepsilon_k = 1$ and, hence, the automorphisms in the presentation of f'' are pairwise mutually outer. Moreover, $e(\mu_1) \ldots e(\mu_t) = 1$ and, hence, $e(\mu) = 1$ for all the derivations in the presentation of f''. Let

$$f''(x) = \sum_{i=1}^{m} \sum_{j=1}^{n(i)} \sum_{k=1}^{s(i,\,j)} a_{ij}^{(k)} x^{g_i \Delta_j} b_{ij}^{(k)},$$

then, according to theorem **2.2.2**, in the tensor product $I_F \otimes I_F = R_F \otimes R_F$ the following equalities are valid:

$$\sum_k a_{ij}^{(k)} \otimes b_{ij}^{(k)} = 0$$

for all $i, j, 1 \leq i \leq m, 1 \leq j \leq n(i)$.

Hence, at any $z_{ij} \in R_F$ we have $\sum_k a_{ij}^{(k)} z_{ij} b_{ij} = 0$, and, consequently,

$$F''(z_{ij}) = \sum_{i=1}^{m} \sum_{j=1}^{n(i)} \sum_{k=1}^{s(i,\,j)} a_{ij}^{(k)} z_{ij} b_{ij}^{(k)} = 0,$$

which is the required proof.

2.3.2. Corollary. *If a multilinear differential identity with automorphisms is fulfilled on a ring* R, *then it is also fulfilled on the ring of quotients* R_F.

Indeed, all identities of type 1. - 10. are valid for any $x, y \in R_F$, i.e., by theorem **2.1.7**, it is sufficient to establish a corollary for reduced identities. As in theorem **2.3.1**, z_{ijk} can assume any values from R_F, the corollary is proved.

2.3.3. Corollary. *Let* $\Delta = \delta_1 .. \delta_n$ *be a correct word of strongly sets of derivations. In this case* $e(\Delta) = e(\delta_1) .. e(\delta_n)$.

Proof. It should be recalled that the support of Δ is, by definition, equal to the support of the image $e(R^{\Delta})$. Let $f = 1 - e(\Delta)$. Then $fx^{\Delta} = 0$ is a reduced identity, i.e., by theorem **2.3.1**, we have $fze(\delta_1) .. e(\delta_n) = 0$ and, hence, $e(\delta_1) .. e(\delta_n) \le e(\Delta)$. On the other hand,

$$x^{\Delta} = e(\delta_n) \cdot (\ldots(e(\delta_2)(e(\delta_1)x^{\delta_1})^{\delta_2})\ldots)^{\delta_n} = e(\delta_1)\ldots e(\delta_n)x^{\Delta},$$

and, therefore, $e(\Delta) \le e(\delta_1) .. e(\delta_n)$. The corollary is proved.

Example. In theorem **2.3.1** the condition of multilinearity of an identity cannot be omitted. Let R be a perfect field of a characteristic $p > 0$. Let us consider an automorphism $g: x \to x^p$ and let $W(x, y) = x^p - y$. Then $W(x, x^g) = 0$, while $W(x, y) \ne 0$ on R.

For the case of the zero characteristic the condition of multilinearity in theorem **2.3.1** is, naturally, inessential, since then any identity is equivalent to a system of complete linearizations of its homogenous components. Nevertheless, in the case of an arbitrary characteristic the presence of a (non-multilinear) nontrivial identity with automorphisms also gives enough information on the structure of the ring R. For instance, using operators Δ_x^y arising in the process of proving the Martindale theorem, one can obtain a nontrivial polylinear identity for a prime ring. The theorem proved above shows that in this case the initial ring obeys a nontrivial generalized identity, i.e., by the Martindale theorem, its central closure is quite primitive, while its sfield is finite-dimensional over the center.

In the next two paragraphs we shall see that for differential identities (without automorphisms) the condition of multilinearity in the theorem on independence excessive.

2.4. Differential Identities of Prime Rings

In this paragraph R is assumed to be a prime ring.

2.4.1. Theorem. *Let* $f(x_1, \ldots, x_n) = 0$ *be a differential identity of a prime ring* R, *where* $f(x_i) \equiv F(x_i^{\Delta_j})$ *is a reduced D-polynomial.*

Then $F(z_{ij})=0$ *is a generalized identity on* R_F.

Proof. If the characteristic of a field C equals zero, then any identity is equivalent to a system of complete linearizations of its homogenous components. *Therefore, we assume* $p>0$.

Since operators Δ_x^y (see formula (23) in **1.13**) do not violate reducibility (if from the onset we replace $(x+y)^\Delta$ with $x^\Delta + y^\Delta$), then the complete linearization $\psi(y_1,\dots,y_n)$ of the polynomial $f(x_1,\dots,x_n)$ will be a reduced identity and, hence, in the ring R the generalized identity $\Psi(z_{ij})=0$, where $\psi(y_1,\dots)=\Psi(y_1^{\Delta^j})$, holds.

Let us assume that this identity is trivial, i.e., it follows from identities of type $cx=xc$, where $c\in C$. Identities of this type do not change the order of sequence of the variables. It implies that if in the expression $\Psi(z_{ij})$ we collect all the terms with the order of sequence of the variables fixed, then their sum will be also a trivial identity. Identifying the variables in this sum in an appropriate way, we get a senior homogenous part of $F(z_{ij})$, i.e., by induction, $F(z_{ij})=0$ is a trivial identity in R and, hence, in R_F as well.

Let us, finally, assume that $\psi(z_{ij})=0$ is a nontrivial identity. According to the Martindale theorem, RC is then a primitive ring with a nonzero socle H, the sfield of which, T, is finite-dimensional over the center C.

2.4.2.Lemma. *Let a prime ring* R *satisfy the nontrivial generalized identity. In this case any derivation acting in a trivial way on a generalized centroid* C *will be inner for* Q.

Proof. Let e be a primitive idempotent from RC. Then the sfield T is isomorphic to $eRCe$ and is finite-dimensional over the center $ec\cong C$. By lemmas **1.13.5** and **1.13.6**, the algebra of multiplications of the sfield T is equal to a complete ring of endomorphisms $End_C T$. In particular, a projection $l: T \overset{on}{\to} C$ has the form $l(x)=\sum a_i x b_i$, where $a_i, b_i\in T$.

Let now μ be a derivation acting trivially on C, $x, y\in R$. We have $l(exe)=c_1e$, $l(eye)=c_2e$, $c_1, c_2\in C$. Hence, $l(exe)^\mu= c_1e^\mu$, $l(eye)^\mu= c_2e^\mu$. This affords

$$l(exe)l(eye)^\mu = l(eye)l(exe)^\mu = c_1c_2ee^\mu,$$

i.e., we have obtained a *polylinear* differential identity.

Let $\quad 1(exe)^{\mu} = \sum[(a_i e)^{\mu} xeb_i + a_i ex(eb_i)^{\mu}] + 1(ex^{\mu}e) \overset{df}{=}$ $A(x) + 1(ex^{\mu}e)$. Then, if μ is an outer derivation, then we get a multilinear reduced identity

$$1(exe)[A(y) + 1(ey^{\mu}e)] = 1(eye)[A(x) + 1(ex^{\mu}e)]$$

and, hence, by theorem **2.3.1**, in the ring R the following identity is valid:

$$1(exe)[A(y) + 1(eze)] = 1(eye)[A(x) + 1(ez_2e)].$$

Substituting into it $x=0$, we get $0 = 1(eye)1(ez_2e)$, which is impossible, since 1 is an epimorphism onto the center. The lemma is proved.

Let us continue proving the theorem. It follows from lemma **2.4.2** that if two derivations coincide on C, then they are equal modulo D_{in}.

In particular, *the derivations induced on* C *by strongly independent derivations are linearly independent over* C. The inverse statement is also valid: if the induced derivations are linearly independent, then the initial ones are strongly independent (it should be recalled that now C is a sfield).

If the field C is finite, then $c \to c^p$ is its automorphism and $(c^p)^{\mu} = pc^{p-1}c^{\mu} = 0$, i.e., all derivations are trivial on C. It means that all derivations of the ring R are inner and, hence, the reduced identity does not include in its presentation derivations and $f \equiv F$, and it now remains to show that $F=0$ is an identity on R_F. We know (theorem **1.15.1**) that $R_F = \mathbb{L}$ is a complete ring of linear transformations of a space over the sfield, in which case RC is dense in $\mathbb{L}$. Since C is finite, one can find a nonzero ideal I of R, such that $IC \subseteq R$ and, hence, F is an identity on IC. However, IC contains the socle of the ring RC and, therefore, IC is also dense in $\mathbb{L}$. Since ring operations are continuous in a finite topology, the identity F is transferred from IC to $\mathbb{L}$, which is the required proof.

Therefore, the only case left to be considered is that if an *infinite generalized centroid* C.

Step 1. Let us first show that our theorem is valid when $R = C$ is an infinite field of a characteristic $p > 0$.

Since the field C is infinite, then infinite is the fields C^P which consists, as we have seen above, of the constants of all derivations, and, therefore, one can assume that $f(x_1,\dots, x_n)$ is a homogenous (over each of the variables) identities (which does not of course imply that F is a homogenous polynomial over each of the variables).

It should be further remarked that it would be sufficient for us to consider the identities depending on one variable. Indeed, if f depends on several variables, then one can single out one of them and consider the rest variables to be coefficients. In this case the coefficients prove to be differential polynomials which depend on a less number of variables and, therefore, allowing for the fact that a polynomial over an infinite field is uniquely determined by its values, one can carry out an evident induction.

Let us begin with the simplest case.

Let $f(x)=\sum_{i=1}^{m}\gamma_i(x^{\delta_i})^{p^n}$, where $\gamma_i\in C, \{\delta_i\}$ are deriva-tions which are linearly independent over C. Then, if $f(x)=0$ is an identity of C, then for any $c\in C$ we have an equation that has a nonzero solution $z_i=\gamma_i$ in C:

$$\sum_{i=1}^{m} z_i(c^{\delta_i})^{p^n}=0.$$

If c runs through the whole field C, then we get an infinite system of equations, each having a zero solution in C, which fact implies that the rank of every finite subsystem is strictly less than m, and since all the coefficients lie in the field C^{p^n}, then every finite subsystem has its solution in C^{p^n}. Choosing now a subsystem in such a way that the space of its solution over C^{p^n} had a maximum possible dimensionality, we come to the conclusion that the solutions of this subsystem will be those of the whole infinite system. Therefore, there can be found elements $c_1^{p^n},\dots, c_m^{p^n}\in C^{p^n}$, which are nonzero and such that

$$0=\sum c_i^{p^n}(x^{\delta_i})^{p^n}=(\sum c_i x^{\delta_i})^{p^n}.$$

It affords $\sum\delta_i c_i=0$, which contradicts linear independence of the derivations $\delta_1,\dots, \delta_m$.

Let now $f(x)=\sum\gamma_i(x^{\Delta_i})^{p^n}$, where Δ_i are pairwise correct words from a linearly independent set of derivations. Let us carry out induction over a senior word Δ_0. There are no empty words among Δ_i, since the substitution $x=1$ shows the corresponding coefficient to equal zero. If the senior word is as long as 1, then the prove is given above.

Let us consider the expression

$$G(x, y)=F(x, y)-x^{p^n}F(y)-F(x)y^{p^n}.$$

All the terms containing the word Δ_0 in this expression are cancelled. Let $\Delta_0 = \delta^s\Delta'$, where Δ' does not begin with δ and $s < p$. Then the word $\Delta = \delta^{s-1}\Delta'$ is nonempty and the terms containing this word in their presentation will have the form

$$s^{p^n}\gamma_0(x^\delta y^\Delta + x^\Delta y^\delta)^{p^n} + \sum \gamma_k(x^{\delta_k} y^\Delta + x^\Delta y^{\delta_k})^{p^n} =$$
$$= [s^{p^n}\gamma_0(x^\delta)^{p^n} + \sum \gamma_k(x^{\delta_k})^{p^n}](y^\Delta)^{p^n} + \dots \quad ,$$

where summation is carried out by those indices $k \neq 0$, for which the word Δ_k has the form $\delta_k\Delta$ (it should be recalled that all the words are correct and $\Delta_k < \Delta_0$). Therefore, according to the inductive supposition and the remark made at the beginning of this step, we have

$$s^{p^n}\gamma_0(x^\delta)^{p^n} + \sum \gamma_k(x^{\delta_k})^{p^n} = 0,$$

which is impossible due to the case considered above.

An arbitrary homogenous differential expression can be reduced to the form

$$\sum_i \gamma_i (x^{\Delta_1})^{n_1^{(i)}} \dots (x^{\Delta_m})^{n_m^{(i)}} \overset{df}{=} \sum \gamma_i V_i(x),$$

where $n_1^{(i)} + \dots + n_m^{(i)} = n = const$, $\Delta_1 < \Delta_2 < \dots < \Delta_m$ are diffe-rent correct words from a certain finite set of derivations.

Let us expand the indices $n_k^{(i)}$ in the powers of p:

$$n_k^{(i)} = r_{0,k}^{(i)} + r_{1,k}^{(i)}p + r_{2,k}^{(i)}p^2 + \dots \quad , 0 \leq r_{j,k}^{(i)} < p.$$

Let us then transform our identity to the form

$$\sum_i \gamma_i \prod_{k=1}^{m} [x^{\Delta_k}]^{r_{0,k}^{(i)}} [(x^{\Delta_k})^p]^{r_{1,k}^{(i)}} [(x^{\Delta_k})^{p^2}]^{r_{2,k}^{(i)}} \dots = 0. \tag{15}$$

Of importance for us here is the fact that the expressions in square brackets are additive homomorphisms, while the powers over them are strictly less than p.

Let us carry out induction over a number $q = \max\{q_i = \sum_{j,k} r^{(i)}_{j,k}\}$. If $q = 1$, then, due to homogeneity, the expression assumes the form considered above. Let us consider the expression

$$G(x, y) = F(x + y) - F(x) - F(y).$$

If this expression is reduced to form (15), assuming that γ_i depend on x, then we see that the number q for the expression $G_x(y) = G(x, y)$ is reduced. The same is valid for the variable y. Moreover, the terms $G(x, y)$ for which $q_i = q - 1$ will emerge only from the terms $F(x)$, for which $q_i = q$, in which case, according to Newton binomial, the corresponding coefficients will be further multiplied by some of the numbers $r^{(i)}_{j,k}$.

Let us now put to a lexicographical order all the m-tuples of the numbers $(n^{(i)}_m, \ldots, n^{(i)}_1)$, for which $q_i = q$, $\gamma_i \neq 0$, and let $(n^{(a)}_m, \ldots, n^{(a)}_1)$ be a maximal m-tuple, with $n^{(a)}_s$ being the first on the right nonzero number and $r = r^{(a)}_{t,s}$ being a nonzero number with the least possible number t. Then in $G_x(y)$ the coefficient at

$$T(y) \overset{df}{=} [(y^{\Delta_s})^{p^t}]^{r-1}[(y^{\Delta_s})^{p^{t+1}}]^{r^{(a)}_{t+1,s}} \cdots$$
$$\cdots \prod_{k>s}[y^{\Delta_k}]^{r^{(a)}_{0,k}} \cdots$$

will be equal to

$$r\gamma_a(x^{\Delta_s})^{p^t} + \sum_{\substack{1 \le j < s \\ 0 \le l}} \gamma_{i(j,l)}(x^{\Delta_j})^{p^l} + \sum_{0 \le l < t} \gamma_{i(s,l)}(x^{\Delta_s})^{p^l},$$

where $i(j, l)$ is such an index i, that $V_i = (x^{\Delta_j})^{p^l} T(x)$.

A homogenous component of the power p^t of this coefficient has the form

$$r\gamma_a(x^{\Delta_s})^{p^t} + \sum_{1 \le j < s} \gamma_{i(j,t)}(x^{\Delta_j})^{p^t}.$$

Therefore, since the field of constants is infinite and allowing for the inductive supposition and the case considered above, we come to the conclusion that $\gamma_a = 0$. This is a contradiction showing that all coefficients γ_i equal zero.

Step 2. Let us now assume that the coefficients of $f(X)$ lie in a wider ring L containing C as a subring of the center of L, while the derivations are determined only on C. In this case we, naturally, assume that the variables are commuting with L, i.e., $F \in L \otimes_C C[z_{ij}]$.

Let us show that if $f(x_1,..., x_n)$ turn to zero on C, then $F(z_{ij})$ is a zero polynomial. Let us present F as a sum $F = \sum l_k \otimes F^{(k)}(z_{ij})$, where l_k are elements of the ring L which are linearly independent over C. Then, by the condition, $\sum l_k F^{(k)}(c_i^{\Delta_j}) = 0$ for $c_i \in C$, i.e., since the elements l_k are linearly independent, we see that $F^{(k)}(x_i^{\Delta_j}) = 0$, and, therefore, $F^{(k)}(z_{ij})$ are zero polynomials. Hence, F is also zero.

Let us go on considering a general case.

Step 3. Let us prove that $f(x_1,... x_n)$ turn to zero on H.

Let us choose a variable x_1 and expand $f(x_1,... x_n)$ into a sum of components homogenous over x_1

$$f(x_1) = f_0(x_1) + \ldots + f_m(x_1).$$

Since the field C^P is infinite, one can choose different central constants $\xi_0,..., \xi_d$, where d is a general degree of f. Let us choose an ideal $I \in F$, such that $I\xi_j^i \subseteq R$ $0 \leq i, j \leq d$. Then for any $r \in I$ we have

$$f(r\zeta_j) = f_0(r) + f_1(r)\zeta_j + \ldots + f_m(r)\zeta_j^m = 0.$$

As these equalities are valid for all j and the Wandermont's determinant, $\det\|\xi_j^i\| = \prod_{j>i} (\xi_j - \xi_i) \neq 0$, then we have the following equalities:

$$f_0(r) = 0,\ f_1(r) = 0, \ldots, f_m(r) = 0.$$

Extending these considerations onto the rest of the variables $x_2, \ldots, x_n$ we come to the conclusion that all multihomogènous components $f_k(x_1, \ldots, x_n)$ of the polynomial $f(x_1, \ldots, x_n)$ turn to zero on I.

If $k = (k_1, \ldots k_n)$ is a set of the degrees of the component $f_k(x_1, \ldots, x_n)$ over $x_1, \ldots, x_n$, respectively, then, substituting an element of type $c_i^p x_i$ instead of x_i, we get

$$f_k(c_i^p x_i) = \prod_i (c_i^p)^{k_i} \cdot f_k(x_1, \ldots, x_n)$$

i.e., all multihomogenous components and, hence, $f(X)$ turns to zero on the elements of type $c_i^p r_i$, where $c_i \in C$, $r_i \in I$.

Then, instead of x_i let us substitute into the D-polynomial $f(X)$ the sums $\sum_{j=1}^{n} x_i^{(j)}$. In this case the multihomogenous components $f^{(k)}(x_i^{(j)})$ of the polynomial $f(\sum x_i^{(j)})$ will be identities on I. It implies that $f(\sum x_i^{(j)})$ will also turn to zero at $x_i^{(j)} = r_i^{(j)} c_j^p$, where $c_j \in C$, $r_i^{(j)} \in I$. Therefore, $f(\sum x_i^{(j)} c_j^p) = 0$ is an identity on I and, hence, $f(x_1, \ldots, x_n) = 0$ is an identity on IC^p.

Let us prove, finally, that $IC^p \supseteq H$. To this end, it suffices to prove that $H \subseteq RC^p$. Indeed, in this case IC^p is a two-sided ideal in RC^p, while the socle is a simple ring and the least ideal.

Let e be a primitive idempotent, $0 \neq I_1$ be an ideal in R, such that $eI_1, I_1 e \subseteq R$. Then $A = eI_1^2 C^p e$ is a subring of a finite-dimensional (over the center) sfield $T = eRCe$, in which case AC is an ideal in T and, hence, $AC = T$. The ring A has no zero divisors and obeys a generalized identity (even a polynomial one, as a subring of a finite-dimensional sfield). Hence, according to corollary **1.13.7**, there is a nonzero element $ce \in A \cap Ce$. We have $e = (ce)^p (c^{-1})^p \in A$. Therefore, *$RC^p$ contains all primitive idempotents.*

If e is a primitive idempotent, $x \in RC$, then $f = e + + ext(1 - e)$ is also a primitive idempotent, as $\varphi\colon t \to tf$ is a homomorphism from T

on $fRCf$. Consequently, $RC^P \supseteq eRC(1-e)$. Analogously, $RC^P \supseteq (1-e)RCe$. As RC^P is a subring, then it also contains the product $eRC(1-e)\ (1-e)RCe = eRCe$. Hence, $eRC = eRCe + eRC(1-e) \subseteq RC^P$, and, by analogy, $RCe \subseteq RC^P$. And, finally, $H = RCe \cdot eRC \subseteq RC^P$.

Step 4. Let us now finish proving the theorem. A standard proof is reduced to the case of one variable: if $0 = F(x_1^{\Delta_j}, \dots, x_n^{\Delta_j})$, then, giving concrete values to all the variables but the first one, we find an identity $F(z_{1j}, x_2^{\Delta_j}, \dots)$; the next step may be fixing the values of all the variables but the second one, etc. .

So, let $f(x) = F(x^{\Delta_j}) = 0$ be an identity on H. Let us arrange all correct words from the derivations in the order of diminishing $\Delta_1 > \Delta_2 > \dots > \Delta_n$ and put in correspondence to each of them its own variable $\Delta_j \leftrightarrow z_j$. The monomials m_1, m_2 of $z_1, \dots, z_n$ will be called *equivalent* provided their degrees over each of the variables are equal. So, $F(z_j)$ falls into a sum of addends, in each of them the equivalent monomials are assembled in a pairwise fashion (i.e., into a sum of multihomogenous components)

$$F(z_j) = F^{(1)}(z_j) + \dots + F^{(k)}(z_j). \tag{16}$$

Each class of equivalence of monomials is determined by a set of degrees over the variables. For definiteness, let us assume that the monomials from $F^{(1)}(z_j)$ are corresponded to by the largest in the lexicographical sense set $(l_1, \dots, l_n)$. Let us show that $F^{(1)}(z_j)$ is an identity on H.

Let m be an arbitrary natural number and let $h_1, \dots, h_m$ be any elements from H. Let us consider m commutative variables $\xi_1, \dots, \xi_m$. Then the following identity (see step 3) is fulfilled on C:

$$g(\xi_1, \dots, \xi_m) \equiv f(h_1\xi_1 + \dots + h_m\xi_m) = 0.$$

If we apply step 2 to this identity, then we will see that the polynomial $G(\xi_{1j}, \dots, \xi_{mj})$ equals zero as an element of the tensor product

$L \otimes_C C[\xi_{1j},\dots,\xi_{mj}]$, where $g(\xi_1,\dots,\xi_m) = G(\xi_1^{\Delta_j},\dots,\xi_m^{\Delta_j})$. In particular, zero is the sum $G^{(1)}(\xi_{1j},\dots,\xi_{mj})$ of all monomials of G which have a total degree l_1 over the variables $\xi_{11},\dots,\xi_{m1}$; a total degree l_2 over the variables $\xi_{12},\dots,\xi_{m2},\dots$; a total degree l_n over the variables $\xi_{1n},\dots\xi_{mn}$.

Now we have to remark that

$$G^{(1)}(\xi_{1j},\dots,\xi_{mj}) \equiv F^{(1)}(\sum_i h_i\xi_{i1},\dots,\sum_i h_i\xi_{in}). \tag{17}$$

Indeed,

$$(\sum_i h_i\xi_i)^{\Delta_j} = \sum_i h_i\xi_i^{\Delta_j} + \dots, \tag{18}$$

where dots denote sums of the terms in which ξ_i is included with a word less than Δ_j. Therefore, terms of the degree l_1 over $\xi_1^{\Delta_1},\dots,\xi_m^{\Delta_1}$ can arise only from monomials of the polynomial $f(X)$ of the degree l_1 over x^{Δ_1}, since Δ_1 the most senior word, while l_1 is the largest of the occurrences x^{Δ_1} in the monomials from $f(X)$.

Of these monomials, only those with the degree l_2 over x^{Δ_2} give both the degree l_1 over $\xi_1^{\Delta_1},\dots,\xi_m^{\Delta_1}$ and the degree l_2 over $\xi_1^{\Delta_2},\dots,\xi_m^{\Delta_2}$ (at the substitution $x = \sum h_i\xi_i$).

By analogous considerations for words Δ_3, Δ_4,etc., one can show that the terms of the necessary degree are obtained only from the monomials $F^{(1)}(x^{\Delta_j})$. Formula (18) shows now the validity of equality (17).

Since the left-hand part of equality (17) turns to zero at all the values of variables, then $F^{(1)}(z_j) = 0$ at all $z_j \in h_1C + \dots + h_mC$. As the elements $h_1,\dots h_m$ were chosen arbitrarily, then $F^{(1)} = 0$ is an identity on H.

If we now apply the same considerations to the difference

$F(x^{\Delta_j}) - F^{(1)}(x^{\Delta_j})$, we will see that $F^{(2)}(z_j) = 0$ is an identity on H. Continuing this process, we see that all $F^{(k)}(z_j)$, and, hence, $F(z_j)$ are identities on H.

Since ring operations are continuous in a finite topology, and H is dense in L, then $F(z_j) = 0$ is an identity on $L = (RC)_F \supseteq R_F$. The theorem is proved.

2.5 Differential Identities of Semiprime Rings

2.5.1. Theorem. *Let* $f(x_1,.. x_n) = 0$ *be a differential identity of a semiprime ring of a characteristic* $p \geq 0$, *where* $f(x_1,..., x_n) \equiv F(x_i^{\Delta_j})$ *is a reduced polynomial. Then* $F(z_{ij} e_j) = 0$ *is a generalized identity of the ring* R_F, *where* $e_j = e(\Delta_j)$.

Let us first make some remarks.

2.5.2. *Any central idempotent is a constant of every derivation.*

If $e^2 = e$, then $(e^2)^{\mu} = 2ee^{\mu} = e^{\mu}$, wherefrom we get $2e^2e^{\mu} = ee^{\mu}$, and, hence, $ee^{\mu} = 0$, i.e., $e^{\mu} = 2ee^{\mu} = 0$.

2.5.3. *Any ideal of a generalized centroid* C *is differential.*

Indeed, since C is regular, any ideal is generated by its idempotents which, as has been shown above, prove to be constants.

2.5.4. *Let* μ *be an arbitrary derivation from* $D(R)$. *This derivation is induced on on stalks* $\Gamma_p(Q)$, *the induced derivation* $\bar{\mu}$ *belonging to* $D(\Gamma_p(R))$.

Indeed, if $a \in \ker \rho_p$, then $e(a) \in p$ and, therefore, $a^{\mu} = (ae(a))^{\mu} = a^{\mu}e(a) \in \ker \rho_p$, i.e., the formula $\rho_p(x)^{\bar{\mu}} = = \rho_p(x^{\mu})$ determines the derivation $\bar{\mu}$ correctly. Moreover, if $I^{\mu} \subseteq R$, then $(IE)^{\mu} \subseteq RE$ and, since the derivations are continuous, we have $(\widehat{RE})^{\mu} \subseteq \widehat{RE}$ and, hence, $\rho_p(\widehat{IE})^{\mu} \subseteq \Gamma_p$, i.e., by **1.12.5**, $\bar{\mu} \in D(\Gamma_p)$.

2.5.5. Now the derivation $\mu \in D(R)$ can be viewed as a unary operation. *This operation is strictly sheaf.*

Conditions (a) and (b) of strict sheafness (see **1.11.7**) are obvious, condition (b) is checked directly: if $\bar{x}^{\mu} = \bar{y}$, then $p \ni e(x^{\mu} - y)$ and $f = 1 - e(x^{\mu} - y) \notin p$, i.e., $p \in U(f)$, in which case $(x^{\mu} - y)f = 0$ or $(xf)^{\mu} = yf$, i.e., the equality $x^{\mu} = y$ is valid on $\Gamma(U(f))$, which is the required proof.

2.5.6. *Let* $\mu_1,.., \mu_n$ *be a strongly independent set of derivations from* $D(R)$, p *be a point of the spectrum, such that* $p \in \bigcap_{i=1}^{n} U(e(\mu_i))$. *Then* $\bar{\mu}_1,.. \bar{\mu}_n$ *is a strongly independent set of derivations of the stalk* Γ_p.

Let, on the contrary, $\sum_{i\geq 1} \xi_i x^{\bar{\mu}_i} - \xi_0 x + x\xi_0 = 0$ for all $x \in \Gamma_p$, and suitable ξ_i, ξ_0 from the ring $Q(\Gamma_p)$, in which case $\xi_1 \neq 0$. We can find an element a of the stalk, such that $\bar{a}\xi_i \in \Gamma_p$, $i = 0, 1,.., n$; $\bar{a}\xi_1 \neq 0$. Let $\bar{a}\xi_i = \bar{b}_i$. Now we get

$$\sum_{i\geq 1} \bar{a}\xi_i(x\bar{a})^{\bar{\mu}_i} - \bar{a}\xi_0 x\bar{a} + \bar{a}x\bar{a}\xi_0 = 0.$$

It means that on the stalk the following $n+2$-ary predicate is true:

$$\forall x(\sum_{i\geq 1} \bar{b}_i x^{\bar{\mu}_i}\bar{a} + \sum \bar{b}_i x\bar{a}^{\bar{\mu}_i} - \bar{b}_0 x\bar{a} + \bar{a}x\bar{b}_0 = 0).$$

Since the operations of derivations are strictly sheaf, then this predicate is also strictly sheaf and, hence, for a certain $e \notin p$ and suitable preimages $a, b_i \in \hat{RE}$ we have the identity

$$e \cdot (\sum_{i\geq 1} b_i x^{\mu_i} a + \sum b_i x a^{\mu_i} - b_0 xa + axb_0) = 0$$

in which case $e \cdot e(b_1) \notin p$, as $\bar{b}_1 \neq 0$. If we apply theorem **3.3.1**, we get $e \cdot b_1 z_1 a\, e(\mu_1) = 0$, i.e., $e \cdot e(b_1)e(a) \cdot e(\mu_1) = 0$. However, not a single idempotent in this product belongs to p and, hence, it cannot turn to zero, which is a contradiction.

Proof of theorem **2.5.1.** Let us consider a Boolean subring in E, generated by all supports of elements and derivations of $f(x_1,\dots,x_n)$. Let $e_1,\dots,e_m$ be a set of all minimal idempotents of this ring. Then $\{e_k\}$ are pairwise orthogonal idempotents and $f(X)=\sum e_k f(X)=\sum f^{(k)}(X)$, where $f^{(k)}(X)$ stands for the expression obtained from f by way of substituting all coefficients t for $e_k t$ and all derivations δ for δe_k. It is also evident that $F(z_{ij}e_j)=\sum F^{(k)}(z_{ij}e_j)$. Therefore, if we prove the theorem for the identities $f^{(k)}(x_1,\dots x_n)=0$, then it will be proved for the general case as well. Going over to considering a ring e_kR, one *can assume that* $e(t)=e(\delta)=1$ *for all* t,δ *which enter the presentation* of $f(x_1,\dots,x_n)$.

As any idempotent from E is a constant, then the identity $f(x_1,\dots x_n)$ is also valid on the ring RE. Indeed, let $f(a_1,\dots,a_n)\neq 0$, where $a_i=\sum b_{ij}e_j, b_{ij}\in R, e_{ij}\in E$. In a Boolean subring, generated by the idempotents e_{ij}, $e(f(a_1,\dots,a_n))$, let us choose a minimal nonzero idempotents $e_0\leq e(f(a_1,\dots,a_n))$ Then for any e_{ij} we have either $e_{ij}e_0=e_0$ or $e_{ij}e_0=0$, and, hence,

$$0\neq e_0 f(a_1,\dots,a_n)=f(\dots,\sum_j b_{ij}e_{ij}e_0,\dots)=f(\sum_{j\in A_i} b_{ij})e_0=0.\cdots$$

In the latter sums j runs through the set A_i of those indices for which $e_{ij}e_0=e_0$. By lemma **1.6.6**, $f(x_1,\dots,x_n)$ is an identity of the closure $\widehat{RE}$ of the ring RE in Q.

According to proposition **1.12.3**, $\Gamma_p(R_F)\subseteq(\Gamma_p(R))_F$ and, hence, applying the homomorphism ρ_p, we see that on the stalk Γ_p valid is the identity $\bar{f}(x_1,\dots,x_n)=0$ obtained from f by substituting all coefficients with their images and by substituting the derivations μ with the induced ones, $\bar{\mu}$. As $e(\mu_i)=1$, then, by remark **2.5.6**, $\bar{f}(x_1,\dots,x_n)$ is a reduced polynomial, i.e., by theorem **2.4.1**, $\bar{F}(z_{ij})=0$ is an identity on the stalk $\Gamma_p(R_F)$. It implies that the domain of values of $F(z_{ij})$, where the variables run through R_F, is contained in the intersection of all kernels ρ_p, which equals zero. The theorem is proved.

2.5.7. Corollary. *Any differential identity fulfilled on a semiprime ring R will also hold on a ring R_F.*

Proof. Let $e_0, e_2, e_3,\dots, e_p,\dots$ be pairwise orthogonal idempo-tents, such that $\sup_{p\geq 0} e_p = 1$ and the ring $e_p R_F$ has a characteristic p (see 2.1.1). The identity $f(X)=0$ holds on R_F iff every identity $e_p f(X)=0$ is valid on $e_p R_F$. Taking into account the fact that $(e_p R)_F = e_p R_F$, it would be sufficient to establish the validity of the corollary for the case when the ring R has a characteristic $p\geq 0$. By theorems **2.1.7** and **2.5.1**, we have to remark that the trivial identities 4. - 9. are fulfilled on R_F in an obvious way. So, the theorem is proved.

The theorem on the algebraic independence of derivations proves to be sometimes useful in a somewhat different formulation.

2.5.8. Corollary. *Let $\Delta_1,\dots,\Delta_m$ be different correct words of a strongly independent set of derivations of a semiprime ring. Then, if the identity of the following type is valid on R*

$$f_1^{\Delta_1}+\ldots+f_m^{\Delta_m}=0,$$

where $f_1,\dots, f_m$ are multilinear generalized polynomials, then the identities $e(\Delta_1)f_1=0,\dots, e(\Delta_m)f_m=0$ hold on R_F.

Proof. Let us choose a variable x_1. As the identity $F(x_1,\dots)=0$ is equivalent to two identities: $F(0,\dots)=0$ and $F(x_1,\dots)-F(0,\dots)=0$, then, with no generality violated, one can assume that x_1 is encountered in all polynomials $f_1,\dots, f_m$. Let us give concrete values to all the variables but x_1. Let us reduce the initial identity to $F(x_1^{\Delta_i})=0$, where $\{\Delta_i \mid 1\leq i\leq n\}$ are all the subwords of the words $\Delta_1,\dots\Delta_m$. Then, by theorem **2.5.1**, the identity $F(z_{1i})=0$ is valid on R. Let Δ_1 be a senior word among $\Delta_1,\dots\Delta_n$. Applying the Leibnitz formula to the expression $(a_1 x_1 a_2)^{\Delta_1}$, we see when reducing, that the expression $x_1^{\Delta_1}$ arises only from the monomials of the polynomial f_1. Therefore, we have $F(z_{11},0,\dots,0)=f_1(z_{11})=0$, i.e.,

$f(z_{11}) = 0$ is an identity on R. The rest of the proof is completed through induction (and corollary **2.5.7**).

2.5.9. Remark. The above corollary does not always hold for non-multilinear identities. For instance, in the field of the characteristic $p>0$ the identity $(x^p)^{\mu} = 0$ holds for any derivation μ, while $x^p \neq 0$.

Nonetheless, considerations from the proof of corollary **2.5.8** show that in an arbitrary case partial linearizations of the identities $f_i = 0$ over every variable will be fulfilled.

2.6 Essential Identities

In this paragraph we shall consider polylinear differential identities with automorphisms of arbitrary (not necessarily semiprime) rings with a unit.

Let $f(x_1, \ldots, x_n)$ be an arbitrary polylinear differential polynomial with automorphisms and coefficients from a given ring R:

$$f(X) = \sum_{\substack{\pi \in S_n, \\ 1 \leq j \leq m, \\ 1 \leq i \leq k, \\ 1 \leq r \leq l.}} a^{(0)}_{ijr,\pi} x_{\pi(1)}^{g_{j,1}\Delta_{i,1}} a^{(1)}_{ijr,\pi} x_{ijr,\pi}^{g_{j,2}\Delta_{i,2}} \cdot \ldots \cdot a^{(n)}_{ijr,\pi} \qquad (19)$$

2.6.1. Definition. *A generalized monomial* f_π of the polynomial $f(X)$ corresponding to a permutation π is a sum of all the monomials occurring in $f(X)$ in whose presentation the order of the letters x_i is fixed and equal to $x_{\pi(1)}, \ldots, x_{\pi(n)}$:

$$f_\pi = \sum_{i,j,k} a^{(0)}_{ijr,\pi} x_{\pi(1)}^{g_{j,1}\Delta_{i,1}} a^{(1)}_{ijr,\pi} x_{ijr,\pi}^{g_{j,2}\Delta_{i,2}} \cdot \ldots \cdot x_{\pi(n)}^{g_{j,n}\Delta_{i,n}} a^{(n)}_{ijr,\pi}$$

It is evident that $f(X)$ falls into a sum of its generalized monomials.

For semiprime rings the process of reducing a DA-polynomial takes place in every generalized monomial individually, since not a single of identities 1. - 10. changes the order of variables. If in the process of reducing all the generalized monomials turn to zero, then, naturally, f will be a zero DA-polynomial. On the other hand, one often can, using a given (generally speaking, non-reduced) identity, easily discover the nonzero values of its generalized monomials under concrete values of the variables. It immediately gives information on $f(X)$ being a nonzero DA-polynomial, in which case there is no need to reduce it. By the same reason it proves to be useful to study the identities for which there is enough information on the values of generalized monomials.

2.6.2. Definition. Let $f(X)$ be an arbitrary multilinear differential polynomial with automorphisms and coefficients from a given ring R. Let us call *a stempel* of f a two-sided ideal I_F of the ring R generated by all the values of all generalized monomials.

Analogously, a stempel of a set $\Gamma = \{ f_\lambda(X) \}$ will be an ideal $I_\Gamma = \sum_\lambda I_f$.

2.6.3. Definition. The identity $f(X) = 0$ of a ring R is called *essential* if $I_f = R$.

Analogously, a system of identities is called *essential* if its stempel coincides with R.

2.6.4. Theorem. *If a ring R with a unit satisfies an essential system of polylinear identities with automorphisms, then R has a polynomial identity*

$$\sum_{\pi \in S_n} \alpha_\pi x_{\pi(1)} x_{\pi(2)} \cdots x_{\pi(n)} = 0,$$

where the coefficients λ_π can be chosen equal to ± 1.

Let us first prove this theorem for a prime R, and then, using a canonical sheaf, transfer it to semiprime rings (making meanwhile sure that the theorem is also valid for differential identities with automorphisms of a semiprime ring) and, finally, employing the Amitsur's method, we can obtain it in a general form.

2.6.5. Proposition. *If a semiprime ring with a unit, R, obeys an essential system of DA-identities, then a certain standard identity holds in R:*

$$\sum_{\pi \in S_n} \alpha_\pi y_{\pi(1)} y_{\pi(2)} \cdots y_{\pi(n)} = 0.$$

The case of a prime ring. Let us transform all identities to a reduced form $F^{(\lambda)}(x_i^{g_{ik}\Delta_j})$. In this case the stempel of the system does not change. By theorem **2.3.1**, the system of identities $F^{(\lambda)}(z_{ikj}) = 0$ is valid in the ring $R_{\mathcal{F}}$. If the left part of any of these identities is zero as a generalized polynomial, then its generalized monomials are also zero. On the other hand, the generalized monomials of polynomial $F^{(\lambda)}(x^{g_{ik}\Delta_j})$ are the sums of generalized monomials of the polynomials $F^{(\lambda)}(z_{ijk})$, in which substitutions $z_{ijk} = x_i^{g_{ik}\Delta_j}$ have been made. By the same reason all the generalized monomials of the corresponding DA-polynomial will assume nonzero values on R.

Therefore, one of the monomials $F^{(\lambda)}(z_{ijk})$ is nontrivial and we can make use of the Martindale theorem.

According to this theorem and theorem **1.15.1**, we see that $\Gamma' = \{F^{(\lambda)}(z_{ijk})\}$ is an essential system of identities of the ring $R_{\mathcal{F}} = L$ of all linear transformations of a space V over a sfield Δ which is finitely-dimensional over C.

Let β be the dimensionality of V over Δ. Let us denote by P a set of all linear transformations of the space V , the rank of which is strictly less than β.

Let us show that if β is infinite, then in the factor-ring $\bar{L} = L/P$ the condition of minimality for one-sided ideals does not hold. Let $\{e_\gamma, \gamma \in A\}$ be a basis of the space V and $A = A_1 \supset A_2 \supset \dots \supset A_n \supset \dots$ be an infinitely descending chain of subsets in A, such that the powers of $A_i \setminus A_{i+1}$ are equal to β, and let

$$I_n = \{\varphi \in L \mid e_\gamma \varphi = 0 \text{ for all } \gamma \in A_1 \setminus A_n\}.$$

Then

$$I_1 + P/P \supset I_2 + P/P \supset \dots \supset I_n + P/P \supset \dots$$

is an infinitely strictly descending chain of the right ideals of the ring L. Indeed, if $I_n + P = I_{n+1} + P$, then for a transformation v, such that

$$e_\gamma v = \begin{cases} e_\gamma, & \gamma \in I_n \setminus I_{n+1} \\ 0, & \gamma \notin I_n \setminus I_{n+1} \end{cases},$$

we should get a presentation $v = a + p$, where $a \in I_{n+1}$, $p \in P$. Let V_1 be a subspace generated by $\{e_\gamma | \gamma \in I_n \setminus I_{n+1}\}$. Then

$$V_1 = V_1 v \subseteq V_1 a + V_1 p = V_1 p$$

However, $\dim V_1 = \beta$, while $\dim V_1 p \leq \dim V_p < \beta$, which is a contradiction.

It should be further remarked that $\bar{L}$ is a simple ring with a unit. For this purpose it is sufficient to show that P is a maximal ideal. If $l \notin P$, then the dimensionality of Vl equals β. Let $W = \ker l$ and W_1 is a complement of W to V, i.e., $V = W \oplus W_1$. Then l maps W_l on Vl in a one-to-one manner and the dimensionality of W_l equals β. Therefore, one can find a one-to-one linear transformation $l_1 : V \to W_1 \subset V$, in which case $l_1 l : V \to Vl$ is also one-to-one , i.e., there is an inverse mapping $l_2 : Vl \to V$. Defining this mapping further on an extension to Vl one can assume $l_2 \in L$. Thus, if $l \notin P$, then $l_1 l l_2 = 1$ for suitable $l_1, l_2 \in L$, i.e., P is a maximal ideal.

Since $\bar{L}$ is a simple ring with a unit, it coincides with its Martindale ring of quotients, and a system of identities Γ'', obtained by substituting all coefficients with their images under a natural homomorphism $L \to \bar{L}$, is fulfilled on it. The stempel of the system Γ'' is equal to a homomorphic image of the stempel I_Γ and, hence, there are nontrivial identities among Γ''. By the Martindale theorem, $\bar{L}$ is again a primitive ring with a nonzero socle and, moreover, as $\bar{L}_F = \bar{L}$, then $\bar{L}$ is a ring of all linear transformations of a vector space $\bar{V}$ over a sfield $\bar{\Delta}$. Since $\bar{L}$ is a simple ring, then the dimension of $\bar{V}$ over $\bar{\Delta}$ must be finite (when the dimension is infinite, the transformations of a strictly less than the dimension rank form an ideal) and, in particular, the condition of minimality for one-sided ideals is fulfilled in $\bar{L}$. As has been shown above, it is impossible for an infinite β.

Therefore, we come to the conclusion that $R_F = L$ is a ring of all linear transformations of the finite-dimension space $\bar{V}$ over the sfield Δ, in which case the dimension of Δ over C is also finite. Hence, L is finite-dimensional over C and, in particular, in the ring L and, hence, in R as well, a certain polynomial identity is fulfilled. For instance, if m is the dimension of L over C, then a standard identity of the power $m + 1$ holds:

$$S_{m+1}(X) \equiv \sum_\pi (-1)^\pi x_{\pi(1)} \cdot \ldots \cdot x_{\pi(m+1)} = 0$$

The case of a semiprime ring. Let p be an arbitrary point of the spectrum. It suffices to show that on the stalk over this point valid is a system of identities $\bar{\Gamma}_p$, the stempel of which contains the unit. Indeed, in this case on the stalk, according to the above proved, a certain predicate is true:

$$\forall\, x_1 \dots x_n S_n(x_1, \dots, x_n) = 0, \text{ where } n = n(p)$$

Since it is a strictly sheaf predicate (**1.11.16**), then there can be found a neighborhood U_p of a point p, such that this predicate is true on the ring of sections over U_p. Since the spectrum forms a compact space (**1.9.13**) and it has an open covering $X = \cup\, U_p$, then it is possible to find a finite subcovering $X = \bigcup_k U_{p_k}$ as well. Let $r = \max_k n(p_k)$. Then on the ring of global sections the standard identity $S_r(x_1, \dots, x_r)$ holds, which is the required proof (see also corollary **1.11.20**).

In order to construct the system of identities $\bar{\Gamma}_p$, let us transform all identities from Γ to a reduced form $F^{(\lambda)}(x_i^{g_{ik}\Delta_j}\varepsilon_{ik})$. In this case the stempel of the system is not subject to any changes. By theorem **2.3.1**, in the ring R_F valid is the system of identities $\Gamma_p = \{F^{(\lambda)}(z_{ikj}\varepsilon_{ik}e_j) = 0\}$, where $e_j = e(\Delta_j)$. Generalized monomials of the polynomial $F^{(\lambda)}(x_i^{g_{ik}\Delta_j}\varepsilon_{ik})$ are sums of generalized monomials of $F^{(\lambda)}(z_{ikj}\varepsilon_{ik}e_j)$, with the substitutions $z_{ikj} = x_i^{g_{ik}\Delta_j}$ (it should be recalled that e_j can be determined as the least idempotent, such that $e_j y^{\Delta_j} = y^{\Delta_j}$ for all y). It, in particular, implies that the stempel of the system Γ_p contains the stempel of the system Γ, i.e., Γ_p is also an essential system. In this case the system $\bar{\Gamma}_p$ obtained from Γ_p by substituting all coefficients with their images under a natural homomorphism is fulfilled on every stalk. Now the stempel of the system obtained is the image of the stempel of Γ_p, i.e., it contains the unit. Therefore, $\bar{\Gamma}_p$ is the sought essential system. The proposition is proved.

Now we can complete the proof of theorem **2.6.4**, using the Amitsur method.

Let us consider a direct product $\bar{R} = \prod_{\lambda \in A} R_{\lambda}$, where A is a set of countable sequences of the elements belonging to R, $R_{\lambda} = R$. If $\varphi \in \bar{R}$, then we set $\varphi^{g}(\lambda) = (\varphi(\lambda))^{g}$, where g is an automorphism of R. In the ring $\bar{R}$ valid is a system of identities Γ' obtained from Γ by substituting coefficients a with $\prod_{\lambda \in A} a_{\lambda}$, where $a_{\lambda} = a$. In this case the unit of the ring $\bar{R}$ lies in the ideal $I_{\Gamma'}$, i.e., Γ' is an essential system of identities of $\bar{R}$.

Let L be a Baer radical of the ring $\bar{R}$. By theorem **1.1.4 (b)**, L is the least semiprime ideal. However, L^{g} is also a semiprime ideal, and, hence, $L^{g} \subseteq L$. Applying the automorphism g^{-1} to this inclusion, we get $L \supseteq L^{g-1}$, i.e., $L = L^{g-1}$ again, due to the minimality. Therefore, the Baer radical is invariant with respect to all automorphisms of the ring $\bar{R}$, i.e., the action of all automorphisms is induced on the factor-ring $\bar{R} / L$.

In this case on the factor-ring valid is an essential system of identities Γ'' obtained by substituting the coefficients with their images under the natural homomorphism, and by substituting the automorphisms with induced ones. By proposition **2.6.5**, the factor-ring obeys a standard identity S_m.

Let us now choose elements $\varphi_1, \ldots, \varphi_m$ in such a way that $\varphi_i(\lambda) = \lambda(i)$. We have $S_m(\varphi_1, \ldots, \varphi_m) \in L$. Since L consists of nilpotent elements (see **1.1.8**), then for a certain k we have $S_m^k(\varphi_1, \ldots, \varphi_m) = 0$. Choosing now an arbitrary $\lambda = (x_1, \ldots, x_m, \ldots)$, we get

$$0 = S_m^k(\varphi_1, \ldots, \varphi_m)(\alpha) = S_m^k(x_1, \ldots, x_m),$$

i.e., a certain degree of a standard identity holds in the ring $\bar{R}$. The theorem is proved.

2.6.6. Remark. The only reason for eliminating derivations in the formulation of theorem **2.6.4** was that a Baer radical can prove to be no differential ideal, which makes direct application of the Amitsur method impossible. It would be of interest to overcome this obstacle in some way.

The situation becomes clear when R has no additive torsion.

2.6.7. Lemma. *If a ring R has no additive torsion, then a Baer radical is a differential ideal.*

Proof. By theorem **1.1.4** **(a)**, it is sufficient to show that the sum of all nilpotent ideals N_1 is a differential ideal and the factor-ring R / N_1 has no additive torsion. The second statement is obvious: if $nx \in N_1$, then the ideal $n(x_1)$ generated by nx is nilpotent, $n^m (x)^m = 0$ and, hence, $(x)^m = 0$, $x \in N_1$.

Let μ be a derivation, $a \in N_1$. Then the ideal (a) generated by the element a , is nilpotent, $(a)^m = 0$. In particular, for any $x_1, \ldots, x_m \in R \cup \{1\}$, the following equality holds

$$(x_1 a)(x_2 a)\ldots(x_m a) = 0$$

Let us differentiate this equality m times:

$$\sum_{\alpha_1 + \ldots + \alpha_m = m} n_\alpha \mu^{\alpha_1}(x_1 a)\mu^{\alpha_2}(x_2 a)\ldots \mu^{\alpha_m}(x_m a) = 0,$$

where n_λ are some integer coefficients. Among all these addends only one, i.e., $n\mu(x_1 a) \ldots \mu(x_m a)$ contains no cofactor of the type xa, i.e. $n\mu(x_1 a) \ldots \mu(x_m a) \in (a)$. Since $\mu(xa) = \mu(x)a + x\mu(a) \equiv x\mu(a) \pmod{(a)}$, then

$$nx_1 \mu(a) x_2 \mu(a) \ldots x_m \mu(a) \in (a)$$

This means that the ideal generated by $n\mu(a)$ is nilpotent by modulo (a), i.e., $n^m(\mu(a))^{m^2} \subseteq (a)^m = 0$. Therefore, $\mu(a) \in N_1$, which is the required proof.

Now, according to remark **2.6.6**, we come to the next theorem.

2.6.8. Theorem. *If a ring R with a unit obeys an essential system of differential identities with automorphisms and there is no additive torsion in R, then a polynomial identity holds in R.*

2.7 Some Applications. Galois extensions of *PI*-rings, algebraic automorphisms and derivations, associative envelopings of Lie ∂-algebras of derivations

Here we shall consider some immediate corollaries from the theorem on algebraic independence of automorphisms and derivations.

It should be recalled that a ring R is called a *PI-ring* if there exists a polynomial $f(X)$ with integer coefficients, one of which at the monomial of the highest degree is equal to one, which identically turns to zero on R

$$\sum_{i=(i_1,\ldots,i_m)} \alpha_i x_{i_1} x_{i_2} \ldots x_{i_m} = 0$$

2.7.1. Theorem. *Let G be a finite group of automorphisms of a ring R which has no additive $|G|$-torsion. Then, if a subring of fixed elements is a PI-ring, then R is also a PI-ring.*

Proof. Using operators Δ_x^y (see formula (22) in **1.13**), one can easily construct a polylinear polynomial which turns to zero on a ring of fixed elements, one of whose coefficients equals the unit:

$$f(x_1,\ldots,x_m) = \sum_{\pi \in S_m} \alpha_\pi x_{\pi(1)} \cdots x_{\pi(m)} = 0 .$$

Let us assume, for definiteness, that the coefficient λ_1 corresponding to an identical permutation, equals the unit.

Let us first assume that the ring R has an element γ, the trace of which, $t(\gamma) = \sum \gamma^g = 1$ (i.e., R has a unit as well).

Since for any $x \in R$ an element $t(x)$ belongs to a ring of fixed elements R^G, then in the ring R the following identity with automorphisms holds

$$\sum \alpha_\pi t(x_{\pi(1)}) t(x_{\pi(2)}) \ldots t(x_{\pi(m)}) = 0 , \qquad (20)$$

the generalized monomial of which, corresponding to a unit permutation, has the form

$$t(x_1) t(x_2) \ldots t(x_m).$$

Substituting into it $x_1 = x_2 = \ldots = x_m = \gamma$ we see that (20) is an essential identity, i.e., by theorem **2.6.4**, R is a PI-ring.

In order to complete the proof of the theorem, it suffices to embed the ring R, which has no additive $|G|$-torsion, into a ring R_1, which contains an element $\frac{1}{|G|}$ (as $t(\frac{1}{|G|}) = 1$), the ring of invariants of which obeys the polynomial identity.

It can is done in a standard way. Let us consider a set of formal expressions of type $\frac{r+m}{n^k}$, where $r \in R$, m, k are integers, $k \geq 0$, $n = |G|$. Let us assume that $\frac{r+m}{n^k} = \frac{r_1 + m_1}{n^{k_1}}$, iff $n^{k_1} r = n^k r_1$, $n^{k_1} m = n^k m_1$ and let us define the operations of addition and multiplication in a standard way. One can easily see that we get a ring, in which case the mapping $r \to \frac{r+0}{1}$ will be an embedding $R \subset R_1$ (here use should be made of the absence of additive n-torsion).

Further on, the action of the group G is extended onto R_1 by the formula $\left(\frac{r+m}{n^k}\right)^g = \frac{r^g + m}{n^k}$. For a fixed element we have $\frac{r+m}{n^k} = \frac{r^g+m}{n^k}$, i.e., $n^k r = n^k r^g$ or $r = r^g$. This, in particular, implies that for fixed $y, z \in R_1$ their commutator, $[y, z] = yz - zy$, lies in the ring of invariants of R and, hence, the following identity holds in the ring R_1^G:

$$f([y_1, x_1], \ldots, [y_m, z_m]) = 0,$$

which is the required proof.

It should be recalled that in the first part of the proof we made use neither of the restriction on the additive structure of the ring R, nor of theorem **2.6.4** in its full volume. Therefore, we can formulate a more general though less subtle statement.

2.7.2. Theorem. *Let G be a finite group of automorphisms of a ring R with a unit. If an ideal $t(R)$ of the ring R^G obeys a polynomial identity of the degree m and $1 \in Rt(R)^m R$, then R is a PI-ring.*

2.7.3. It should be recalled that an automorphism s of the algebra R over a field F is called *algebraic* if there is a polynomial $\varphi(t) \in F[t]$, such that $\varphi(s) = 0$ in the ring $End_F R$ (i.e., s is an algebraic element considered as a linear transformation of the space R over the field F).

The result presented below was obtained by V.E.Barnaumov under somewhat more severe restrictions.

2.7.4. Corollary. *Let s be an algebraic automorphism of the algebra R over a field F, such that the unit is not a multiple root of its minimal polynomial. Then, if the subalgebra of invariants of s obeys a polynomial identity, then R is a PI-algebra.*

Proof. Let $\varphi(t) = \lambda_0 + \lambda_1 t + ... + \lambda_n t^n$ be the minimal polynomial for s. Then

$$\alpha_0 a + \alpha_1 a x^s + \ldots + \alpha_n a x^{s^n} = 0$$

for all $a, x \in R$. It implies that $(\alpha_0 a, ..., \alpha_n a) \in V(s^0, s^1, ..., s^n)$ (see **1.3.4**). Therefore, by lemma **1.3.4**, the ideal $R\varphi(1)$ is nilpotent. If $\varphi(1) \neq 0$, then this ideal coincides with R and there is nothing to prove. Therefore, herefrom one can assume that $\varphi(1) = 0$ and $\varphi(t) = (t-1)\psi(t)$, in which case $\psi(1) \neq 0$.

As $\varphi(s) = 0$, then $\psi(s) \cdot s = \psi(s)$, i.e., $a^{\psi(s)}$ is a fixed element for any $a \in R$.

Let $R^{\#} = R \oplus F$ be an algebra obtained from R by an external addition of a unit. Let us extend the action of s on $R^{\#}$, assuming that $(a+\alpha)^s = a^s + \alpha, a \in R, \alpha \in F$. Let $f(x_1, ... x_n) = 0$ be a polylinear identity fulfilled in the subalgebra of fixed elements. In this case the following polylinear identity is valid in the subalgebra of fixed elements of the algebra $R^{\#}$:

$$g(x_1, y_1, \ldots, x_m, y_m) \equiv f([x_1, y_1], \ldots, [x_m, y_m]) = 0.$$

This means that the following identity with automorphisms holds in $R^{\#}$:

$$h \equiv g(x_1^{\Psi(S)}, y_1^{\Psi(S)}, \ldots, x_m^{\Psi(S)}, y_m^{\Psi(S)}) = 0.$$

Generalized monomials of the polynomial h have the form $\alpha_\pi x_{\pi(1)}^{\psi(s)} y_{\pi(1)}^{\psi(s)} ... x_{\pi(m)}^{\psi(s)} y_{\pi(m)}^{\psi(s)}$. In particular, the stempel $I_h(R^{\#})$ contains elements $\alpha_\pi 1^{\psi(s)} ... 1^{\psi(s)} = \alpha_\pi \psi(1)^{2m} \in F$, and, hence, $1 \in I_h$, i.e., by theorem **2.6.4**, the ring $R^{\#}$ and, therefore, the ring R, obey a polynomial identity, which is the required proof.

2.7.5. Definition. The derivation μ is called *outer*, if

$\mu C \cap D_{in} = 0$, i.e., the product μC can not be a nonzero inner derivation for $c \in C$. This definition is equivalent to the fact that the one-element set $\{\mu\}$ is strongly independent.

2.7.6. Lemma. *Any derivation* μ *falls into a direct sum of inner,* μ_{in}, *and outer,* μ_{out}, *derivations*

$$\mu = \mu_{in} + \mu_{out}, \quad e(\mu_{in}) \cdot e(\mu_{out}) = 0.$$

Proof. Let us consider an exact sequence of the homomor-phisms of C-modules

$$0 \longrightarrow D_{in} \cap \mu C \longrightarrow \mu C \longrightarrow \mu C / D_{in} \cap \mu C \longrightarrow 0.$$

By lemma **1.6.21**, the last but one module in this chain is both projective and injective, i.e., the chain splits and, hence,

$$\mu C = \mu_{in} C \oplus \mu_{out} C.$$

We have $\mu_{out} C \cap D_{in} \in (\mu C \cap D_{in}) \cap \mu_{out} C = 0$. Besides, $\mu_{in} C \cong e(\mu_{in}) C$, $\mu_{out} C \cong e(\mu_{out}) C$, and, hence, $e(\mu_{in}) \cdot . e(\mu_{out}) = 0$. The lemma is proved.

2.7.7. Definition. An endomorphism φ of the abelian group $< Q, + >$ will be called algebraic if there is a polynomial $f(t) = \sum_{i=1}^{n} r_i t^i$, $r_1, \dots, r_n \in Q$, such that $\sum r_i \varphi^i \equiv 0$ and the coefficients $r_1, \dots, r_n$ generate an ideal which has a zero annihilator in Q (i.e., $\sup\{ e(r_1), \dots, e(r_n) \} = 1$).

2.7.8. Theorem. *Any algebraic derivation* μ *of a semiprime ring* R *with no additive torsion is inner for* Q.

Proof. By condition, $\sum r_i \mu^i = 0$ for suitable $r_i \in Q$, i.e., a differential identity $\sum r_i x^{\mu^i} = 0$ holds in the ring R. Let $\mu_{in} = ad\xi$. Then we have the following reduced identity:

$$\sum r_i (x (ad\xi)^i) + \sum r_i x^{\mu_{out}^i} = 0.$$

Therefore, by theorem **2.3.1**, the following identity is valid

$$\sum r_i(x(ad\xi)^i) + \sum r_i e(\mu_{out}) z_i = 0.$$

Substituting into this identity $x = z_j = 0,\ j \neq i$, we get $r_i e(\mu_{out}) = 0$. Therefore, the ideal generated by the elements $r_1, \dots, r_n$ annihilates the idempotent $e(\mu_{out})$, i.e., $e(\mu_{out}) = 0 = \mu_{out}$. It implies that $\mu = \mu_{in}$ is an inner derivation, which is the required proof.

It should be remarked that the theorem allows enhancements. One can, for instance, require that μ be algebraic only on certain subrings (or termal sets).

2.7.9. Corollary. *Let* R *be a prime ring of a zero characteristic and let us assume that the restriction of derivation* μ *on a nonzero left ideal,* Rb, *proves to be algebraic. Then* μ *is an inner derivation for* Q.

Proof. By analogy with the above proof, we get a reduced identity

$$\sum r_i(xb(ad\xi)^i) + \sum r_i x^{\mu^i_{out}} b + \dots = 0,$$

wherefrom we have $r_i zbe(\mu_{out}) = 0$ for all i and, hence, $\mu_{out} = 0$. The corollary is proved.

2.7.10. $D(R)$. By $\Phi(L)$ we shall denote an associative subring in $End(Q,+)$ generated by L and the operators of left multiplications, l_c, $c \in C$. Let L_{in} be a set of all inner derivations from L. $B(L)$ will designate an associative C-subalgebra in Q, generated by the elements corresponding to the derivations from L_{in}. In the considerations to follow we shall assume R to have a characteristic $p \geq 0$.

2.7.11. Corollary. *If a derivation* μ *belongs to* $\Phi(L)$, *then there is an element* $\delta \in L$, *such that* $\mu - \delta$ *is an inner derivation and the corresponding element from* Q *belongs to* $B(L)$.

Proof. By condition, $\mu = \phi(\delta_1, \dots, \delta_n)$, where $\delta_1, \dots, \delta_n \in L$; ϕ is a polynomial with the coefficients from l_C. Let us transform the expression $x^{\phi(\delta_1, \dots, \delta_n)}$ to a reduced form $x^{f(\delta'_1, \dots, \delta'_n)}$. Then f will be a polynomial with the coefficients from $l_{B(L)} r_{B(L)}$ written to the right from the words (by r_b we denote here the operator of right multiplication by b).

Let us consider a C-module M generated by a (strongly independent) set of derivations $\delta'_1,\ldots,\delta'_m$, and let $M' = M + \mu C$. Let, then, M'_{in} be a submodule of inner derivations from M'. In this case $M'_{in} \cap M = 0$, since $\delta'_1,\ldots,\delta'_m$ is a strongly independent set.

By lemmas **1.6.21**, the finitely-generated submodule $A = M' / M'_{in}$ of the module $End Q / D_{in}$ is projective and, therefore, the exact sequence $0 \to M'_{in} \to M' \to A \to 0$ splits. In particular, M'_{in} is a finitely-generated and, hence, an injective module.

Therefore, we see that the injective submodule $M + M'_{in}$ is split from M' by a direct summand $M' = (M + M'_{in}) \oplus \mu' C$. Besides, one can assume that $\mu' \equiv \mu' (\mathrm{mod}\ M + M'_{in})$, i.e., $\mu = \mu' + \delta + ad\xi$, where $\delta = \sum \delta'_i \lambda_i \in L$, $\lambda_i \in C$, $\xi \in Q$.

The set $\delta'_1,\ldots,\delta'_m, \mu'$ is, by construction, strongly independent. Therefore, the identity $x ad\xi + x^{\mu'} + \sum \lambda_i x^{\delta'_i} - x^{f(\delta'_i)} = 0$ is a reduced one. As in the presentation of this identity the derivation μ' occurs but once, then, by theorem **2.3.1**, the following identity is valid in R:

$$x\, ad\xi + y e(\mu') + \sum \lambda_i e(\delta'_i) z_i + F(z_{ij}, x) = 0. \tag{21}$$

Substituting $x = z_i = z_{ij} = 0$ into it, we get $e(\mu') = 0$, i.e., $\mu' = 0$. Thus, $\mu - \delta = ad\xi$. Then, substituting $z_i = z_{ij} = 0$ into identity (21), we obtain an equality $x ad\xi = x \sum l_{b_i} r_{d_i}$, where $b_i, d_i \in B(L)$.

Thus, $l^{\xi} - r^{\xi} = \sum l_{b_i} r_{d_i}$. Allowing for the fact that $l_Q r_Q \cong l_Q \otimes_C r_Q$, we obtain an equality $l_{\xi} = \sum c_i l_{b_i}$, where $c_i \in C$. Hence, $\xi = \sum c_i b_i \in B(L)$.

The corollary is proved.

2.7.12. Corollary. *Let L be a differential Lie C-subalgebra in $D(R)$, where R is a semiprime ring of a characteristic $p \geq 0$, and C is its generalized centroid. Then, if L contains no nonzero inner derivations for Q, then $\Phi(L)$ is isomorphic to the universal enveloping for L.*

Proof. Let T be a tensor algebra of a space $\hat{L} = L + l_C$ over a field F of central constants. The identical mapping $L \to L$ determines the epimorphisms of the associative F-algebras $\psi: T \to \Phi(L)$. Let A be an associative enveloping for L. Then we have a C-linear homomorphism of (restricted) Lie algebras $\xi_A: L \to A^{(-)}$, such that the action of L on C coincides with that of $\xi_A(L)$, and the set $\xi_A(L)$, together with the space $1 \cdot C$, generates A as an associative algebra. Let us extend the mapping ξ_A to $\hat{L}$, setting $\xi_A(1_C) = 1 \cdot C$. It is sufficient now to prove that the kernel of ξ_A is contained in the kernel of ψ.

If $f \in T$, then one can consider f as a formal associative polynomial of derivations with non-commutative coefficients from 1_C. In this case the condition $f \in \ker\xi$ implies that the linear differential identity $x^f = 0$ holds in the ring R. By theorems **2.1.7** and **2.3.1**, this identity results from identities of type 4. - 8. (It should be recalled that there exist no nontrivial linear generalized identities, which can, for instance, be seen from theorem **2.2.2**). Therefore, the task now is to show that the elements T corresponding to identities 4. - 10., belong to the kernel of ψ.

The elements corresponding to identities 1. - 7., 9. lie in the kernel of ψ because ξ_A is a C-linear homomorphism of restricted Lie algebras. Corresponding to identity 4. are the elements of the type $1_C\mu - \mu 1_C = 1_{C^\mu}$ which are transformed, under the action of ξ_A to the elements of the type $(1_A \cdot c)(\mu\xi_A) - (\mu\xi_A)c - 1_A \cdot c^\mu = c^{(\mu\xi_A)} - c^\mu = 0$. Identities of type 8 are absent as, by condition, L contains no nonzero derivations which are inner for Q. The corollary is proved.

2.7.13. Corollary. *Let G be a finite group of automorphisms of a semiprime ring R and let μ be a derivation of R acting in a trivial way on a fixed ring R^G. Then, if $e(t(R)) = 1$, then μ is an inner derivation for Q. Here $t(x) = \sum_{g \in G} x^g$.*

Proof. Let p be an arbitrary simple number or zero, e_p be a corresponding idempotent (see **2.1.1**). Let us transform the expression $e_p t(x) = \sum_{g \in G} e_p x^g$ to a reduced form $\sum_{i,j} a_{ij} x^{g_i} b_{ij} \varepsilon_i$. Then in the ring $e_p R$ the following reduced identity is valid:

$$e_p t(x)^\mu \equiv \sum_{i,j} (a_{ij}^\mu x^{g_i} b_{ij} + a_{ij} x^{g_i} b_{ij}^\mu)\varepsilon_i +$$

$$+ \sum_{i,j} a_{ij}[x^{g_i} ad\xi]\varepsilon_i b_{ij} + \sum_{i,j} a_{ij} x^{g_i \mu_{out}} b_{ij}\varepsilon_i = 0.$$

By theorem **2.3.1**, we get the following identities in $e_p R$:

$$\sum_j a_{ij}\, y b_{ij}\varepsilon_i e(\mu_{out}) = 0$$

Substituting into these identities the elements x^{g_i} instead of y and summing over al i, we get:

$$0 = \sum_{i,j} a_{ij} x^{g_i} b_{ij} e(\mu_{out})\varepsilon_i = e_p t(x) e(\mu_{out}).$$

As $e(t(R)) = 1$, then $e_p e(\mu_{out}) = 0$ and, hence, $\mu_{out} = 0$ because $\sup\{e_p\} = 0$. The corollary is proved.

It should be remarked that the condition $e(t(R)) = 1$ is equivalent to the fact that the annihilator of the ideal $Rt(R)R$ in the ring R equals zero. This condition is, for instance, fulfilled if R has no additive $|G|$-torsion, in which case it is the annihilator of the set $t(R)$ that equals zero (see lemma **1.3.4**).

REFERENCES

S.Amitsur [5,7];

V.E.Barbaumov [12];

K.I.Beidar [14];

K.I.Beidar, A,V,Mikhalev [15];

V.K.Kharchenko [71,73,74,75,78,81,84];

E.Posner [138];.

L.Rowen [142];

A.M.Slin'ko [145].

CHAPTER 3
THE GALOIS THEORY OF PRIME RINGS

The contribution of the Galois theory for a class of rings is commonly understood as a proof of the principal correspondence theorem of definite types of finite (or reduced-finite) groups of automorphisms and those of subrings from a given class.

Let R be a ring, S a subring and G a group of automorphisms (i.e., a subgroup of the group of all automorphisms) of the ring R. An element $a \in R$ is called *G-invariant* if $a^g = a$ for every $g \in G$. A set of all G-invariants is denoted by R^G or by $\mathbb{I}(G)$. A set of all automorphisms for which the elements of S serve as invariants is denoted by $A(S)$. It is evident that $\mathbb{I}(G)$ is a subring of the ring R, and $A(S)$ is a group of automorphisms.

As was the case in the classical theory of fields, the group $A(S)$ is called a *Galois group of the ring* R over S. A subring S is called a *Galois subring* of R, and R is, respectively, a *Galois extension of the ring* S, if $S = \mathbb{I}(G)$ for a group of automorphisms G.

The correspondences $G \to \mathbb{I}(G)$ and $S \to A(S)$ invert the inclusion relations, i.e., if $G_1 \subseteq G_2$, then $\mathbb{I}(G_1) \supseteq \mathbb{I}(G_2)$ and if $S_1 \subseteq S_2$, then $A(S_1) \supseteq A(S_2)$. We also have

$$A(\mathbb{I}(G)) \supseteq G\ ,\ \mathbb{I}(A(S)) \supseteq S\ ,$$

which immediately yield

$$\mathbb{I}(A(\mathbb{I}(G))) = \mathbb{I}(G)\ ,\ A(\mathbb{I}(A(S))) = A(S).$$

Therefore, the mappings under discussion set a one-to-one correspondence between Galois groups and Galois subrings. And, hence, the groups $A(S)$ and the subrings $\mathbb{I}(G)$ are of primary interest in the Galois theory.

In order to prove the correspondence theorem in a class of rings $\aleph$, it is necessary (and sufficient) to answer the following questions:

(1) Under what conditions does a subring of fixed elements for a group G of automorphisms of a ring $R \in \aleph$ belong to $\aleph$?

(2) When will a group G which obeys condition (1), be a Galois group ?

(3) Under what conditions is an intermediate ring $S \in \aleph$, $I(G) \subseteq S \subseteq R$ a Galois subring ?

An analogous approach is also possible for studying derivations. In this case the role of Galois objects is played by differential (restricted) Lie Z-algebras and the subrings of constants of such algebras. In this case the same problems as for groups, i.e., (1) - (3) arise.

In the three chapters to follow we are going to develop the Galois theory for automorphisms and derivations in classes of prime and semiprime rings. As above, we shall consider a somewhat more general situation, assuming that the automorphisms lie in $A(R)$, while derivations in $D(R)$ (see **1.7**).

3.1. Basic Notions

Let R be a prime ring, G be a group of automorphisms. It should be recalled that by Φ_g we denote a set of all elements $\varphi \in R_F$ such that $x\varphi = \varphi x^g$ for all $x \in R_F$ (see **1.7.5**, **1.7.6**). For these sets valid are the relations $\Phi_g \Phi_h \subseteq \Phi_{gh}$ (see formula (8) in **1.7**) which, in particular, afford that Φ_g is a linear space over a generalized centroid of the ring R. Moreover, corollary **1.7.9** states that Φ_g will be nonzero iff g is an inner automorphism for Q.

3.1.1. Definition. *The algebra of a group of automorphisms* G is a C-algebra $B(G) = \sum_{g \in G} \Phi_g$.

Therefore, the algebra of the group has a basis of elements which correspond to the automorphisms which are inner for Q, i.e., it is an inner part of the group in a ring form.

If G is a finite group, then its algebra $B(G)$ will be finite-dimensional over C. Of a finite order will also be a factor-group G / G_{in}, where G_{in} is a normal subgroup of all inner for Q automorphisms.

3.1.2. Definition. A group G is called reduced-finite if its algebra $B(G)$ is finite-dimensional, while the factor-group G / G_{in} is finite. In this case the number $\dim_C B(G) \cdot |G / G_{in}|$ is called a *reduced order* of the group G.

3.1.3. Definition. A reduced-finite group G is called an *M-group* (*a Maschke group*) if its algebra is semiprime (or, which is equivalent, semisimple).

3.1.4. Definition. A group G is called an *N-group* (*a Noether group*) if any inner automorphism of the ring Q corresponding to an invertible element from $B(G)$ belongs to G.

3.1.5. Lemma. *Any Galois group is a Noether group.*

Proof. Let $G = A(S)$. If $s \in S$, then for all $g \in G, \varphi \in \Phi_g$ we have $s\varphi = \varphi s^g = \varphi s$, i.e., Φ_g and, hence, $B(G)$ as well, are element-wise commutative with the subring S. In particular, if b is an invertible element from $B(G)$ then the inner automorphism $\tilde{b}(x) = b^{-1}xb$ acts trivially on S, i.e., $\tilde{b} \in G$, which is the required proof.

The value of Maschke groups will be discussed later (see **3.6**), and now we go over to the central notion.

3.1.6. Definition. A Maschke group G is called *regular* if it is also a Noether group.

3.1.7. Definition. A regular group G is called *quite regular* if its algebra is simple.

Let us pay attention to the fact that all these notions (but that of a reduced order) refer only to the inner part of the group G. Hence, any finite group of automorphisms which contains no non-trivial automorphisms inner for Q, will be quite regular. Such a group will be naturally called a (*finite*) *group of outer automorphisms*.

The basic notions referring to subrings will be introduced later. Here we are going to dwell only on the finiteness of the ideal over a subring. The point is that in the classical Galois theory of importance is the fact that the field has a finite dimension over a Galois subfield. Under general conditions this is not the case, while a certain relation of finiteness (developed by A.I.Shirshov when studying rings with polynomial identities) is valid and even plays an important part in a number of cases.

3.1.8. Definition. A subring A of the ring R is called *right-Shirshov finite* over a subring $D \in R$, if there are elements $r_1, \ldots, r_n \in R$, such that $A \subseteq r_1 D + \ldots + r_n D$.

3.1.9. Definition. A two-sided ideal I of the ring R is called *right-locally finite over a subring* $D \subseteq R$ if any finitely-generated right ideal of the ring R lying in I is a right-Shirshov finite subring over D.

3.2 Some Properties of Finite Groups of Outer Automorphisms

This paragraph is of an auxiliary character and all its statements will be further significantly generalized. However, the information accumulated in the preceding chapters makes it possible to obtain these particular cases, which could be of great help for theoretical endeavours, with no special effort at all.

3.2.1. Theorem. *Let G be a finite group of outer (for Q) automorphisms of a prime ring R. Then the centralizer of a ring of invariants R^G in the ring of quotients R_F is equal to the generalized centroid C.*

Proof. Let an element a commute with all fixed elements of the ring R. Then the following identity with automorphisms is valid

$$a \sum_{g\in G} x^g - \sum_{g\in G} x^g \, a = 0 \tag{1}$$

under all $x \in R$, such that $x^g \in R$ for any $g \in G$. As $G \subseteq A(R)$ and G is finite, then there exists a two-sided ideal $I \neq 0$, such that $I^g \subseteq R$ for all $g \in G$, i.e., (1) is an identity on I. By theorem **2.2.2**, we have $a \otimes 1 - 1 \otimes a = 0$, i.e., $a \in C$, which is the required proof.

3.2.2. Theorem. *Any finite group of outer automorphisms of a prime ring is a Galois group.*

Proof. Let h be an arbitrary automorphism , $h \in A(R)$, which acts trivially on R^G, i.e., $h \in A\!\!I(G)$. Then, as was the case in the preceding theorem, we can find an ideal I on which the following identity with automorphisms is fulfilled:

$$\sum_{g\in G} x^{gh} - \sum_{g\in G} x^g = 0$$

This identity cannot be a reduced one, since theorem **2.2.2** would immediately yield the equality $1 \otimes 1 = 0$. The irreducibility of this identity implies that one of the automorphisms $(g_i h)(g_j h)^{-1}$, $(g_i h)g_k^{-1}$, $g_i \neq g_j \in G$ is inner. Let $g_i h g_k^{-1} = \tilde{b}$. Then $b \in AI(G)$ and, hence, b commutes with all fixed elements, i.e., by the preceding theorem, $\tilde{b} = 1$. Thus, $h = g_k g_i^{-1} \in G$, which is the required proof.

3.2.3. Theorem. *A fixed ring of a finite group of outer automorphisms of a prime ring is prime.*

Proof. Let $aR^G b = 0$. Then we have an identity on an ideal $I \neq 0$:

$$a \sum_{g \in G} x^g b = 0.$$

By theorem **2.2.2**, we have $a \otimes b = 0$ and, hence, either $a = 0$ or $b = 0$, which is the required proof.

3.2.4. Theorem. *Let G be a finite group of outer automorphisms of a prime ring R. Then there exists a nonzero ideal of the ring R which is right-locally finite over R^G.*

Proof. By theorem **1.7.10**, there are elements t_j, v_j, such that $a = \sum_j t_j v_j \neq 0$; $\sum_j t_j v_j^g = 0$ for all $g \in G$, $g \neq 1$. Setting $\tau(x) = \sum_{g \in G} x^g$, we get

$$ax = \sum_j t_j \tau(v_j x).$$

If now x runs through the ideal I, such that $I^g \subseteq R$ for all $g \in G$, then we see that $aI \subseteq \sum t_j R^G$. It remains to notice that the ideal RaI is locally-finite over R^G, since $\sum_i r_i a I \subseteq \sum_{i,j} r_i t_j R^G$. The theorem is proved.

3.2.5. Theorem. *Let G be a finite group of outer automorphisms and let $R \supseteq S \supseteq R^G$. Then any automorphism $\varphi\colon S \to R$, which is trivial on R^G, is extended to an automorphisms from G*

Both this theorem and somewhat more complex next ones will be deduced below from general results (see **3.9** and **3.11**).

3.2.6. Definition. A subring S is called *rationally complete* in a prime ring R if the condition $Ax \subseteq S$ for $x \in R$, and a nonzero ideal A of S imply $x \in S$. For a prime subring S this condition is seen to be equivalent to $S_F \cap R = S$.

3.2.7. Theorem. *Let G be a finite group of automorphisms $R \supseteq S \supseteq R^G$. Then S will be a Galois subring of R iff it is rationally complete.*

The above theorems allow the reader to easily formulate the correspondence theorem for outer groups, and show the situation to be convenient for this case.

Let us now try to elucidate , in general terms, another extreme situation , that of inner automorphisms, from the point of view of the notions proved earlier. Since a fixed ring of a reduced-finite group of inner automorphisms coincides with a centralizer of the finite-dimensional algebra $\mathcal{B}(G)$, then we come to the necessity to study (from a general standpoint) centralizers of finite-dimensional algebras in prime rings. It is still more important since practically any finite-dimensional algebra with a unit is generated by its invertible elements and can, therefore, serve as an algebra of a reduced-finite group (with the only exception for algebras over the two-element field $GF(2)$, the factor-algebras by the radical of which have two or more direct addends isomorphic to $GF(2)$).

3.3 Centralizers of Finite-Dimensional Algebras

3.3.1. Definitions. A finite-dimensional algebra with a unit, B, over a field C is called *centralizable* if for any prime ring R with a generalized centralizer C, such that $B \subseteq Q(R)$, the centralizer of B in R is nonzero.

It should be recalled that a finite-dimensional algebra with a unit is called *Frobenius* if it has a non-degenerate associative bilinear form. We will call the algebra B *partially Frobenius* (p.F.) if it has a partially non-degenerate associative bilinear form, i.e., if there exists a nonzero right, ρ, and a nonzero left, λ, ideals, and a bilinear form $\rho \times \lambda \to C$, such that $(rb, l) = (r, bl)$ for all $r \in \rho$, $l \in \lambda$, $b \in B$, and if the equality $(r, \lambda) = 0$ is valid only at $r = 0$, while the equality $(\rho, \lambda) = 0$ is valid only at $l = 0$.

These notions can be also formulated in terms of conjugated modules. It should be recalled that for the left B-module, λ, we denote by λ^* the conjugated right B-module, $\lambda^* = Hom_C(\lambda, C)$, with the module structure given by the formula $(\varphi b)(l) = \varphi(bl)$. Accordingly, for the right B-module,

ρ, by ρ^* we shall denote the conjugated module $\rho^* = Hom_C(\rho, C)$, with the module structure given by the formula $(b\psi)r = \psi(rb)$. The latter formula can be more conveniently presented as $r(b\psi) = (rb)\psi$, by locating the argument to the left from the function.

On conjugated modules, there naturally arises a non-degenerate associative bilinear form $\lambda^* \times \lambda \to C: (\varphi, 1) = \varphi(1)$ and, respectively, $\rho \times \rho^* \to C: (r, \psi) = r\psi$. In the finite-dimensional case, the inverse is also valid. If for a couple of modules, ρ and λ, right and left, respectively, there is a non-degenerate bilinear associative form $\rho \times \lambda \to C$, then $\lambda \cong \rho^*$ and $\rho \cong \lambda^*$.

The definition of a Frobenius algebra can be now given by the formula $B^* \cong B$, where, on the one hand, B is a left regular module, and on the other hand, it is a right module. Being partially Frobenius means the existence of a nonzero one-sided ideal ρ, such that its conjugated module is isomorphic to another one-sided ideal.

It would now be natural to give one more definition, that of algebras which are *"Frobenius by parts"*, i.e., such that the sum of all right ideals conjugated with the left ones coincides with the whole algebra B. Such algebras have been studied elsewhere and are called *quasi-Frobenius* (*QF*).

Both Frobenius and quasi-Frobenius algebras originated in the theory of presentation of groups and finite-dimensional algebras. The two following theorems characterizing these algebras are well known, and we refer the reader to see the proof elsewhere ([36]. theorems **61.2** and **61.3**).

3.3.2. Theorem. *The following statements on a finite-dimensional algebra with a unit, B, are equivalent:*

(1) *the algebra B is Frobenius;*

(2) *there is a linear form $\varepsilon \in B^*$, the kernel of which contains no nonzero one-sided ideals of the algebra B;*

(3) *for all left ideals λ and right ideals ρ of the algebra B the following equalities are valid:*

$$\dim r(\lambda) + \dim \lambda = \dim B,$$
$$\dim 1(\rho) + \dim \rho = \dim B,$$

where $r(\lambda) = \{ b \in B \mid \lambda b = 0 \}$, $1(\rho) = \{ b \in B \mid b\rho = 0$ are the right and left, respectively, annihilators in B (sometimes they are denoted by $ann_B^r(\lambda)$ and $ann_B^1(\rho)$, respectively).

3.3.3. Theorem. *The following statements on a finite-dimensional algebra with a unit, B, are equivalent:*

(1) *the algebra B is quasi-Frobenius;*

(2) *the left* B*-module* B *is injective;*
(3) *the right* B*-module* B *is injective;*
(4) *for every left ideal* λ *and right ideal* ρ *of the algebra* B *the following equalities are valid:*

$$l(r(\lambda)) = \lambda, \quad r(l(\rho)) = \rho$$

3.3.4. Examples of Frobenius algebras.

(a) *A group algebra of a finite group is Frobenius.*

If $B = C[G]$ and $b = \sum c_g g$ is an arbitrary element of a group algebra, then we set $\varepsilon(b) = c_e$, where e is a unit element of the group G. Then ε is a linear form, the kernel of which contains no nonzero one-sided ideals.

(b) *The universal p-enveloping of a finite-dimensional Lie p-algebra is Frobenius.*

If $\mu_1, \dots, \mu_n$ is the basis of a Lie p-algebra L, then the basis of its p-enveloping are the words of the type $\mu_1^{m_1}, \dots, \mu_n^{m_n}$, where $0 \le m_i < p$. Let us denote by $\varepsilon(b)$ the coefficient at $\mu_1^{p-1} \dots \mu_n^{p-1}$ in the decomposition of the element $b \in u(L)$ over this basis. Then ε is the sought linear form.

(c) *A semi-simple algebra is Frobenius.*

It suffices to remark that all irreducible left (right) modules of a simple algebra are mutually isomorphic and any module is a direct sum of irreducible modules. Hence, from dimensional considerations, $({}_B B^*) \cong B_B$ (provided B is simple). For the semi-simple algebra B we have a decomposition into a sum of simple modules, $B = B_1 \oplus \dots \oplus B_n$. Therefore,

$$B^* = B_1^* \oplus \dots \oplus B_n^* \cong B_1 \oplus \dots \oplus B_n \cong B.$$

3.3.5. Lemma. *In a quasi-Frobenius algebra the annihilator of the intersection of one-sided ideals is equal to the sum of their annihilators:*

$$l(\rho_1 \cap \rho_2) = l(\rho_1) + l(\rho_2),$$
$$r(\lambda_1 \cap \lambda_2) = r(\lambda_1) + r(\lambda_2).$$

Proof. Let us, for instance, check the first equality. According to statement (4) of theorem **3.3.3**, it is sufficient to prove that the right annihilators of both parts of the equality coincide, i.e., that $\rho_1 \cap \rho_2 =$

$r(l(\rho_1)+l(\rho_2)$. However, $r(l(\rho_1)+l(\rho_2)=rl(\rho_1)\cap rl(\rho_2)=\rho_1\cap\rho_2$, which is the required proof. Thus, the lemma is proved.

3.3.6. In linear algebras well known is the fact that the conjugated modules ρ, λ have *mutually dual bases*, i.e., the bases $\{a_i\}_{i=1}^n$, $\{b_i\}_{i=1}^n$ over C, such that

$$(a_i, b_j)=\begin{cases}1, & \text{if } i=j,\\ 0, & \text{if } i\neq j,\end{cases}$$

where (ρ, λ) is the corresponding form.

Moreover, for any basis $\{a_i\}$ of the module ρ there is a dual basis $\{b_i\}$ of the conjugated module ρ^*.

The most important property of the operation of module conjugation is that it is an exact contravariant functor, i.e., for any exact sequence of the homomorphisms of modules

$$0\longrightarrow E\longrightarrow F\longrightarrow D\longrightarrow 0$$

we have an exact sequence

$$0\longrightarrow D^*\longrightarrow F^*\longrightarrow E^*\longrightarrow 0$$

and, besides, the conjugation is involutory, i.e., $(A^*)^*\cong A$ for a finite-dimensional module A.

We shall leave these facts as exercises for the reader (if he is not familiar with them) and go over to the basic topic.

3.3.7. Theorem. *An algebra* B *is centralizable iff it is partially Frobenius.*

Let us first prove two lemmas.

3.3.8. Lemma. *If the dimension of the algebra* B *is greater than one, then the free product* $B *_C C[X] \overset{df}{=} B\langle X\rangle$ *is prime and its generalized centroid equals* C.

Proof. The basis of the algebra $B\langle X\rangle$ consists of the words of the type $\beta_{i1}x\beta_{i2}x\ldots x\beta_{in}$, $n\geq 1$, where β_{ij} are elements of a fixed basis of B. Let us order the basis of B in a standard way and extend this order to the basis $B\langle X\rangle$ (i.e., we assume that a longer word is bigger than a shorter one, while the words of the same length are compared lexicographically). Then if u, v are elements from $B\langle X\rangle$, then the senior word of the product uxv is $\bar{u}x\bar{v}$, where $\bar{u}$, $\bar{v}$ are the senior words of u and v, respectively. Therefore, $uB\langle X\rangle v\neq 0$ and $B\langle X\rangle$ is a prime ring.

Let z be an element of a generalized centroid, $z\notin C$. We can find a nonzero element $f\in B\langle X\rangle$, such that $h=fz\in B\langle X\rangle$. Let us assume that the element $z\notin C$ is chosen in such a way that the senior word of h be the smallest word possible. If $h=0$, then $B\langle X\rangle fB\langle X\rangle z=0$ and, hence, $z=0\in C$, which is impossible. Let u be an arbitrary element from $B\langle X\rangle$. Then $fuhz=fzuh=huh=hufz$. Therefore, $fuh=huf$.

Let us show that the latter equality implies the equality of the senior words $\bar{f}$ and $\bar{h}$. If the degrees of the words $\bar{f}$ and $\bar{h}$ over x coincide, then these words are of the same length and $\bar{f}x\bar{h}=\overline{fxh}=\overline{hxf}=\bar{h}x\bar{f}$, which affords $\bar{f}=\bar{h}$. Let, for instance, $\deg_x\bar{f}<\deg_x\bar{h}$. Then, since the dimension of B is greater than one, one can find an element b_1, such that $\bar{f}xb_1$ is not the initial subword for $\bar{h}$. If we assume $u=xb_1x$, then $\bar{f}xb_1x\bar{h}=\overline{fxb_1xh}=\overline{huf}=\bar{h}xb_1x\bar{h}$, which means that $\bar{f}xb_1$ is the initial subword of the word $\bar{h}$, which is a contradiction.

Therefore, $\bar{f}=\bar{h}$, i.e., one can choose $\lambda\in C$ in such a way that the senior word of the element $h-\lambda f$ be smaller than $\bar{h}$. Then $f(z-\lambda)=h-\lambda f$, in which case $z-\lambda\notin C$. This contradicts the choice of the elements z and h and the lemma is thus proved.

3.3.9. Lemma. *Let B be a subalgebra of an algebra B' over a field C. Then the centralizer of B in the algebra $B'\langle X\rangle$ is generated by the centralizer of B in B' and by the elements of type $\sum a_iub_i$, where $\{a_i\},\{b_i\}$ are dual bases of the conjugated left B-submodule λ in B' and the right B-submodule ρ in B', respectively.*

Proof. Let $\{a_i\},\{b_i\}$ be dual bases of conjugated B-submodules in B'. If $b\in B$, then $ba_i\in\lambda$, $b_ib\in\rho$ and, hence, $ba_k=\sum_i\lambda_{ki}a_i$, $b_ib=\sum_k\lambda'_{ik}b_k$. Since the corresponding linear form is associative, valid are the equalities $\lambda'_{ik}=(b_ib,a_k)=(b_i,ba_R)=\lambda_{ki}$, and, hence,

$$\sum_i(a_iub_i)b=\sum_{i,k}\alpha'_{ik}a_iub_k=\sum_{i,k}\alpha_{ki}a_iub_k=b\sum_k a_kub_k,$$

i.e., the elements of the algebra $B'\langle X\rangle$, used in the lemma, lie in the centralizer of the subalgebra B.

Let us consider the inverse case. Let us fix a basis $\{\beta_i\}$ of the algebra B. Then any element f of the free product is presented as

$$b_0 + \sum_j \sum_i a_i^{(j)} u_j b_i^{(j)},$$

where u_j are mutually different subwords of the basis words, which begin and end with x. If the element f lies in the centralizer of B, then for any $b \in B$ we have

$$[b b_0 - b_0 b] + \sum_j [\sum_i b a_i^{(j)} u_j b_i^{(j)} - a_i^{(j)} u_j b_i^{(j)} b] = 0$$

This implies that if all the elements $ba_i^{(j)}$, $a_i^{(j)}$, $b_i^{(j)} b$, $b_i^{(j)}$ are written over the basis $\{\beta_i\}$, then in the obtained linear combination of basis words of the type $\beta_{i1} x \beta_{i2} x \ldots x \beta_{im}$ all the terms are cancelled. Since all u_j are mutually different, then only the terms arising in the same square bracket can be cancelled. It means that the sums in the square brackets are zero.

Therefore, b_0 belongs to the centralizer of B in B', and all the sums $\sum_i a_i^{(j)} u_j b_i^{(j)}$ commute with B, so the index j can be henceforth omitted.

Let $f = \sum_{i=1}^{n} a_i u b_i$, assuming the minimal possible number n in this presentation of the element f. In this case the elements $\{a_i\}$ and $\{b_i\}$ will be linearly independent over C.

Let $\rho = \sum b_i C$, $\lambda = \sum a_i C$. Let us define a non-degenerate bilinear form on $\rho \times \lambda$, assuming that $(b_i, a_j) = \delta_{ij}$ is a Kronecker symbol. The task now is to show that ρ is the right, and λ is the left B-submodule, and that the form is associative.

Let $b\in B$. Let us extend $\{a_k\}_{k=1}^{n}$ to the basis $\{a_k\}_{k\in I}$ of the algebra B' and write the products $ba_i=\sum_{k=1}^{n}\alpha_{ik}a_k$, $1\le i\le n$ in this basis. In the same way let us extend $\{b_i\}_{i=1}^{n}$ to the basis $\{b_i\}_{i\in I}$ and write the products $b_kb=\sum_{i=1}^{m}\alpha'_{ki}b_i$, $1\le k\le n$. Then

$$\begin{aligned} bf-fb &= \sum_{i=1}^{n}\sum_{k=1}^{m}\alpha_{ik}a_kub_i-\sum_{k=1}^{n}\sum_{i=1}^{m}\alpha'_{ki}a_kub_i= \\ &= \sum_{i,k=1}^{n}(\alpha_{ik}-\alpha'_{ki})a_kub_i+\sum_{i=1}^{n}\sum_{k=n+1}^{m}\alpha_{ik}a_kub_i- \\ &-\sum_{k=1}^{n}\sum_{i=n+1}^{m}\alpha'_{ki}a_kub_i=0 \end{aligned} \tag{2}$$

This means that if we decompose the left and write coefficients at u over the basis $\{\beta_i\}$, then, after cancelling the similar terms, all the terms should be cancelled, i.e., substituting in a formal way the letter u by the sign $\otimes$ in equality (2), we get a correct equality in the tensor product $B'\otimes B'$. However, the elements $a_k\otimes b_i$, $k, i\in I$ form a basis of this tensor product, i.e., equality (2) shows that $\alpha'_{ik}=\alpha_{ki}$ at $1\le k, i\le n$ and $\lambda_{ik}=0$ at $k>n$; $\lambda_{ki}=0$ at $i>n$. The latter two conditions are equivalent to the fact that ρ is the right, and λ is the left B-submodule in B', while the former is equivalent to associativity of the form

$$\begin{aligned} (b_{k_0}b, a_{i_0}) &= \Big(\sum_{i=1}^{n}\alpha'_{k_0i}b_i, a_{i_0}\Big)=\alpha'_{k_0i_0}=\alpha_{i_0k_0}= \\ &= \Big(b_{k_0}, \sum_{k=1}^{n}\alpha_{i_0k}a_k\Big)=(b_{k_0}, ba_{i_0}). \end{aligned}$$

The lemma is proved.

Proof of theorem **3.3.7**. Let B be a centralizable algebra. In the free product $B\langle X\rangle$ let us consider an ideal R generated by the basis words which contain a variable x (i.e., R are polynomials without free terms with the coefficients from B of the noncommutable variable x). In this case R is a prime ring as an ideal of the prime $B\langle X\rangle$. Besides, $Q(R) = Q(B\langle X\rangle)$ and, therefore, the generalized centroid is equal to C, and $Q(R) \supseteq B$.

If B has no nonzero conjugated one-sided ideals, then, by lemma **3.3.9**, the centralizer of B in $B\langle X\rangle$ is equal to the center of B which does not intersect with R, i.e., $Z_R(B) = 0$, which is a contradiction.

Inversely, let B be a partially Frobenius algebra, and ρ, λ be its conjugated right and left ideals, respectively. In these ideals let us choose conjugated bases $\rho = \sum_{i=1}^{m} b_i C$, $\lambda = \sum_{i=1}^{m} a_i C$. Then, by lemma **3.3.9**, the element $\sum_{i=1}^{m} a_i x b_i$ commutes with the elements of B in $B\langle X\rangle$.

If now $B \subseteq Q(R)$ for a prime ring R with a generalized centroid C, then for any $r \in R$ the mapping $x \to r$ is extended to the homomorphism $B\langle X\rangle \to Q$, which is identical on B, i.e., the element $f(r) = \sum_{i=1}^{m} a_i r b_i$ lies in $Z_Q(B)$.

Let us choose an ideal $I \neq 0$ of the ring R, such that $a_i I \subseteq R$, $I b_i \subseteq R$. Then $f(r) \in Z_R(B)$ at $r \in I^2$. Let us, therefore, assume $f(I^2) = 0$. In this case, by corollary **1.7.14**, one can choose elements $t_j, v_j \in I^2$, such that

$$a_0 = \sum_j t_j a_1 v_j \neq 0; \quad \sum_j t_j a_k v_j = 0 \text{ at } k \neq 1.$$

Hence, $a_0 r b_1 = \sum t_j f(v_j r) = 0$, i.e., $a_0 R b_1 = 0$, which contradicts the primarity of R. The theorem is proved.

3.3.10. Theorem. *Let B be a non-centralizable algebra, S be an arbitrary algebra over C. Then there is a prime ring R with a generalized centroid C, such that $B \subseteq Q(R)$ and $Z_R(B) \cong S$.*

Proof. Let us consider an algebra $B' = B \otimes_C S'$, where S' is obtained from S by an external joining of a unit. Let us show that B' cannot contain conjugated nonzero finite-dimensional B-submodules.

Since C is a field, then S' is a free C-module and, hence, B' is the free left and free right B-module. If now ρ, λ are a couple of nonzero conjugated B-submodules in B', then we can construct a homomorphism of right B-modules, $\varphi\colon B' \to B$, such that $\varphi(\rho) \neq 0$. For the epimorphism $\rho \to \varphi(\rho) \to 0$ we have a conjugated monomorphism $0 \to \varphi(\rho)^* \to \rho^* \cong \lambda$, i.e., $\varphi(\rho)^*$ can be considered to be a submodule of the module λ. In the same way we can construct a homomorphism of left B-modules, $\psi\colon B' \to B$, such that $\psi(\varphi(\rho)^*) \neq 0$. In this case the module $[\psi(\varphi(\rho)^*)]^*$ can be identified with the submodule of the module $\varphi(\rho) \cong (\varphi(\rho)^*)^*$, and we can find in B a pair of nonzero conjugated ideals $\psi(\varphi(\rho)^*)$, $[\psi(\varphi(\rho)^*)]^*$, which contradicts the non-centralizability of the algebra B, according to theorem **3.3.7**.

Therefore, by lemma **3.3.9**, the centralizer of B in the free product $B'\langle X \rangle$ is equal to that in B', i.e., it is equal to $Z_B(B) \otimes_C S'$. Let now a subring R to be generated by $1 \otimes S$ and by all the elements which are essentially dependent on x. Then $R \cap (Z_B(B) \otimes S') = 1 \otimes S$ and, hence, $Z_B(B) \cong S$. It is now left to remark that R contains the nonzero ideal of the ring $B'\langle X \rangle$ and make use of lemma **3.3.8**. The theorem is proved.

The class of centralizable algebras is sufficiently broad. Alongside with quasi-Frobenius algebras, it also includes all finite-dimensional commutative and all finite-dimensional matrix-local algebras. Moreover, if B is centralizable and B_1 is arbitrary, then the direct sum $B \oplus B_1$ is also centralizable. The latter peculiarity, together with theorem **3.3.10**, give rise to pessimistic prospects as to the possibilities of studying partially Frobenius algebras from general positions. The minimal restriction which now becomes evident is that the algebra B is Frobenius "by parts". This intuition is confirmed by the following two theorems.

3.3.11. Definition. An algebra B over a field C will be called *quite centralizable* if for any prime ring R with a generalized centroid C and such that $B \subseteq Q(R)$, one can find a nonzero ideal I in R which is right-locally-finite over the centralizer $Z_R(B)$.

3.3.12. Theorem. *An algebra B is quite centralizable iff it is quasi-Frobenius.*

Proof. Let B be quite centralizable. As R, let us consider a ring $B\langle X \rangle = B *_C C\langle X \rangle$, where $C\langle X \rangle$ is a free associative algebra of a countable rank. Since $B\langle X \rangle = B\langle X \setminus \{x\} \rangle *_C C\langle x \rangle$, where x is an arbitrary element of X, then by lemma **3.3.8**, $B\langle X \rangle$ is a prime algebra with a generalized centroid C. Let I be a locally-finite ideal over

$Z_R(B), 0 \neq w \in I$. Then, by definition, one can find elements $w_1, \dots, w_n \in R$, such that $wR \subseteq \sum w_i Z_R(B)$. Let x be a letter from X which is not encountered in the presentation of $w, w_1, \dots, w_n$. Then

$$wx = \sum_{i=1}^{n} w_i \tau_i(x), \tag{3}$$

where $\tau_i(x) \in Z_R(B)$ in which case the degree of $\tau_i(x)$ over x can be assumed to be equal to one. By lemma **3.3.9**, the centralizer of B in the ideal of the ring R is generated by the elements of the type $\sum_{i=1}^{m} a_j u b_j$, where $\lambda = \sum a_j C$ and $\rho = \sum b_j C$ are conjugated left and right, respectively, ideals of the algebra B. Then equality (3) assumes the form

$$wx = \sum_{i=1}^{n} w'_i \sum_{j=1}^{m} a_j^{(i)} u^{(i)}(x) b_j^{(i)}, \tag{4}$$

where w'_i are still independent of x. In the right-hand part of equality (4) one can neglect all the terms in which the last letter of $u^{(i)}(x)$ is different from x, all the addends on the left end in x. Allowing for the fact that the free product $B\langle X \rangle$ can be naturally presented as $(B *_C C\langle X \setminus \{x\} \rangle) *_C C[x]$, we see that in equality (4) the right coefficient at x in the left-hand part (i.e., the unit) must be linearly expressed by the right coefficients in the right-hand part. Hence, one can find a finite number of finite-dimensional ideals $\rho_1, \dots, \rho_n$, such that $1 \in \rho_1 + \dots + \rho_n$, in which case $\rho^*_i \cong \lambda_i <_l B$. Therefore, $B = \rho_1 + \dots + \rho_n$, i.e., B is a finite-dimensional algebra.

The epimorphism of the right B-modules $\sum \oplus \rho_i \xrightarrow{\varphi} \sum \rho_i = B$ determines the embedding of the conjugated left modules $B^* \xrightarrow{\varphi^*} (\sum \oplus \rho_i) = \sum \oplus \lambda_i \subseteq \sum \oplus_B B$. However, $(B_B)^*$ is an injective module, since it is conjugated with the free one. Hence, this module is singled out from its extension $\sum \oplus_B B$ by a direct addend and is, therefore, projective. Thus, the conjugated module $(B^*)^* = B_B$ is injective and, by theorem **3.3.3**, B is a Frobenius algebra.

The inverse statement will be obtained in corollary **3.5.7**.

3.3.13. Theorem. *Let R be a prime ring, B be a quasi-Frobenius subalgebra of $Q(R)$. Then the centralizer of the ring $Z_R(B)$ in the ring R_F (and, moreover, in $Q(R)$) is equal to B*

Proof. Let B be a quasi-Frobenius algebra, $B \subseteq Q(R)$ and let z be an element of the ring R_F commutable with $Z_R(B)$. Let λ, ρ be a couple of conjugated ideals, $a_1, \dots, a_n$; $b_1, \dots b_n$ be their dual bases. Let us extend the basis of the space λ to the basis $a_1, \dots, a_n$; $z a_1, \dots, z a_k$ of the space $\lambda + z\lambda$. Then, by lemma **3.3.9**, one can find a nonzero ideal I of the ring R, such that $\sum a_i x b_i \in Z_R(B)$ for all $x \in I$ and, hence,

$$z(\sum a_i x b_i) - \sum_{i=1}^{n} (a_i x b_i) z = 0 \tag{5}$$

According to corollary **2.2.11**, we have

$$\sum_{i=1}^{n} z a_i \otimes b_i - \sum_{i=0}^{n} a_i \otimes b_i z = 0. \tag{6}$$

In the factor-space

$$(\lambda + z\lambda) \otimes (\rho + \rho z) / \lambda \otimes (\rho + \rho z) \cong (z\lambda / z\lambda \cap \lambda) \otimes (\rho + \rho z),$$

equality (6) is transformed into $\sum_{i=1}^{k} z a_i \otimes (b_i + \sum_{j>k} c_j b_j) = 0$. It is, however, impossible at $k \geq 1$, as $\{z a_i\}_{i=1}^{k}$ are linearly independent elements in the factor-space $z\lambda / (z\lambda \cap \lambda)$. Therefore, $k = 0$, i.e., $z \cdot \lambda \subseteq \lambda$.

Since the choice of λ is z-independent, and in a quasi-Frobenius algebra the sum of all the left ideals conjugated with the right ones contains the unit, then

$$z \cdot 1 \in z \sum_{\lambda^* \cong \rho <_r B} \lambda \subseteq \sum_{\lambda^* \cong \rho <_r B} \lambda = B,$$

which is the required proof.

3.4 Trace Forms

Let G be a reduced-finite group of automorphisms of a prime ring R, $G \subseteq A(R)$.

3.4.1. Lemma. *Let* λ, ρ *be nonzero conjugated left and right, respectively, ideals of an algebra* $B(G)$, *and let* $a_1, \dots, a_n; b_1, \dots, b_n$ *be their dual bases. Let us choose a system of representatives* $1 = g_1, g_2, \dots, g_m$ *of cosets of a normal subgroup* H *of inner for* Q *automorphisms. Then for any* $x \in Q$ *the following inclusion is valid*

$$\tau_{\lambda,\rho}(x) \overset{df}{=} \sum_{j=1}^{m} \Big(\sum_{i=1}^{n} a_i x b_i \Big)^{g_j} \in Q^G. \tag{7}$$

In this case there exists a nonzero ideal I of the ring R, such that $0 \neq \tau_{\lambda,\rho}(I) \subseteq R^G$.

Proof. By lemma **3.3.9**, we have $v(x) = \sum_{i=1}^{n} a_i x b_i \in Z_Q(B(G)) = Q^H$. For any $g \in G$ the equalities $g_j g = h_j g_{\pi(j)}$ are valid, where π is a permutation of the set $1, \dots, m$ which is g-dependent; $h_j \in H$. Therefore, $v(x)^{g_j g} = v(x)^{g_{\pi(j)}}$. Summing these equalities over all j, we get $\tau(x)^g = \tau(x)$.

Let us now choose a nonzero ideal I of the ring R in such a way that $a_i I \subseteq R$, $I b_i \subseteq R$ and $I^{g_j} \subseteq R$ for all i, j. Then $\tau(x) \in R^G$ for all $x \in I^3$. The lemma is proved.

3.4.2. Definition. We shall call a *trace form* the form $\tau_{\lambda,\rho}$ constructed in lemma **3.4.1**. It is evident that if the algebra of a group G is not centralizable, then the group has no trace forms at all.

3.4.3. Proposition. *If the form* $\tau(x) = \sum a_i x^{g_i} b_i$ *has only invariant values,* $a_i, b_i \in B(G)$, $g_i \in G$, *then* $\tau(x)$ *is a trace form, i.e., there can be found left and right conjugated ideals* λ, ρ of $B(G)$ *such that on* R_F *we have* $\sum a_i x^{g_i} b_i \equiv \tau_{\lambda,\rho}(x)$.

Proof. As in lemma **3.4.1**, let $1 = g_1, \dots, g_m$ be a system of representatives of cosets of H. Let us rewrite the form $\tau(x)$ as

$$\tau(x) = \sum_{j=1}^{m} \Big(\sum_i a_{ij} x b_{ij}\Big)^{g_j}.$$

As the values of $\tau(x)$ commute with the elements from $\mathcal{B}(G)$, then for any $b \in \mathcal{B}$ we have a reduced identity $\tau(x)b - b\tau(x) = 0$. By **2.2.2**, it means that $v_j(x) = \sum_i a_{ij} x b_{ij}$ is a linear form commutative with $\mathcal{B}(G)$. Viewing this form as an element of the free product $\mathcal{B}(G) * C\langle X\rangle$ we can, by lemma **3.3.9**, find conjugated ideals λ_j, ρ_j, such that $v_j(x) = v_{\lambda_j, \rho_j}(x)$. The task now is to show that $v_j(x) \equiv v_1(x)$. For this purpose let us consider an identity $\tau(x) = \tau(x)^{g_j}$. Considering the part of this identity which is free from automorphisms (see **2.2.2**), we see that $v_1(x) \equiv v_j(x)$. The lemma is proved.

3.4.4. Definition. If the algebra of the group G is Frobenius, then the form $\tau_{\mathcal{B},\mathcal{B}}(x)$ is called a *principal trace form.*

3.4.5. Remark. *Let* $\mathcal{B}$ *be a Frobenius algebra. Then the trace form* $\tau(x)$ *is principal iff one of the following equivalent conditions is met:*

(a) *for any nonzero* $b \in \mathcal{B}$ *the form* $\tau(bx)$ *is nonzero;*
(b) *for any nonzero* $b \in R_F$ *the form* $\tau(xb)$ *is nonzero.*

Proof. Let λ be a proper left ideal. Then, since the algebra is Frobenius, its right annihilator is nonzero: $\lambda b = 0$, $b \neq 0$. We have $\tau_{\lambda,\rho}(bx) \equiv 0$.

Inversely, if $\tau_{\mathcal{B},\mathcal{B}}(bx) \equiv 0$, then $\sum a_i b \otimes a^*_i = 0$, where $\{a_1, \dots a_n\}$ is a basis of $\mathcal{B}$, $\{a^*_1, \dots, a^*_n\}$ is its dual basis. Since a^*_i are linearly independent, then $a_i b = 0$ for all i. Hence, $\mathcal{B}\, b = 0$ and, consequently, $b = 0$, since $\mathcal{B}$ contain the unit. The remark is proved.

3.4.6. Definition. A subring $S \subseteq R$ is called *almost intermediate,* if for any trace form $\tau(x)$ there is a nonzero ideal I of the ring R, such that $\tau(I) \subseteq S$.

3.4.7. Lemma. *Let the algebra of a reduced-finite group G be semisimple (i.e., G is an M-group, see 3.1.3), then a subring S will be almost intermediate iff for the principal form $\tau(x)$ there exists a nonzero ideal I, such that $\tau(I) \subseteq S$.*

The proof easily follows from the fact that any one-sided ideal of a semisimple finite-dimensional algebra is generated by an idempotent: $\lambda = \mathbf{B}e$, $\rho = f\mathbf{B}$ and, hence, $\tau_{\lambda,\rho}(x) = \tau_{\mathbf{B},\mathbf{B}}(exf)$.

The notion of an almost intermediate ring is so far not going to be of any importance. Of importance for future proofs is the fact that not all fixed elements but only the values of trace forms are used in proofs. Throughout the present chapter the reader can make no distinction between intermediate ($R \supseteq S \supseteq R^G$) and almost intermediate rings.

3.4.8. Lemma. *A principal trace form is uniquely defined by a nondegenerate associative bilinear form on $\mathbf{B}(G)$.*

Proof. When constructing a principal trace form the choice of a basis $a_1, \ldots, a_n$ and representatives $1 = g_1, \ldots g_m$ of cosets of H can be arbitrary. The form $v(x)$ is independent of the choice of the basis $a_1, \ldots, a_n$. It can be easily seen, since under elementary transformations of the basis $a_i \to \lambda a_i$ or $a_i \to a_i + \beta a_j$, the dual basis is transformed by an inverse mapping $b_i \to \lambda^{-1} b_i$ or $b_j \to b_j - \beta b_i$, respectively, and, hence, the form v does not undergo changes: $\lambda a_i x \lambda^{-1} b_i = a_i x b_i$ and, respectively, $(a_i + \beta a_j) x b_i + a_j x (b_j - \beta b_i) = a_i x b_i + a_j x b_j$.

If we choose another system of representatives, $f_1, f_2, \ldots, f_m$, then $f_i = h_i g_i$ for suitable $h_i \in H$. Therefore, $v(x)^{g_i} = v(x)^{h_i g_i} = v(x)^{g_i}$ and the form $\tau(x)$ is independent of a concrete choice of representatives. The lemma is proved.

3.4.9. Theorem. *A principal trace form is uniquely determined to the accuracy of a substituting variable $x_1 = bx$, where b is an invertible element from $\mathbf{B}(G)$.*

Proof. A nondegenerate bilinear form f on $\mathbf{B}(G)$ sets an isomorphism $\varphi_f : {}_B B \to ({}_B B)^*$. If g is another form, then $\varphi_f \varphi_g^{-1}$ is an isomorphism of the left module ${}_B B$ onto itself. As $Hom({}_B B, {}_B B) \cong B$, then $\varphi_f \varphi_g^{-1}$ is determined by the right multiplication by an invertible element b. Thus, $x\varphi_f \varphi_g^{-1} = xb$ or $x\varphi_f = (xb)\varphi_g$ and, since $(x\varphi_f)(t) = f(t, x)$, then $f(t, x) = g(t, xb)$. If now $a_1, \dots, a_n$ is the basis of B and $a_1^*, \dots a_n^*$ is its dual basis with respect to f, then $a_1^* b, \dots a_n^* b$ will be the dual basis for the form g and, hence, $v_f(x) = v_g(bx)$. The theorem is proved.

3.5 Galois Groups

The results obtained in the preceding paragraph more of less convincingly demonstrate that we should concentrate our attention on studying the groups of automorphisms which have quasi-Frobenius algebras.

3.5.1. Theorem. *Let G be a reduced-finite group of automorphisms , the algebra of which, $B(G)$, is quasi-Frobenius. Then in R_F the centralizer of a fixed ring $I(G) = R^G$ is equal to $B(G)$ (while the centralizer of any almost intermediate ring is contained in $B(G)$).*

Proof. Let λ, ρ be arbitrary conjugated left and right, respectively, ideals of the algebra of a group G, and $\tau_{\lambda,\rho}(x)$ be a trace form determined by them (see **3.4.1**). Let us choose a nonzero ideal I, so that $\tau(I) \subseteq R$. In this case, if z belongs to the centralizer of R^G in R_F, then the following identity can be written:

$$z\tau(x) - \tau(x)z = 0.$$

This is a reduced identity, i.e., by theorem **2.2.2**, we have the relation

$$\sum z a_i \otimes b - \sum a_i \otimes b_i z = 0.$$

It is exactly relation (6) from the proof of theorem **3.3.13**. As λ, ρ is an arbitrary pair of conjugated ideals of a quasi-Frobenius algebra, then, as was the case in theorem **3.3.13**, we have $z \in B(G)$, which is the required proof.

3.5.2. Theorem. *Let G be a reduced-finite N-group of automorphisms, the algebra of which is quasi-Frobenius, and let S be an almost intermediate ring. Then $A(S) \subseteq G$. In particular, G is a Galois group.*

Proof. We are to show that any automorphism $h \in A(R)$, which leaves the elements of the subring S fixed, lies in G.

Let us choose arbitrary conjugated nonzero ideals λ, ρ of the algebra $B(G)$ and let $a_1,..., a_n; b_1,..., b_n$ be their dual bases. Let $1 = g_1,..., g_m$ be a system of representatives of cosets of a normal group G_{in}.

Let us choose a nonzero ideal I, so that $\tau(I) \subseteq S$. Now we can write the identity which is fulfilled on I:

$$\sum_j v(x)^{g_j h} - \sum_j v(x)^{g_j} = 0,$$

where $v(x) = \sum_{i=1}^{n} a_i x b_i$.

If this is a reduced identity, then, by theorem **2.2.2**, we have a contradiction. Irreducibility implies that one of the automorphisms $(g_j h)(g_i h)^{-1}, (g_j h) g_i^{-1}, i \neq j$ is inner. Let $g_i h g_j^{-1} = \tilde{b}$. Then $\tilde{b} \in A(S)$ and, hence, b commutes with all the elements from S, i.e., by the previous theorem, $b \in B(G)$. Since G is an N-group, then $\tilde{b} \in G$, i.e., $h = \tilde{b} g_k g_i^{-1} \in G$, which is the required proof.

3.5.3. Theorem. *Let G be a reduced-finite group of automorphisms, its algebra being quasi-Frobenius. Then in R there is a nonzero local-finite over R^G ideal of the ring R.*

This theorem will result from the following proposition.

3.5.4. Proposition. *Under the condition of the preceding theorem, there exist elements $a \neq 0$, $r_1,..., r_t \in R$, such that at all $x \in R$ we have*

$$ax = \sum_{i=1}^{t} r_i \tau_i(x), \tag{8}$$

where τ_i are certain mappings from R to R^G.

Let us show how theorem **3.5.3** is deduced from this proposition. Let us consider a set I of elements a, for which there exist elements $r_1,\dots, r_{m(a)} \in R$, such that at all $x \in R$ equality (8) holds. It is evident that IR is a locally-finite over R^G ideal of the ring R. It is the proposition that yields $I \neq 0$.

Proof of proposition **3.5.4**. Let λ, ρ be arbitrary conjugated left and right ideals of the algebra of the group G. By lemma **3.5.1**, we can find a form $\tau_{\lambda,\rho}$ and an ideal I, such that at $x \in I$

$$\tau(x) = \sum_{j=1}^{m} \left(\sum_{i=1}^{n} a_i x b_i \right)^{g_j} \in R^G. \tag{9}$$

holds.

Let us fix an index i_0. Since the elements $a_1,\dots, a_n$ are linearly independent over C, and the automorphisms g_i are mutually outer (i.e., $\Phi_{gh^{-1}} = 0$ at $g \neq h$), then, by proposition **1.7.10**, there are elements $v_k, w_k \in I$, such that $\sum_k v_k a_i w_k = 0$ at $i \neq i_0$; $\sum v_k a_{i_0} w_k = a_0 \neq 0$; $\sum v_k a_i^{g_j} w_k = 0$ at all i and at $j \neq 1$.

Let us multiply equality (9) from the left by v_k and substitute xw_k instead of x. Summing over all k, we get

$$a_0 x b_{i_0} = \sum_k v_k \tau(w_k x). \tag{10}$$

Let us now make use of the fact that the conjugated ideals λ, ρ, as well as the elements i_0 have been chosen arbitrarily. Since the algebra is quasi-Frobenius, one can find a system of elements b_i, such that $\sum c_i b_i = 1$ for suitable $c_i \in C$ and for every b_i the equality of type (10) holds:

$$a_i x b_i = \sum_k v_k^{(i)} \tau(w_k^{(i)} x). \tag{11}$$

Since the ring R is prime, one can find elements $y_1,\dots, y_{m-1}$, such that $a = a_1 y_1 a_2 y_2 \cdots y_{m-1} a_m \neq 0$. Multiplying equality (11) from the left by $a_1 y_1 \cdots a_{i-1} y_{i-1}$ and substituting $x = y_i a_{i+1} \cdots y_{m-1} a_m$ we get the following equality

$$axb_i = \sum_k r_k^{(i)} \tau_k^{(i)}(x).$$

Let us multiply this equality by c_i and sum it over all i, getting thus an equality of the required type:

$$ax = \sum_k r_k \tau_k(x).$$

In this equality the elements r_k are obtained by multiplying $r_k^{(i)}$ by the elements c_i from the generalized centroid and, therefore, they can lie not in R. To eliminate this difficulty, it is sufficient to multiply the latter identity from the left by an element $y \in R$, such that $ya \neq 0$, $yc_i \in R$. This element does exist, as $Ja \neq 0$, where J is an ideal, such that $Jc_i \subseteq R$. The proposition is proved.

3.5.6. Corollary. *Let G be a reduced-finite group of automorphisms of a simple ring R with a unit. If the algebra of the group G is quasi-Frobenius, then R is a finitely generated module over R^G.*

Proof. Let I be a locally-finite ideal over R^G. By condition, we have $1 \in I = R$ and, hence, $1 \cdot R$ is contained in a finitely generated right R^G-submodule of the module R, which is the required proof.

3.5.7. Corollary. *Let B be a quasi-Frobenius C-subalgebra in $\mathfrak{A}(R)$. Then the ring R contains a nonzero locally-finite ideal over the centralizer of B in R.*

Proof. If B is generated by its invertible elements, then we can formally apply theorem **3.5.3**. If this is not the case, then for arbitrary conjugated ideals λ, ρ in B and their dual bases $a_1, \dots a_n$; $b_1, \dots, b_n$, the form $\sum a_i x b_i$ assumes the values in the centralizer of B and one can repeat the considerations of the proof of proposition **3.5.4**, starting with formula (9) and assuming $m = 1$, $g_1 = 1$.

3.6 Maschke Groups. Prime Dimensions

We should not forget that our aim is to develop the Galois theory in the class of semiprime rings and, therefore, we should elucidate under what conditions a fixed ring is semiprime.

3.6.1. Theorem. *Let G be a reduced-finite group of automorphisms of a prime ring R, with its algebra, $\mathcal{B}(G)$, being quasi-Frobenius. In this case a fixed ring R^G is semiprime iff $\mathcal{B}(G)$ is semi-simple.*

Proof. Let $\mathcal{B}(G)$ be not semi-simple. Let us show that we can find nonzero conjugated ideals λ, ρ, such that $\rho^g\lambda^h=0$ for all $g, h\in G$.

Let J be a radical of the algebra $\mathcal{B}(G)$. This is the largest nilpotent ideal, and, in particular, $J^g\subseteq J$ for all $g\in G$. Let n be the largest number, such that $A=J^n\neq 0$. Then $A^2=0$, and the ideal A is invariant under the action of G: $A^g\subseteq A$, $g\in G$.

Since the sum of all right ideals conjugated with the left ones contains the unit, then $A\rho_1\neq 0$ for one of the ideals, ρ_1. Let $a\rho_1\neq 0$, $a\in A$. Let us consider the epimorphism of the right B-modules $\rho_1\to a\rho_1\to 0$. A conjugated exact sequence has the form: $0\to(a\rho_1)^*\to(\rho_1)^*\cong\lambda_1$, i.e., the right ideal, $\rho=a\rho_1$ is conjugated with a left ideal, λ, contained in λ_1. Since ρ is contained in the invariant ideal A, then $\rho\cdot\rho^{gh^{-1}}=0$ for all $h, g\in G$. Since bilinear forms are associative, we get $0=(\rho\,\rho^{gh^{-1}},\ \lambda)=(\rho,\rho^{gh^{-1}}\lambda)$. Since the form is nondegenerate, we get the equality $\rho^{gh^{-1}}\lambda=0$ or $\rho^g\lambda^h=0$.

For the ideals ρ, λ let us construct a $\tau_{\lambda,\rho}: I\to R^G$, where I is a nonzero ideal of R. Then $\tau_{\rho,\lambda}(I)$ is a nonzero ideal of R^G, in which case the equalities $\rho^g\lambda^h=0$ show its square to be equal zero. Thus, R^G is not semiprime.

Inversely, let $\mathcal{B}(G)$ be a semisimple algebra and let us assume that $tR^Gt=0$ for a nonzero $t\in R^G$. Since $\mathcal{B}(G)$ is generated by invertible elements corresponding to inner automorphisms from G, and t commutes with these elements, then t commutes with all elements from $\mathcal{B}(G)$. In particular, in the algebra $\mathcal{B}$ the annihilator of t is a proper two-sided ideal, and we can find a non-zero simple addend S of the algebra $\mathcal{B}$, which does not intersect the annihilator of t. It means that if $s_1,\dots, s_n$ are linearly independent over C elements from S, then $ts_1,\dots, ts_n$ will also be linearly independent.

Let λ, ρ be conjugated right and left ideals of S. Using lemma

3.4.1, we can construct a trace form $\tau_{\lambda,\rho}: I \to R^G$, where $0 \neq I$. Then, by condition, on the ideal I the following identity holds

$$t\tau_{\lambda,\rho}(x)t = \sum_{j=1}^{m} t(\sum_{i=1}^{n} a_i x b_i)^{g_j} t = 0.$$

This beng a reduced identity, by theorem **2.2.2**, we get the equality

$$\sum_{i=1}^{n} ta_i \otimes b_i t = 0,$$

which contradicts the fact that $ta_1, \dots, ta_n$ and $b_1 t, \dots, b_n t$ are linearly independent systems of elements. The theorem is proved.

We have already discussed the case when the ring of invariants is semiprime (corollary **1.3.7**). Therefore, the following theorem, which is a variation of the Maschke theorem, seems to be quite natural.

3.6.2. Theorem. *If G is a finite group of automorphisms of a prime ring R, and R has no additive $|G|$-torsion, then $\mathbf{B}(G)$ is semiprime.*

Proof. This theorem allows a standard proof in the spirit of the proof of the Maschke theorem. In order not to give here quite well-known considerations, let us show how this theorem results from the preceding theorem and corollary **1.3.7**.

Let H be a subgroup of inner for Q automorphisms, with its order dividing that of the group G and being, thus, co-prime with the characteristic of C. If $h \in H$, then let us fix an element $\varphi_h \in \mathbf{B}(G) = \mathbf{B}(H)$, such that $h = \tilde{\varphi}_h$. In this case the elements φ_h generate $\mathbf{B}$, and $\varphi_h \varphi_g = c(h, g)\varphi_{hg}$ for all $g, h \in H$, where $c(h, g)$ are uniquely determined nonzero elements from C.

Let us consider an algebra B', the dimension of which is equal to the order of the group H, and basis elements u_h are multiplied by the formula $u_h u_g = c(h, g) u_{hg}$. Since the coefficients $c(h, g)$ are uniquely defined, this is an associative algebra and the mapping $u_h \to \varphi_h$ determines the homomorphism $B' \to \mathbf{B}$. The task now is to show that B' is a semisimple algebra.

It should be remarked that B' is a Frobenius algebra. It has a non-degenerate bilinear form

$$(u_h, u_g) = \begin{cases} c(h, g), & g = h^{-1} \\ 0, & g \neq h^{-1} \end{cases}$$

The associativity of this form results from that of the algebra $\mathcal{B}$: if $fhg = 1$, then $(u_f u_h, u_g) = c(f, h)c(fh, g)$ and $(u_f, u_h u_g) = c(h,g)c(f,hg)$, in which case $(\varphi_f \varphi_h)\varphi_g = c(f,h)\varphi_{fk}\varphi_g = c(f,h)c(fh,g)\varphi_{fhg}$ and $\varphi_f(\varphi_h \varphi_g) = c(h, g)\, c(f, hg)\varphi_{fgh}$, wherefrom we get the required equality.

Let us embed the algebra B' in a prime ring R_0 with a generalized centroid C (for instance, $R_0 = B'\langle x \rangle\rangle$). Then the action of H on R_0 can be governed by the formula $x^h = u_h^{-1} x u_h$. By corollary **1.3.7**, the subring of invariants of H is semiprime and by theorem **3.6.1**, the algebra of the group H, which is equal to B', is also semiprime, which is the required proof.

3.6.3. Remark. Constructions from the preceding proof are applicable to any finite group of automorphisms (without any restriction on the additive group of the ring). In particular, *the algebra of any finite group is a homomorphic image of the Frobenius algebra* B'. If the order of a finite group of automorphisms coincides with the reduced order, then this homomorphism is an isomorphism. Applying theorem **3.5.2**, we come to the conclusion that in this case *a Galois group generated by* G *is obtained from* G *by adding an inner automorphism corresponding to the elements from* $\mathcal{B}(G)$ (whose number, obviously, can be infinite).

Theorem **3.6.1** shows that of prime interest for us are groups with a semisimple algebra, and theorem **3.6.2** prompts definition **3.1.3** of Maschke groups.

3.6.4. Definition. The *prime dimension of the ring* R is the largest number n, such that R contains a direct sum of n nonzero two-sided ideals.

The prime dimension of a prime ring is equal to one. A semiprime ring can have either finite or infinite prime dimension. However, if this dimension equals one, then the ring is prime: if $IJ = 0$, then in a semiprime ring we have $I \cap J = 0$ and, hence, the sum $I + J$ is direct.

3.6.5. By analogy, the *invariant prime dimension of the ring* B on which the group G acts, is the largest number n, such that B contains a direct sum of n nonzero invariant two-sided ideals. If the invariant prime dimension is equal to one, then the ring is called *G-prime*. It is evident that if B is a semisimple finite-dimensional algebra, then its G-primeness implies that B has no proper invariant ideals, i.e., it is *G-simple*.

3.6.6. Lemma. *The following statements are equivalent for a semiprime ring* R.

(1) *The prime dimension of the ring* R *is* n.

(2) *The ring* R *contains an essential* (see **1.4.4**) *direct sum of ideals* $I_1 \oplus ... \oplus I_n$, *each of which is a nonzero prime subring.*

(3) *The ring of quotients* $R_{\mathcal{F}}$ *is decomposed into a direct sum of* n *nonzero prime rings.*

(4) *The ring of quotients* $Q(R)$ *is decomposed into a direct sum of* n *nonzero prime rings.*

(5) *The generalized centroid of* R *is isomorphic to a direct sum of* n *fields.*

(6) *The ring* $R_{\mathcal{F}}$ *has exactly* 2^n *different central idempotents.*

Proof will be carried out going along a circle.

(1) ⇒ (2). Let $A = I_1 \oplus ... \oplus I_n$ be a direct sum of nonzero ideals. Then, by lemma **1.4.3.**, $A + ann\, A$ is a direct sum of nonzero ideals and, since n is a prime dimension, $A \in \mathcal{F}$. If one of the ideals, say. I_1, is not a prime ring, i.e., $EF = 0$ for nonzero ideals E and F of the ring R, then $(I_1 E I_1)(I_1 F I_1) = 0$, in which case $E_1 = I_1 E I_1$, $F_1 = I_1 F I_1$ are now the ideals of R, and we thus find a direct sum $E_1 \oplus F_1 \oplus I_2 \oplus ... \oplus I_n$, with its length greater than n.

(2) ⇒ (3). By proposition **1.4.12.**, we have

$$R_{\mathcal{F}} = (I_1 \oplus \ldots \oplus I_n)_{\mathcal{F}} = (I_1)_{\mathcal{F}} \oplus \ldots \oplus (I_n)_{\mathcal{F}}.$$

(3) ⇒ (4). Let $R_{\mathcal{F}} = R_1 \oplus ... \oplus R_n$, and $1 = e_1 + ... + e_n$ be the corresponding decomposition of the unit. Then $Q = Qe_1 + ... + Qe_n$ is the sought direct decomposition of Q, since e_i are central orthogonal idempotents and if, for instance, $qQe_1r = 0$ for $q, r \in Qe_1$, then, by lemma **1.5.6.**, we have a product $e(q)e(r) = 0$, which yields $e(q)R_1e(r) = 0$, i.e., $e(q) = 0$ or $e(r) = 0$. According to the definition of a support (see **1.5**) we, thus, get either $q = 0$ or $r = 0$.

(4) ⇒ (5). Let $Q = Q_1 \oplus ... \oplus Q_n$. Since ideals Q_i annihilate each other, then the center of Q is equal to a direct sum of the centers of Q_i.

Each of these centers has no zero divisors, i.e., it is a field (see lemma **1.5.3**).

(5) ⇒ (6). One should prove that a direct sum of n fields has exactly 2^n idempotents. It is exactly the sums of the units of addends over all possible subsets of these units, and the number of these subsets is exactly 2^n.

(6) ⇒ (1). Let $e_1, \dots, e_m$ be all nonzero minimal central idempotents. In this case these idempotents are mutually orthogonal. Since the number of central idempotents is finite, any such idempotent is greater than a minimal one (see **1.5.4** for the definition of an order). In particular, the idempotent $1-(e_1+\dots+e_m)$ cannot be nonzero, since its least minimal idempotent annihilates all e_i. Then for any central idempotent e we have $e = e\cdot 1 = ee_1 + \dots + ee_m = \sum e_i$, $i \in A \subseteq \{1, \dots, m\}$, since the product ee_i is equal either to 0 or e_i. Moreover, the two sums of the minimal idempotents over different subsets are different and, hence, $2^m = 2^n$, i.e., $m = n$.

By the definition of a ring of quotients, one can find an essential ideal I, such that $Ie_i \subseteq R$ for all i. Then $Ie_1 + \dots + Ie_n$ is a direct sum of nonzero ideals in R, which implies that the prime dimension of R is not less than n.

If $I_1 \oplus \dots \oplus I_l$ is a direct sum of nonzero ideals of the ring R, then the supports $f_k = e(I_k)$ are mutually orthogonal and one can find 2^l different central idempotent-sums $\sum_{k\in A} f_k$ over the subsets $A \subseteq \{1, \dots, l\}$, i.e., $2^l \leq 2^n$ and $l \leq n$. The lemma is proved.

3.6.7. Theorem. *Let G be a Maschke group of automorphisms of a prime ring R. Then the prime dimension of a fixed ring is equal to the invariant prime dimension of the algebra $\mathcal{B}(G)$.*

Proof. Let $\mathcal{B}(G) = B_1 \oplus \dots \oplus B_n$ be a decomposition of a semisimple algebra $\mathcal{B}(G)$ into a direct sum on nonzero invariant G-simple ideals. In each of the rings B_i let us choose conjugated left and right ideals λ_i, ρ_i, respectively. These ideals annihilate all the other addends B_j, $j \neq i$ and they are, therefore, also conjugated as $\mathcal{B}(G)$-modules.

Using lemma **3.4.1**, let us construct trace forms $\tau_{\lambda_i, \rho_i}: I \to R^G$, where I is a nonzero ideal of the ring R. Let I_i be the image of the corres-ponding mapping. Then I_i is an ideal of R^G and the ideals $I_i, 1 \leq i \leq n$ annihilate each other, i.e., $I_1 + \ldots + I_n$ is a direct sum in R^G.

Let us, for instance, show that I_1 is a prime ring. Let $0 \neq a \in I_1$. Then the element a, as the one lying in R^G, commutes with the elements from $B(G)$ and, hence, its annihilator in $B(G)$ is a two-sided invariant ideal. On the other hand, by construction, I_1 annihilates the sum $B_2 + \ldots + B_n$, i.e., $ann\, a \supseteq B_2 + \ldots + B_n$. The intersection $ann\, a \cap B_1$ is an invariant ideal in B_1, in which case this ideal is not equal to B_1, since in the opposite case a would annihilate the whole algebra $B(G)$. Therefore, $ann\, a \cap B_1 = 0$, as B_1 is G-simple.

Let $0 \neq v, w \in I_1$; $a_1, \ldots, a_m$; $b_1, \ldots, b_m$ be dual bases of the ideals λ_1 and ρ_1. As the annihilators of v and w have a zero intersection with B_1, then $va_1, \ldots, va_m$, as well as $b_1 w, \ldots b_m w$ are linearly independent sets.

If now $vI_1 w = 0$, then we have a reduced identity on I: $v\tau_{\lambda_1, \rho_1}(x)w = 0$, which yields, by theorem **2.2.2**, $\sum_{i=1}^{m} va_i \otimes b_i w = 0$, which contradicts the linear independence of the above sets.

According to the previous lemma, the prime dimension of R^G is n, which is the required proof.

3.6.8. Theorem. *Let G be a reduced-finite group of automorphisms of a prime ring R, the algebra of which is quasi-Frobenius. A fixed ring R^G is prime iff $B(G)$ contains no proper trace ideals (i.e., it is G-simple).*

The proof results immediately from theorems **3.6.2** and **3.6.7**. As a somewhat more complex corollary, we get the following result.

3.6.9. Theorem. *Let G be a finite group of automorphisms of a simple ring R which has no additive $|G|$-torsion. Then a fixed ring R^G is isomorphic to a direct sum by not more than $|G|$ simple rings.*

Proof. By theorem **3.6.2**, the group G is a Maschke group. The invariant prime dimension m of the algebra of this group is certainly less than its dimension which is not greater than the order n of the group G. By theorem **3.6.7**, the prime dimension of R^G is $m \leq n$.

Let now $I = I_1 + \ldots + I_m$ be an arbitrary direct sum of m nonzero ideals of the ring R^G. Let us show that $I = R^G$.

For this purpose let us first remark that the left annihilator of I in

R is equal to zero. Indeed, in the opposite case $ann_l I$ is the left invariant not nilpotent ideal of the ring R which has, by the Bergman-Isaaks theorem, a nonzero intersection with R^G and, therefore, the sum $I + (ann_l I \cap R^G)$ will be direct.

It should be then remarked that IR is a prime subring of R. If $aIRb = 0$, $b \neq 0$, then $aI = 0$ (since R is prime) and, hence, $a \in ann_l I = 0$.

Therefore, IR is an invariant prime subring of R. Applying to it proposition **3.5.4**, we can find a nonzero element a, such that $aIR \subseteq IR(IR)^G$.

It should be also remarked that $n(IR)^G \subseteq I$. Indeed, if $v = \sum i_\alpha r_\alpha$, $i_\alpha \in I$, $r_\alpha \in R$, then $nv = t(v) = \sum i_\alpha t(r_\alpha) \in I$, where $t(x) = \sum_{g \in G} x^g$ is a trace of the element x.

Now we have a chain of inclusions

$$R = R(na)IR \subseteq R \cdot IRn(IR^G) \subseteq RIR \cdot I = RI.$$

Hence $nR^G \subseteq n(RI)^G \subseteq I$ (here we make use of the right analog of the latest remark). Since R has no additive n-torsion, then $nR = R$ and $nR^G = R^G$. Thus, $R^G = nR^G \subseteq I$.

We are now to show that each of the ideals I_k is a simple ring. If A is a proper ideal in, for instance, I_1, then $A_1 = I_1 A I_1$ is an ideal of R^G, in which case $0 \neq A_1 \neq I_1$. However, we have already proved that the direct sum $A_1 + I_2 + \ldots + I_m$ is also equal to R^G, i.e., $A_1 = I_1$. The theorem is proved.

3.7 Bimodule Properties of Fixed Rings

In this paragraph we shall, to a certain degree of accuracy, describe the (R, R^G)- and (R^G, R)-subbimodules in R_F for a Maschke group G.

3.7.1. Theorem. *Let G be an M-group of automorphisms of a prime ring R, and let V be an (R, S)-submodule in R_F, where S is an almost intermediate ring. In this case there exists an idempotent $e \in \mathcal{B}(G)$*

and a nonzero ideal I *of the ring* R, *such that* $Ie \subseteq V \subseteq R_F e$.

This theorem will be deduced from the following proposition. It should be recalled that for the linear form $f(x)$ and an element $\beta \in R \otimes R^{op}$, $\beta = \sum r_k \otimes t_k$ we have determined $f(x) \cdot \beta = \sum t_k f(r_k x)$. For an element $a \in R_F$ we have assumed $a \cdot \beta = \sum t_k a r_k$ and if $g \in G$, then $\beta^g = \sum r_k^g \otimes t_k$.

3.7.2. Proposition. *Let* V *be a nonempty subset in* R_F. *Then there are elements* $a \in R$, $a \neq 0$, $v_1, \ldots, v_m \in V$, $\beta_1, \ldots, \beta_m \in R \otimes R^{op}$, *such that for any* $x \in R_F$ *the following equality holds:*

$$axe = \sum_{i=1}^{m} (v_i \tau(x)) \cdot \beta_i, \tag{12}$$

where e *is an idempotent from* $\mathbf{B}(G)$ *determined by the condition* $(1 - e)\mathbf{B}(G) = ann_{\mathbf{B}}^{(r)} V$; $\tau(x)$ *being the principal trace form.*

Let us show in what way theorem **3.7.1** can be deduced from here. Let us choose a nonzero ideal J of R, so that $\tau(J) \subseteq S$. Then the right part of equality (12) will be contained in V at any $x \in J$, i.e., $aJe \subseteq V$. Hence, we have $Ie \subseteq V$, where $I = RaJ$. As $V(1 - e) = 0$ by the definition of e, then $V = Ve \subseteq R_F e$, which fact proves the theorem.

Proof of proposition **3.7.2.** Let v be an arbitrary element from V, and let $(1 - \rho_v)\mathbf{B}$ be its right annihilator in $\mathbf{B}$, where ρ_v is an idemopotent. Let us choose a basis $a_1, \ldots, a_k$ over C of this annihilator and extend it to the basis $a_1, \ldots, a_n$ of the algebra $\mathbf{B}$. Let $a_1^*, \ldots, a_n^*$ be the dual basis of the algebra $\mathbf{B}$.

It should be remarked that the elements $va_{k+1}, \ldots, va_n$ are linearly independent over C: if $\sum c_i va_i = 0$, then $v \sum c_i a_i = 0$, i.e., $\sum_{i>k} c_i a_i \in \sum_{i \leq k} Ca_i$, which is a contradiction.

Let us write the principal trace form in the basis $a_1, \ldots, a_n$, and multiply it from the left by v. Thus, we get

$$va_{k+1} xa_{k+1}^* + \ldots + va_n xa_n^* + v \sum_{ij} (a_i xa_i^*)^{g_i} = v\tau(x)$$

By theorem **1.7.10**, for every s, $k < s \leq n$ one can find an element $\beta_s \in R \otimes R^{op}$, such that $d_s = (va_s) \cdot \beta_s \neq 0$ and $(va_i) \cdot \beta_s = 0$ at $i \neq s$; $(va_i^{g_j}) \cdot \beta_s^{g_j} = 0$ at $j \neq 1$ and any i. Hence, we have

$$d_s x a_s^* = (v\tau(x))\beta_s, \quad k < s \leq n. \tag{13}$$

It should be now remarked that $a_{k+1}^*, \ldots a_n^*$ *forms a basis of the left ideal* $\mathcal{B}\rho$. Indeed, the linear hull of these elements consists of all the elements which are orthogonal to $a_1, \ldots a_k$ with respect to the linear form, i.e., $Ca_{k+1}^* + \ldots + Ca_n^* = \{b \in \mathcal{B} \mid (b, (1-\rho)\mathcal{B}) = 0$. Since the bilinear form is associative and nondegenerate, our remark is proved.

Then, the right annihilator of V in the algebra $\mathcal{B}$ is equal to the intersection of all right annihilators of the elements from V, i.e., $(1-e)\mathcal{B} = \bigcap_{v \in V} (1-\rho_v)\mathcal{B}$. Since the algebra $\mathcal{B}$ is finite-dimensional, we can assume that the latter intersection is taken over a finite set of elements from V. Going now over to the left annihilators in $\mathcal{B}$, we get $\mathcal{B}e =$

$= \sum_{v \in \{v_1, \ldots, v_m\}} \mathcal{B}\rho_v$. Let us now write formula (13) for every $v_i, 1 \leq i \leq m$. We thus obtain a system of equalities:

$$d_{s,i} x a_{s,i}^* = (v_i \tau(x)) \cdot \beta_{s,i}, \; k_i < s \leq n, 1 \leq i \leq m,$$

in which case (this being of prime importance) $e = \sum_{s,i} a_{s,i}^* c_{s,i}$ for suitable $c_{s,i} \in C$.

Since the ring R is prime, the intersection of all (R, R)-bimodules, generated by the elements $d_{s,i}$, $k_i < s \leq n, 1 \leq i \leq m$, respectively, is nonzero. It implies that there are elements $\beta'_{s,i} \in R \otimes R^{op}$ and a nonzero element $d \in R$, such that $d = d_{s,i} \cdot \beta'_{s,i}$ for all s, i, $k_i < s \leq n$, $1 \leq i \leq m$. Moreover, we can find an element $u \in R$, such that $du \neq 0$ and $uc_{s,i} \in R$ for all s, i. Now we get a system of equalities

$$dxa_{s,i}^* = (v_i \tau(x))\beta_{s,i} \cdot \beta'_{s,i} \text{ and}$$

$$duxe = \sum_{s,\,i} (v_i\tau(x))\,\beta_{s,\,i}\cdot\beta'_{s,\,i}\cdot(1\otimes uc_{s,\,i}).$$

Setting $a = du$, $\beta_i = \sum_s \beta_{s,\,i}\cdot\beta'_{s,\,i}(1\otimes uc_{s,\,i})$, we see that the proposition is proved.

3.7.3. Theorem. *Let G be an M-group of automorphisms of a prime ring R, and W be a (R^G, R)-submodule in R_F. Then $eI \subseteq W \subseteq eR_F$ for a certain nonzero ideal I of R and an idempotent $e \in B(G)$..*

The proof is symmetric to the previous one. Here, as above, the ring R^G can be substituted for by any almost intermediate one.

3.7.4. Corollary. *If the right annihilator of the (R, R^G)-submodule , $V \subseteq R_F$, in the algebra B is zero, then V contains a nonzero ideal of the ring R. Analogously, if the left annihilator of the (R^G, R)-submodule, $W \subseteq R_F$, in the algebra $B(G)$ is zero, then W contains a nonzero ideal of the ring R.*

It is evident that the left analog of proposition **3.7.2** is also valid. Let us formulate it in a somewhat different form.

3.7.5. Proposition. Let W be a nonempty subset in R_F. Then there are elements $a \in R$, $w_1, \ldots, w_n \in W, t_i, r_i \in R, 1 \le i \le n$, such that for any $x \in R_F$ the following identity holds:

$$exa = \sum_i \tau(xr_i)w_it_i, \tag{14}$$

where $B(G)(1-e) = ann_B^{(1)} W$, and τ is the principal trace form.

3.7.6. Proposition. *Let G be an M-group, S be its almost intermediate ring, a be its idempotent from R_F, such that $sa = asa$ for any $s \in S$. Then there exists an idempotent $\rho \in B(G)$, such that $a\rho = \rho$, $\rho a = a$. Inversely, if $\rho a = a$ and $a\rho = \rho$ for a certain $\rho \in B(G)$, then $sa = asa$ for any $s \in R^G$.*

Proof. If $\rho a = a$, $a\rho = \rho$, $s \in R^G$, then $sa = s\rho a = \rho sa = a\rho sa = as\rho a = asa$.

Let now $sa = asa$ for all $s \in S$. Let us consider the right ideal $V = aR_F$. If $s \in S$, then $saR_F = asaR_F \subseteq aR_F$. Therefore, V is a (S, R)-submodule and, by theorem **3.6.5**, we have $\rho I \subseteq V \subseteq \rho R_F$, where ρ is an idempotent from $B(G)$. The second inclusion shows that $\rho a = a$. If $i \in I$, then $\rho i = ar$ and, hence, $a\rho i = a^2 r = ar = \rho i$.

Therefore, $(a\rho - \rho)I = 0$, i.e., $a\rho = \rho$. The proposition is proved.

3.8 A Ring of Quotients of a Fixed Ring

In this paragraph we shall calculate the Martindale ring of quotients of a fixed ring of a Maschke group.

3.8.1. Lemma. *Let G be an M-group. Then for the left (right) ideal A and any nonzero ideal I we have $\tau(IA) \neq 0$ (and, respectively, $\tau(AI) \neq 0$~), where τ is the principal trace form.*

Proof. Let us consider the (R, R^G)-bimodule AR^G. By theorem **3.7.1**, we can find a nonzero ideal J and an idempotent $e \in B(G)$, such that $Je \subseteq AR^G \subseteq R_F e$. If $\tau(IA) = 0$, then

$$\tau(IJe) \subseteq \tau(IAR^G) \subseteq \tau(IA)R^G = 0.$$

Herefrom by theorem **2.2.2**, we get $\sum a_i \otimes eb_i = 0$, where $\{a_i\}, \{b_i\}$ are dual bases of the algebra of the group G. This equality is possible only if $eb_i = 0$ for all i, i.e., $eB(G) = 0$, which is a contradiction. The lemma is proved.

3.8.2. Theorem. *Let G be an M-group of automorphisms of a prime ring R. Then we have*

$$(R_F)^G = (R^G)_{F_1},$$

where F_1 is a set of all essential ideals of R^G.

3.8.3. Lemma. *Let A be an essential ideal of a fixed ring. Then the annihilators (both left and right) of A in R_F are equal to zero and each of the sets RA, AR contains a nonzero two-sided ideal of the ring R.*

Proof. If $qA=0$ and $Iq \subseteq R$, where I is a nonzero ideal of R, then $(Iq)A \subseteq (0)$ and, therefore, the left annihilator L of the set A in R is nonzero. Let us choose an ideal I in such a way that $\tau(I) \subseteq R$ for the principal trace form τ (see **3.4.1**). By lemma **3.8.1**, we have $\tau(IL) \neq 0$, but on the other hand, $\tau(IL)A = \tau(ILA) = 0$ and $\tau(IL) = 0$, since A is an essential ideal.

Thus, the annihilator of A in $B(G)$ is zero (both left and right, since A commutes with elements from $B(G)$). By corollary **3.7.4**, the left ideal RA contains a nonzero two-sided ideal of the ring R. In particular, the right annihilator of the set RA and, hence of the set A as well, is zero in R_F. Corollary **3.7.4** also implies that AR contains a nonzero two-sided ideal of the ring R. The lemma is proved.

3.8.4. Lemma. *If in the formulation of the theorem we have $I \in F$, then $I \cap R^G \in F_1$.*

Proof. It is evident that $I \cap R^G$ is a two-sided ideal of R^G. If $a(I \cap R^G)=0$ for $0 \neq a \in R^G$, then by lemma **3.8.1**, we have $\tau(aIJ) \neq 0$ for any nonzero ideal J of R. Moreover, the ideal J can be chosen in such a way that $\tau(J) \subseteq I$ and, hence, $0 \neq \tau(aIJ) = a\tau(IJ) \subseteq a(I \cap R^G) = 0$, which is a contradiction.

Proof of theorem **3.8.2.** Let us first show that $(R^G)_F$ is naturally embedded into R_F. Let $\xi \in Hom(A, R^G)$, $A \in F_1$. Let us determine the correspondence $\xi^h: RA \to R$ by the formula

$$(\sum r_\alpha a_\alpha)\xi^h = \sum r_\alpha(a_\alpha \xi).$$

Let us show that ξ^h is a mapping. Let $V = \{\sum_\alpha r_\alpha(a_\alpha\xi) \mid \sum_\alpha r_\alpha a_\alpha = 0\}$. Then V is the left ideal and for any ideal I, such that $\tau(I) \subseteq R^G$, we have $\tau(\sum_\alpha i r_\alpha(a_\alpha\xi)) = \sum \tau(ir_\alpha)(a_\alpha\xi) = [\sum \tau(ir_\alpha)a_\alpha]\xi = [\tau(\sum ir_\alpha a_\alpha)]\xi = 0$, where i is any element from I. Therefore, by lemma **3.8.1**, we have $V = 0$. It means that ξ^h is a homomorphism of left R-modules.

By lemma **3.8.3**, its domain of definition, RA, contains a nonzero ideal of the ring R, and, hence, ξ^h determines an element from R_F, which will be also designated by ξ^h. It is now evident that the mapping $h: \xi \to \xi^h$ is an embedding of $(R^G)_{F_1}$ in R_F.

We are now to show that the image of h coincides with $(R_F)^G$. If $\xi \in (R^G)_{F_1}$, then

$$x(\xi^h)^g = [(\sum r_\alpha a_\alpha)^{g^{-1}}\xi]^g = [\sum r_\alpha^{g^{-1}}(a_\alpha \xi)]^g = (\sum r_\alpha a_\alpha)\xi^h = x\xi^h.$$

Hence, the image of h is contained in $(R_F)^G$. Inversely, let $\varphi \in (R_F)^G$, $\varphi: I \to R$. Let us consider a restriction of φ on R^G. By lemma **3.8.4**, the domain of the definition of this restriction belongs to F_1. Then, $(I \cap R^G)\varphi \subseteq (R_F)^G \cap R = R^G$. Therefore, the restriction of ξ belongs to $(R^G)_{F_1}$, in which case $\xi^h = \varphi$, since $R(I \cap R^G) \supseteq J \in F$ and $J(\xi^h - \varphi) = 0$. The theorem is proved.

3.8.5. Corollary. *Under the conditions of theorem 3.8.2, the equality* $Q(R)^G = Q(R^G)$ *is valid.*

The proof results immediately from theorem **3.8.2** by way of applying lemma **3.8.4.**

3.9 Galois Subrings for M-groups

In this paragraph we shall elucidate under what conditions the intermediate ring S, $R \supseteq S \supseteq R^G$ will be a Galois subring of an M-group.

Let G be an M-group of automorphisms of a prime ring R ($G \subseteq A(R)$, see **1.7**), S be an (almost) intermediate ring. By theorem **3.5.1**, a centralizer of S in the ring R_F is contained in the algebra $B = B(G)$ of the group G. *This centralizer will be henceforth designated* *by* Z.

Let us introduce the following conditions for the ring S.

BM (*Bimodule property*). Let e be an idempotent from $B(G)$, such that $se = ese$ for all $s \in S$. Then there exists (an idempotent) $f \in Z$, such that $ef = f$, $fe = e$.

RC (*Rational completeness*). If A is an essential ideal of S and $Ar \subseteq S$ for a certain $r \in R$, then $r \in S$.

SI (*Sufficiency of invertible elements*).The ring Z is generated by its invertible elements and if for an automorphism $g \in G$ there is a nonzero element $b \in B(G)$, such that $sb = bs^g$ for all $s \in S$, then there exists an invertible element with the same property.

In relation with the first part of condition **SI**, we should at once remark that if a field C (a generalized centroid of R) contains at least three elements, then any finite-dimensional algebra with a unit over C (and, moreover, Z) is generated by its inverse elements. Therefore, this part of the condition is restrictive only in the case when C is a two-element field.

3.9.1. Theorem. *Any intermediate Galois subring for a Maschke subgroup obeys conditions* **BM**, **RC** *and* **SI**.

Proof. Let $S = R^H$, where H is a subgroup of the group G, the algebra of which, $B(H)$, is semiprime. Then, by theorem **3.5.1**, the centralizer Z of the subring S in $B(G)$ (and even in R_F) is equal to $B(H)$. Now condition **BM** immediately results from proposition **3.7.6**.

Condition **RC** can be checked in the following way: if $h \in H$, then $A(r - r^h) = 0$ and, hence, $RA(r - r^h) = 0$, i.e., lemma **3.8.5** yields $r = r^h$, which is the required proof.

Let us check condition **SI**. The algebra $Z = B(H)$ is, by the definition, generated by invertible elements (see **3.1.1**).

Let $bs = s^g b$ for all $s \in S$. Let τ_H be a principal trace form of the group H. Then for all x from a suitable nonzero ideal we have the equality $b\tau_H(x) = \tau_H(x)^g b$. If this is a reduced identity, then, by theorem **2.2.2**, we have $\sum ba_i \otimes b_i = 0$, where $\{a_i\}, \{b_i\}$ are dual bases of $B(H)$, i.e., $ba_i = 0$ for all i, which is impossible. Therefore, this identity is not reduced, and, hence, $h_1 g \equiv h_2 (\mathrm{mod}\ G_{in})$ for some $h_1, h_2 \in H$, i.e., $g = h\tilde{a}$, where $h \in H$ and a is an invertible element from $B(G)$. We have $a^{-1}s = s^{\tilde{a}}a^{-1} = s^g a^{-1}$, i.e., a^{-1} is the sought element. The theorem is proved.

Now we are to prove the inverse statement.

3.9.2. Theorem. *Let* G *be a regular group of automorphisms of a prime ring* R. *Then if an intermediate subring* S *obeys conditions* **BM**, **RC** *and* **SI**, *then* $S = R^H$ *for a certain M-subgroup* H *of the group* G

In order to elucidate the essence and meaning of the notions and auxiliary lemmas considered below, let us first make a general outline for proving the theorem. First, the group H is immediately calculated: $H = A(S) = \{g \in G \mid s^g = s (\forall\ s \in S)\}$ and one has only to show that its algebra $B(H) = Z$ is semisimple (lemma **3.9.4**). The proof is then reduced to that of the equality $S = R^{A(S)}$. Therefore, for any S (possibly even not obeying conditions **BM**, **RC** and **SI**) let us introduce a Galois closure, $\bar{S} = I(A(S)) = R^{A(S)}$. Condition **RC** implies that for any $\bar{s} \in \bar{S}$ it is sufficient to find a suitable ideal of denominators, $I\bar{s} \subseteq S$. Let us pose (and solve) a more difficult problem: to find a common ideal of denominators, i.e., to show that S contains an essential ideal of the ring $\bar{S}$ (theorem **3.9.16**), which task, certainly, requires some ways of constructing (or finding) elements from S. A certain assistance is given by inclusion $S \supseteq R^G$, i.e., for any trace form τ we have a relation $\tau(x) \in S$, where x runs through a certain nonzero ideal of R. If instead of x we put into this relation, for instance, a product $b_i x$ and multiply it from the left by an element $s_i \in S$, then we get a new form with its values lying in S. The sum $\sum s_i \tau(b_i x)$ of such forms also assumes the values from S. The basic idea of the proof is to obtain, using such transformations and their right analogs, a form $t'(x)$ having the values from S, and such that the mapping $x \to t'(x)$ proves to be $(\bar{S}, \bar{S})$-bimodule, i.e., the factors from $\bar{S}$ can be put beyond the sign t': $t'(\bar{s} x \bar{s}_1) = \bar{s} t'(x) \bar{s}_1$. This can be achieved if such putting beyond the sign becomes possible in every addend $ax^g b$, i .e., $a\bar{s}^g x^g b = \bar{s} a x^g b$ and $ax^g \bar{s}^g b = ax^g b\bar{s}$, or $a\bar{s}^g = \bar{s} a$, $\bar{s}^g b = b\bar{s}$. The latter relations clarify the meaning of the sets $\Phi_g^{(S)}$ introduced in **3.9.5** (see also condition **SI**). In fact, in such a way we shall be able to obtain a principal H-fixed form τ_H.

We should now analyze what happens with the coefficients of the form when applying the above transformations. It is evident that the set of all values which the given (left) coefficient can transform into, forms a (S, R)-submodule and such submodules should be studied (lemma **3.9.3**).

The very transformation of a form can be identified with a linear combination $\sum b_i \otimes s_i \in R \otimes S$, where a tensor product is taken over the ring of integers. The action of an element on the (left) coefficient at x in the addend $(axb)^g$ is implemented by the formula $a \to \sum s_i^{g^{-1}} ab_i$, i.e., $a \to (a^g \cdot \beta^g)^{g^{-1}}$ if we assume that $\beta^g = \sum b_i^g \otimes s_i$, with the dot denoting the action that has already been encountered (see **1.7**). An interrelation of these transformations on finite sets of elements and sets $\Phi_g^{(S)}$ has been studied in lemmas **3.9.10** and **3.9.13**.

Let us get down to a detailed presentation. Let us fix some notations for an M-group G and an (almost) intermediate ring S. Through Z, as has been assumed, we shall denote a centralizer of S in the algebra $\mathcal{B}(G)$.

3.9.3. Lemma. *Let S obey condition **BM** and A be an (S, R)-subbimodule of R_F. Then there is an idempotent $f \in Z$, such that $fR_F \supseteq A \supseteq fI$ for a nonzero ideal I of R. In the same manner, if A is an (R, S)-subbimodule, then $R_F f \supseteq A \supseteq If$ for an idempotent $f \in Z$ and a nonzero ideal I.*

Proof. By theorem **3.7.3**, we get $eR_F \supseteq A \supseteq eI$ for an idempotent $e \in \mathcal{B}(G)$. Since A is a left S-module, then $sei \in A$ for all $s \in S$, $i \in I$, i.e., $sei = er_i$, $r_i \in R_F$. Multiplying it from the left by e, we get $esei = er_i = sei$ and, as i is an arbitrary element of the ideal I, then $se = ese$. Applying property **BM**, we can find an idempotent $f \in Z$, such that $ef = f$, $fe = e$. Let J be a nonzero ideal, such that $fJ \subseteq R$ (it should be recalled that $\mathcal{B}(G) \subseteq Q(R)$). We have: $fJI = e(fJ)I \subseteq eI \subseteq A$. Besides, $fR_F \supseteq f(eR_F) = eR_F \supseteq e(fR_F) = fR_F$.

Let now A be an (R, S)-subbimodule. Then, by theorem **3.7.1**, we have $R_F e \supseteq A \supseteq Ie$ for a certain $e \in \mathcal{B}(G)$, i.e., $ies = r_i e$ for any $s \in S$, $i \in I$ and suitable $r_i \in R_F$. Multiplying this equality from the right by e and taking into account that i is an arbitrary element of the ideal, we see that $ese = es$ or $s(1-e) = (1-e) \cdot s(1-e)$. By condition **B M**, we can find an idempotent $1 - f \in Z$, such that $(1-e)(1-f) = 1-f$, $(1-f)(1-e) = (1-e)$, wherefrom we get $ef = e$, $fe = f$. If now J is a nonzero ideal of the ring R, such that $Jf \subseteq R$, then $IJf \subseteq I(Jf)e \subseteq Ie \subseteq A$. Besides, $R_F f \supseteq (R_F e)f = R_F e$. The lemma is proved.

3.9.4. Lemma. *If an (almost) intermediate ring S obeys property* **BM**, *then it is semiprime, with its centralizer Z being semisimple.*

Proof. If $T < S$ and $T^2 = 0$, then $(T + RT)\cdot(TR + T) = 0$. By lemma **3.9.3**, we can find idempotents $e, f \in Z$, such that $R_F e \supseteq T + RT \supseteq Ie$; $fR_F \supseteq T + TR \supseteq fJ$, where $I, J \in F$. Then for any $t \in T$ we have $te = t$, $ft = t$ and, hence, $tef = tf = ft = t$. However, $IefJ = 0$, i.e., $ef = 0$ and $T = 0$.

To prove that Z is semisimple, it is sufficient to show that any principal right ideal pZ of the algebra Z is generated by an idempotent. As $p \in Z$, then pR is an (S, R)-bimodule and, by lemma **3.9.3**, we have $fQ \supseteq pR \supseteq fI$, where $I \in F$, f is an idempotent from Z. Let us first show that $pB = fB$. For this purpose let us make use of the fact that B is quasi-Frobenius:

$$pB = ann_B^{(r)}(ann_B^{(l)} pB) = ann_B^{(r)}(B(1 - f)) = fB,$$

where the equality $ann_B^{(l)} pB = B(1 - f)$ follows immediately from the chain of inclusions $fQ \supseteq pR \supseteq fI$ and from the fact that f is an idempotent.

Let $f = pb$, where $b \in B$. Then for any $s \in S$ we have $0 = p[b, s]$. Since $ann_Q^{(r)}(p)$ is an (S, R)-bimodule, then by lemma **3.9.3**, one can find an idempotent $e \in Z$, such that $eQ \supseteq ann_Q^{(r)}(p) \supseteq eJ$. In particular, $(1 - e)[b, s] = 0$, i.e., $(1 - e)bs = (1 - e)sb = s(1 - e)b$ and, hence, $(1 - e)b \in Z$. Besides, $pe = 0$, i.e., $p(1 - e) = p$. From here we get $f = pb = p(1 - e)b \in pZ$ and, as $p = fp$, then $pZ = fZ$. The lemma is proved.

3.9.5. Definitions and notations. For every isomorphism $\varphi\colon S \to R_F$ we shall denote by $\Phi_\varphi^{(S)}$ a set of all elements $a \in R_F$, such that $sa = as^\varphi$ for all $s \in S$. Analogously, $\Psi_\varphi^{(S)} = \{a \in R_F \mid \forall\, s \in S\colon as = s^\varphi a\}$. It should be remarked that if φ acts identically on $R^G \cap S$, then $\Phi_\varphi^{(S)} \subseteq B(G)$. Indeed, if $x \in R^G$, then $xa = ax^\varphi = ax$, i.e., by theorem 3.5.1, $a \in B(G)$.

3.9.6. Lemma. *The following relations are valid:*

$$(\Phi_g^{(S)})^h = \Phi_{h^{-1}gh}^{(S^h)};\quad \Phi_g^{(S)}(\Phi_h^{(S)})^g \subseteq \Phi_{gh}.$$

In particular, setting in the second inclusion $h=1$ and $g=1$, by turn, we get

$$\Phi_g^{(S)} Z^g \subseteq \Phi_g^{(S)}, \quad Z\Phi_h^{(S)} \subseteq \Phi_h^{(S)}.$$

Here $g, h \in A(R)$.

Proof. Applying an automorphism h to the equality $sa = as^g$, we get $s^h a^h = a^h s^{gh}$ or $s^h a^h = a^h \cdot (s^h)^{h^{-1}gh}$, which proves the first equality. If $a \in \Phi_g^{(S)}$ and $b \in \Phi_h^{(S)}$, then $sab^g = as^g b^g = a(bs^h)^g = ab^g s^{hg}$. The lemma is proved.

3.9.7. Definitions and notations. Let us denote by $L(S)$ a subring in the tensor product $Q \otimes Q^{op}$ over the ring of integers, which is generated by elements of the kind $r \otimes 1, 1 \otimes s$, where $r \in R$, $s \in S$ and the ring Q^{op} is antiisomorphic to Q, but has the same additive group.

If $\beta = \sum r_i \otimes s_i \in L(S)$, then let us assume $\varphi(\beta) = \sum r_i \otimes s_i^{\varphi}$ (unlike $\beta^{\varphi} = \sum r_i^{\varphi} \otimes s_i$, see **1.7**). If B is a subset in $L(S)$, then we set

$$B^{\perp} = \{ r \in Q \mid \forall \beta \in B \quad r \cdot \beta = 0 \}$$

It should be remarked that $r \cdot \beta = \sum s_i r r_i$, where $\beta = \sum r_i \otimes s_i$. For $r \in Q$ let us define

$$r^{\perp S} = \{ \beta \in L(S) \mid r \cdot \beta = 0 \} = r^{\perp} \cap L(S).$$

3.9.8. Lemma. *Let B be a right ideal of $L(S)$, $g_1, \ldots, g_n$ be isomorphisms of the ring S to $Q(R)$ acting identically on $R^G \cap S$. Let us assume that images S^{g_i} obey condition* **BM**. *Then for any elements $r_1, \ldots, r_n \in Q(R)$ the following formula is valid:*

$$B^{\perp} + \sum_{i=1}^{n} \Phi_{g_i}^{(S)} r_i = \left(\bigcap_{i=1}^{n} g_i^{-1} \left(r_i^{\perp S^{g_i}} \right) \cap B \right)^{\perp}.$$

Proof. Let us first prove that the left part of the equality is contained in the right one. For this purpose it is sufficient to show that if $g_i(\beta)=\sum b_k\otimes s_k^{g_i}\in r_i^{\perp S^{g_i}}$, then $ar_i\cdot\beta=0$ for all $a\in\Phi_g^{(S)}$. We have

$$(ar_i)\cdot\beta=\sum_k s_k ar_i b_k=a\sum_k s_k^{g_i}r_i b_k=a(r_i g_i(\beta))=0$$

Let us prove the inverse inclusion by induction over n.

Let $n=1$, $v\in(g_1^{-1}(r_1^{\perp S^{g_1}})\cap\mathbb{B})^{\perp}$. As $v\in Q(R)$, then $vI\subseteq R$ for a certain $I\in F$. Let $\mathbb{B}_1=\mathbb{B}(I\otimes 1)$, then $v\mathbb{B}_1\subseteq R$ and $\mathbb{B}_1^{\perp}=\mathbb{B}^{\perp}$. It is, therefore, sufficient to show that $v\in\Phi_{g_1}^{(S)}r_1+\mathbb{B}_1^{\perp}$.

If $\beta\in g_1^{-1}(r_1^{\perp S^{g_1}})\cap\mathbb{B}_1$, then $v\cdot g_1(\beta)=0$ and, hence, $\varphi\colon r_1\cdot g_1(\beta)\to v\cdot\beta$ is a correctly defined mapping of the set $A=r_1\cdot g_1(\mathbb{B}_1)$ to R, where β runs through $\mathbb{B}_1$. Moreover, φ is a homomorphism of right R-modules. Indeed,

$$\varphi[(r_1\cdot g_1(\beta))r]=\varphi[r_1\cdot g_1(\beta(r\otimes 1))]=$$
$$=[v\cdot\beta(r\otimes 1)]=(v\cdot\beta)r=[\varphi(r_1\cdot g_1(\beta))]r.$$

Since $\mathbb{B}_1$ is the right ideal in $\mathbb{L}(S)$, then A is an (S^{g_1},R)-bimodule. By lemma **3.9.3**, $A\supseteq\rho J$, $J\in F$, where $\rho^2=\rho$ belongs to the centralizer of S^{g_1} in $\mathbb{B}(G)$ and $\rho a=a$ for all $a\in A$.

Let us extend the action of φ on J by the formula $\bar\varphi(j)=\varphi(\rho j)$. As $\bar\varphi$ is a homomorphism of right R-modules and its domain of definition belongs to F, then $\bar\varphi(j)=\xi j$ for a certain $\xi\in R_F^0$ and for all $j\in J$ (here R_F^0 is the right ring of quotients of R with respect to $\boldsymbol{F}$, see **1.4**).

Let us show that $\xi\in\Phi_{g_1}^{(S)}$. Let $s\in S$. We can find an ideal $I_1\in F$, such that $s^{g_1}I_1\subseteq R$. Let j be an arbitrary element from I_1J, $\rho j=a=r_1\cdot g_1(\beta)\in A$. Then $\rho s^{g_1}=s^{g_1}\rho$ and $s^{g_1}j\in J$ and, hence,

$$\xi s^{g_1}j=\bar\varphi(s^{g_1}j)=\varphi(\rho s^{g_1}j)=\varphi(s^{g_1}\rho j)=$$
$$=\varphi[s^{g_1}(r_1\cdot g_1(\beta))]=\varphi[r_1\cdot g_1(\beta(1\otimes s))]=$$
$$v\cdot\beta(1\otimes s)=s(v\cdot\beta)=s\varphi(\rho j)=s\bar\varphi(j)=s\xi j.$$

This affords $(\xi s^{g_1} - s\xi)I_1 J = 0$ and, hence, $s\xi = \xi s^{g_1}$ for all $s \in S$. In particular, if $s \in R^G$, then $s\xi = \xi s$, i.e., by theorem **3.5.1**, we get $\xi \in B(G) \subseteq Q(R)$. Therefore, $\xi \in \Phi_{g_1}^{(S)}$.

Let us now show that $\xi a = \varphi(a)$ for all $a \in A$. Let us find an ideal $I_a \in F$, such that $I_a \cdot a \subseteq R$. If $l \in JI_a \cap R^G$, then $la \in \rho J$, so that $\rho a = a$ and $l\rho = \rho l$. Hence, $l\xi a = \xi la = \varphi(la) = l\varphi(a)$. Therefore, $l(\xi a - \varphi(a)) = 0$ and, hence, $R(JI_a \cap R^G)(\xi a - \varphi(a)) = 0$. Using lemmas **3.8.3** and **3.8.4**, we get $\xi a = \varphi(a)$.

Finally, for any $\beta \in B_1$ we have

$$(v - \xi r_1) \cdot \beta = v \cdot \beta - \xi r_1 \cdot \beta = v\beta - \xi(r_1 \cdot g_1(\beta)) =$$
$$= v \cdot \beta - \varphi(r_1 \cdot g_1(\beta)) = v \cdot \beta - v \cdot \beta = 0.$$

Consequently, $v \in \Phi_{g_1}^{(S)} r_1 + B_1^{\perp}$. This proves the lemma for the case $n = 1$.

Let us now assume that

$$B^{\perp} + \sum_{i=1}^{n-1} \Phi_{g_i}^{(S)} r_i = \Big(\bigcap_1^{n-1} g_i^{-1}(r_i^{\perp} S^{g_i}) \cap B\Big)^{\perp} \overset{df}{=} B_2^{\perp} \tag{15}$$

where $B_2 = \bigcap_1^{n-1} g_i^{-1}(r_i^{\perp} S^{g_i}) \cap B$. Then B_2 will be a right ideal of $L(S)$. Hence, according to the case $n = 1$, we have

$$B_2^{\perp} + \Phi_{g_n}^{(S)} r_n = (g_n^{-1}(r_n^{\perp} S^{g_n}) \cap B_2)^{\perp},$$

however, $B_2 \cap g_n^{-1}(r_n^{\perp} S^{g_n}) = \bigcap_1^n g_i^{-1}(r_i^{\perp} S^{g_i}) \cap B$ and, by formula (15), we have

$$B^{\perp} + \sum_{i=1}^{n-1} \Phi_{g_i}^{(S)} r_i + \Phi_{g_n}^{(S)} r_n = \Big(\bigcap_1^n g_i^{-1}(r_i^{\perp} S^{g_i}) \cap B\Big)^{\perp}$$

which is the required proof.

3.9.9. Remark. The proved lemma is to be used later in a somewhat more general form. Let $\bar{R} = R \oplus .. \oplus R$ be a direct sum of n copies of the ring R. The group G acts on every addend, and, therefore, $\bar{R}$ is acted upon by a direct power G^n. Besides, a group of permutations S_n acts on $\bar{R}$, rearranging the addends. Let $\bar{G} = G^n \rtimes S_n$ be a group generated by G^n and S_n, while let S be an almost intermediate ring, i.e., a subring of $\bar{R}$ containing all sums $t(x) \oplus .. \oplus t(x)$ at $x \in I$ for an ideal $I \in \mathcal{F}(R)$.

It should be remarked that lemma **3.9.8** is also valid under these conditions. For this purpose it should be noted that the algebra of the group $\bar{G}$ is equal to $\mathcal{B}(G)^n$, the centralizer of S is equal to a direct sum $Z_1 \oplus .. \oplus Z_n$ of the centralizers of S-projections on the addends of $\bar{R}$ and is, therefore, contained in $\mathcal{B}(G)^n$. Besides, any right (left) $\bar{R}$-submodule in $\bar{R}_F = R_F \oplus .. \oplus R_F$ is decomposed into a direct sum of its components. These remarks show lemmas **3.8.3** and **3.8.4**, as well as theorem **3.7.1** and lemma **3.9.3** to remain valid under these conditions as well. Now we are to pay our attention to the fact that in the proof of lemma **3.9.8** presented above we made use of the enumerated statements but not of the primeness of the ring R.

3.9.10. Corollary. *Under the conditions of the previous lemma (or under those of the previous remark) the following equality holds:*

$$\sum_{i=1}^{n} \Phi_{g_1}^{(s)} r_i = \left(\bigcap_{1}^{n} g_i^{-1}\left(r_i^{\perp} S^{g_i} \right) \right)^{\perp}.$$

3.9.11. Definition. Let S be a subring of R . The form $t(x) = \sum a_i x^{g_i} b_i$ is called an *S-form*, provided there is a nonzero ideal I of the ring R, such that $t(a) \in S$ for all $a \in I$. For instance, by the construction of **3.4.1**, any trace form is an R^G-form.

3.9.12. Theorem. *Let S be an (almost) intermediate ring obeying properties **BM** and **SI**, $H = A(S)$. Then H is an M -subgroup of the group G and the principal H-trace form will be an S-form.*

This theorem directly results from the following proposition.

3.9.13. Proposition. *Under the conditions of the preceding theorem, there are elements* $a \in R$, $a \neq 0$, $r_i, v_i \in R$, $s_i \in S$, $1 \leq i \leq m$, *such that for any* $x, y \in R_F$ *the following equality holds:*

$$\tau_H(yax) = \sum_i \tau(yr_i)s_i\tau(v_i x),$$

where τ_H *is the principal* H*-trace form,* τ *is the principal* G*-trace form.*
Let us first prove two lemmas.

3.9.14. Lemma. *If the centralizer* Z *of an almost intermediate ring* S*, obeying condition* ***BM****, is generated by invertible elements, then the group* $A(S)$ *is an* M*-group.*

Proof. As Z is generated by invertible elements, then the algebra of the group $A(S)$ is equal to Z and, by lemma **3.9.4**, is semisimple. It means that $A(S)$ is an M-group.

3.9.15. Lemma. *Under the conditions of theorem* **3.9.12**, *for two principal forms of the groups* G *and* $H = A(S)$ *the following relation is valid:*

$$\tau(x) = \tau_H(dx) + \sum_i (w_i x v_i)^{h_i} + \sum_j (w_j x v_j)^{g_j^{-1}}, \qquad (16)$$

for which the following conditions are met:

(a) $\Phi_{g_j}^{(S)} = 0$ *for all* j;

(b) $h_j \in H$ *for all* i;

(c) $Zd \cap \sum_i Zw_i = 0$;

(d) *the left annihilator of the element* d *in* Z *is zero:* $ann_Z^{(l)} d = 0$.

Proof. Let us consider the algebra B as a right module over Z. Since Z is semisimple, we can find a direct decomposition $B = Z \oplus V$. Let us choose a basis $z_1, ..., z_k$ of the algebra Z over C and extend it with elements $v_{k+1}, ... v_n \in V$ up to the basis of the algebra B. Let $w_1, ..., w_n$ be a dual basis, i.e., $(v_i, w_j) = \delta_{ij}$, where $v_i = z_i$ at $1 \le i \le k$ and the brackets denote an associative bilinear nondegenerate form on B.

Let $A = \sum_{i=1}^{k} w_i C$, $D = \sum_{i>k} w_i C$. Then $A = V^{\perp}$ and $D = Z^{\perp}$. As the form is associative, we have $(V, ZA) = (VZ, A) = 0$, $(Z, ZD) = (Z, D) = 0$, i.e., A and D are left Z-submodules of B, and, since they are not intersecting, then $B = A \oplus D$.

Limiting the bilinear form to a pair of Z-modules (Z, A), we get ${}_Z A \cong ({}_Z Z)^*$, i.e., ${}_Z A \cong {}_Z Z$ since a semisimple algebra is Frobenius. Thus, we can find an element $d \in A$, such that $A = Zd$ and the left annihilator of d in Z is zero (d is an image of the unit at the isomorphism $A \cong Z$). In particular, $w_1 = z_1^* d, \ldots, w_k = z_k^* d$ for some elements $z_1^*, \ldots, z_k^* \in Z$, these elements forming the basis of Z. On Z let us define a bilinear form $[x, y] = (x, yd)$. This form is associative and nondegenerate since $\{z_1^*, \ldots, z_k^*\}$ proves to be a dual basis to $\{z_1, \ldots, z_k\}$.

Now we can construct a principal form which commutes with Z:

$$v_H(x) = z_1^* x z_1 + \ldots + z_k^* x z_k.$$

For the principal form which commutes with B we have:

$$v(x) = v_H(dx) + \sum_{i > k} w_i x v_i ,$$

in which case of importance for us is the fact that $Zd \cap \sum_{i > k} Zw_i = 0$ and that the left annihilator d in Z is zero.

Let us choose a system of representatives of the right cosets of the group $G_{in}H$ over the subgroup G_{in}, which consists of the elements $1 = h_1, h_2, \ldots, h_t$, lying in H, and a system of representatives $1 = g_1, \ldots, g_s$ of the right cosets of the subgroup $G_{in}H$ of the group G. Both these systems are finite, as G is a reduced-finite group.

Now we have a presentation for the principal G-invariant form:

$$\tau(x) = \sum_{j,\, l} v(x)^{h_j g_l} = \sum_{i=1}^{t} (v_H(dx))^{h_j} +$$

$$+ \sum_{j \geq 1} \sum_{i > k} (w_i x v_i)^{h_j} + \sum_{j \geq 1} \sum_{i \geq 1} \sum_{l \geq 2} (w_i x v_i)^{h_j g_l} .$$

It should be remarked that $\Phi_g^{(S)} = 0$ at any $g = (h_j g_l)^{-1}$, where

$i \geq 2$. Indeed, in the opposite case there is, by property **SI**, an invertible element $b \in B$, such that $b^{-1}sb = s^g$ for all $s \in S$, i.e., $bg^{-1} \in A(S) = H$ and, hence, $h_j g_1 \in G_{in} H$. As G_{in} is a normal subgroup, then $g_1 \in G_{in} H$, which contradicts the choice of $g_1, \ldots, g_s$.

Thus, simplifying the notations, we find a correlation of the required type. The lemma is proved.

3.9.16. The proof of proposition 3.9.13. Preserving the notations of the preceding lemma, let us set $\aleph = \bigcap_i w_1^{\perp S} \cap \bigcap_j g_j^{-1}(w_j^{\perp S})^{g_j}$. Then, applying an element $\beta \in \aleph$ to both parts of equality (16), we get:

$$\tau_H((d \cdot \beta)x) = \tau(x) \cdot \beta. \tag{17}$$

Since $\aleph$ is a right ideal in $L(S)$, then $d \cdot \aleph$ is a (S, R)-subbimodule in R_F. By lemma **3.9.3**, we can find an idempotent $f \in Z$, such that $fI \subseteq d\aleph \subseteq fR_F$. Hence, $[(1 - f)d] \cdot \aleph = (1 - f)(d \cdot \aleph) = 0$, i.e., $(1 - f)d \in \aleph^{\perp}$. Applying corollary **3.9.10.** and using conditions (a), (c) and (d) for relation (16), we get $f = 1$.

Now, due to proposition **3.7.5**, one can find a finite set of elements $v_1, \ldots, v_n \in d \cdot \aleph$, such that for suitable $a, r_i, t_i \in R$, $a \neq 0$ and any $y \in R_F$ the following relation is valid:

$$ya = \sum_i \tau(yr_i)v_i t_i.$$

Let $v_i = d \cdot \beta_i$, where $\beta_i \in \aleph$. Then by formula (17) we have $\tau_H(v_i t_i x) = \tau(t_i x)\beta_i$. Herefrom we get

$$\tau_H(yax) = \tau_H(\sum_i \tau(yr_i)v_i t_i x) =$$

$$= \sum_i \tau(yr_i)\tau_H(v_i t_i x) = \sum_i \tau(yr_i)[\tau(t_i x) \cdot \beta_i].$$

Writing now the actions of β_i in detail, we get an equality of the required type. The proposition is proved.

3.9.17. The validity of theorem **3.9.2** is now certain: if S is an intermediate subring, obeying conditions **BM, RC** and **SI**, then one can find a nonzero ideal I of the ring R, such that $\tau_H(I) \subseteq S$, where $H = A(S)$, but $\tau_H(I)$, being an essential two-sided ideal of the ring $R^H = \bar{S}$ and, since S is rationally complete, we get $\bar{S} = S$, which is the required proof.

3.9.18. Corollary. *Let S be an (almost) intermediate ring obeying properties **BM** and **SI**. In this case the ring S contains an essential ideal of the ring $\bar{S} = \mathbb{I}\, A(S)$.*

Indeed, $\tau_H(I) \subseteq S$ for a nonzero ideal I of R and the set $\tau_H(I)$ is an essential ideal of $\bar{S}$.

3.9.19. Finally, it should be remarked that the key statement of this paragraph (proposition **3.9.13.**) remains to be valid under somewhat more general conditions as well (see remark **3.9.9.**). Namely, let $\bar{R} = R \oplus ... \oplus R$, $\bar{G} = G^n \leftthreetimes S_n$ and let the almost intermediate ring S obey condition **BM**, and let us assume $A(S) = (H^m \leftthreetimes S_m) \times G''$, where $H < G$, S_m is a group of permutations of the first m components of the decomposition of $\bar{R}$, while the group G'' acts trivially on these components. Let us denote by e a unit of the first component and assume that $e\Phi^{(S)}_{g\leftthreetimes(1j)}=0$ at all $j > m$ and any $g \in G^n$.

Let us set $\tau_{\bar{G}}(x) = \sum_{i=1}^{n} \tau^{(1i)}(x)$, where $(1i)$ is a transvection from S_n. Then $\tau_{\bar{G}}$ is a trace form for the group $\bar{G}$. Analogously, $\tau_{A(S)} = \sum_{j=1}^{m} \tau_H^{(1j)}(x)$ is a trace form for the group $A(S)$.

3.9.20. Lemma. *There is a nonzero element $a \in R$ (to be more exact, $a \in e\bar{R}$), elements $r_1,..., r_k$, $v_1... v_k \in \bar{R}$ and elements $s_1,..., s_k \in S$, such that for any $x, y \in \bar{R}_F$ the following equality is valid:*

$$\tau_{A(S)}(yax) = \sum_{i=1}^{k} \tau_{\bar{G}}(yr_i)s_i\tau_{\bar{G}}(v_i x).$$

Proof. Using lemma **3.9.15**, let us write the following equality:

$$\tau_{\bar{G}}(x) = \sum_{j=1}^{m} \tau_H^{(1\,j)}(dx) + \sum_{j=1}^{m} \sum_i (w_i x v_i)^{h_i \lambda (1\,j)} +$$
$$+ \sum (w_j x v_j)^{g_j^{-1}} +$$
$$+ \sum_{k=m+1}^{n} (\tau_H(dx) + \sum (w_j x v_j)^{g_j^{-1}})^{(1k)},$$

where the elements d, w and v lie in the first component $e\bar{R}$.

Introducing renotations, we get a presentation

$$\tau_{\bar{G}}(x) = \tau_{A(S)}(dx) + \sum_i (w_i x v_i)^{h_i} + \sum_j (w_j x v_j)^{g_j^{-1}},$$

for which the following conditions are met:

(a) $\Phi_{g_i}^{(S)} = 0$ for all j;

(b) $h_i \in A(S)$ for all i;

(c) $Zd \cap \sum Zw_i = 0$;

(d) the left annihilator of d in Ze is zero: $ann_{Ze}^{(l)} d = 0$.

Now we are to repeat, nearly word per word, the proof of proposition **3.9.13.** (see **3.9.16**), replacing R with $\bar{R}$; τ_H, τ with $\tau_{A(S)}$, $\tau_{\bar{G}}$, respectively, paying a special attention to the fact that $d\aleph$ is contained in the first component and, hence, the idempotent f will be equal to the unit e of the first component. Analogously, proposition **3.7.5** proves the existence of the elements a, r_i, t_i, $a \neq 0$ from the first component for which the following equalities hold:

$$ya = \sum_i \tau_G(yr_i)v_i t_i = \sum_i \tau_{\bar{G}}(yr_i)v_i t_i,$$

since $e\tau_{\bar{G}}(x) = \tau_G(x)$. The lemma is proved.

3.10 Correspondence Theorems

We can now summarize the data obtained in the preceding paragraph in a form traditional for the Galois theory.

3.10.1. Theorem. *Let G be a regular group of automorphisms of a prime ring R. Then the mappings $H \to \mathbb{I}(H)$, $S \to \mathbb{A}(S)$ set a one-to-one correspondence between all regular subgroups of the group G and all intermediate subrings obeying conditions* **BM**, **RC** *and* **SI**.

The proof results immediately from theorems **3.5.2**, **3.9.1** and **3.9.2**.

Since conditions **BM** and **SI** refer only to interactions of the elements of the algebra of the group G with intermediate subrings, then for outer groups we immediately get the following theorem.

3.10.2 Theorem. *Let G be a finite group of outer automorphisms of a prime ring R. Then the mappings $H \to \mathbb{I}(H)$, $S \to \mathbb{A}(S)$ set a one-to-one correspondence between all the subrings of the group G and all intermediate rationally-complete subrings of R.*

Another particular case, when the algebra of a group contains few idempotents and many invertible elements, arises when $\mathbb{B}(G)$ is a sfield. This happens when R contains no zero divisors, i.e., it is a domain. Indeed, then $Q(R)$ also has no zero divisors and the algebra of any reduced-finite group is a sfield, and then we come to the next theorem.

3.10.3. Theorem. *Let G be a reduced-finite N-group of automorphisms of the domain R. Then the mappings $H \to \mathbb{I}(H)$, $S \to \mathbb{A}(S)$ set a one-to-one correspondence between all N-subgroups of the group G and all the intermediate rationally-complete subrings.*

It would be useful to remark here that the condition of rational completeness for intermediate rings in the case of domains is equivalent to a stronger elementary condition: if $rs = s_1$ for some $0 \neq s,\ s_1 \in S,\ r \in R$, then $r \in S$ (such subrings are called *antiideals*). This fact can, for instance, be deduced from theorem **3.10.3** and from the evident fact that a fixed ring of a domain is an antiideal: if $rs = s_1$, then $(rs)^g = s_1^g$, i.e., $r^g s = s_1$ and, hence, $(r - r^g)s = 0$.

The next natural step is to consider groups H the algebras of which, $\mathbb{B}(H)$, are simple (see definition **3.1.7**). This case is quite close to the general situation for M-groups, as studies of an arbitrary M-group can be reduced to such groups to the accuracy of matrix constructions.

3.10.4. Definition. A reduced-finite group of automorphisms, G, of a prime ring R is called an *F-group* if its algebra is simple.

3.10.5. Let us present a general scheme of the above-mentioned reduction. Let $\mathcal{B}(G) = B_1 \oplus ... \oplus B_k$ be a decomposition of the algebra of an M-group G into G-simple components. Let us denote by e_i a unit of the algebra B_i. Then $Q_i = e_i Q e_i$ is an invariant subring and $Q^G = Q_1^G \oplus ... \oplus Q_n^G$, in which case the group G acts on Q in such a way that its algebra is isomorphic to B_i, i.e., it is G-simple.

Let us now consider the case of a G-simple algebra. Let $\mathcal{B}(G) = B_1 \oplus ... \oplus B_n$ be a decomposition into simple components. As the decomposition is unique to the accuracy of permutation of the addends, then for any $g \in G$ there is a permutation π_g of the numbers $1,..., n$, such that $B_k^g = B_{\pi_g(k)}$. Since the algebra is G-simple, for every k there is an automorphism $g_k \in G$, such that $B_1^{g_k} = B_k$ and, in particular, all the components of the decomposition are isomorphic.

Let us denote by G_k a subgroup of all automorphisms for which $\pi_g(k) = k$. In other words, there are isomorphisms which leave the unit e_k in its place. All groups G_k are certain to be mutually conjugated: $G_k = g_k^{-1} G_1 g_k$.

Let us consider a subring $Q_k = e_k Q e_k$. The action of the group G_k is naturally induced on it, and the algebra of the induced group is $B_k = e_k \mathcal{B}(G)$. Let us prove this fact.

Let $\xi \in (Q_k)_F$ and let us assume that $x\xi = \xi x^g$ for all $x \in Q_k$ and a certain $g \in G$. If $g = \tilde{b}$ is an inner automorphism from G_k, then it is inner for $(Q_k)_F$ as well as the element $e_k b$ will correspond to it. We then come to the conclusion that ξ and $e_k b$ are linearly dependent over a generalized centroid of the ring Q_k which is equal to $e_k C$, i.e., $\xi \in B_k$.

If g is not an inner automorphism, then one can choose an element $u \in Q_k$ in such a way that $u\xi \neq 0$, $u\xi \in Q_k$ and for any $y \in Q$ we get the following identity:

$$u e_k y e_k (u\xi) = (u\xi) e_k^g y^g e_k^g u^g,$$

i.e., by proposition **2.2.2**, we get: $ue_k \otimes e_k(u\xi) = 0$, which is impossible, as $ue_k = u$, $e_k(u\xi) = u\xi$.

Let us, finally, show that $Q_k^{G_k} \cong Q^G$, the case $k = 1$ being sufficient for this purpose. Let us construct two mappings in the following way. If $a \in Q^G$, then we set $\varphi(a) = e_1 a$. If $d \in Q_1^{G_1}$, then we set $\psi(d) = d + d^{g_2} + \ldots + d^{g_n}$. Since $e_1^{g_k} = e_k$ and the idempotents $e_1, \ldots, e_n$ are orthogonal, then $\varphi(\psi(d)) = d$. Hence,

$$\psi(\varphi(a)) = e_1 a + e_2 a + \ldots + e_n a = a.$$

The fact that these mappings preserve the operations is also trivial.

Therefore, taking into consideration the fact that the ring Q can be presented as a ring of generalized matrices

$$Q = \| e_k Q e_i \|,$$

studies of an M-group of automorphisms of a prime ring is, in a certain sense, reduced to studies of F-groups.

Let us remark that a regular F-group is called quite regular (see **3.1.7**).

3.10.6. Theorem. *Let G be a regular group of automorphisms of a prime ring R. Then the mappings $H \to I(H)$ and $S \to A(S)$ set a one-to-one correspondence between all quite regular subgroups of the group G and all intermediate (prime) rings, obeying conditions* **BM** *and* **RC** *and having simple centralizers in $\mathbb{B}(G)$.*

To prove it, it suffices to remark that by theorem **3.5.1** a fixed ring of an F-group has a simple centralizer (and is prime, by theorem **3.6.2**), and to show that an intermediate subring under condition **BM** with a simple centralizer obeys condition **SI**. The first part of condition **SI** is fair, as any simple finite-dimensional algebra is generated by its invertible elements. The second part will be proved in a more general form.

3.10.7. Lemma. *Let intermediate rings S and S_1 have simple centralizers and obey condition* **BM**. *Then for any isomorphism $g: S \to S_1$, which acts identically on R^G, either the set $\Phi_g^{(S)}$ contains an invertible element, or it equals zero.*

Proof. Let Z be a centralizer of S in $\mathbb{B}(G)$, and Z_1 be that of S_1. The set $\Phi_g^{(S)}$ is a (Z, Z_1)-bimodule. Let us view this set as a left Z-

module. Let $0 \neq \Phi_g^{(S)} = L_1 \dot{+} \ldots \dot{+} L_m$ be a decomposition of this module into irreducible components. Then $L_i \cong Ze_i$, where e_i is a primitive idempotent from Z, $Z = Ze_1 \dot{+} Ze_2 \dot{+} \ldots \dot{+} Ze_n$.

Let us first assume that $m \geq n$. In this case the left Z-module Z is embeddible into $\Phi_g^{(S)}$ and, hence, $\Phi_g^{(S)}$ contains an element a, the left annihilator of which in Z is zero. As $sa = as^g$ for all $s \in S$, then aR_F is an (S, R)-subbimodule in R_F and, by lemma **3.9.3**, we have $eR_F \supseteq aR_F \supseteq eI$ for some $I \in F$, $e \in Z$. Since $(1-e)a = 0$, then $e = 1$ and, hence, $aR_F \supseteq I$. In particular, the left annihilator of a in the algebra $B(G)$ is zero and, therefore, $\dim_C B(G)a = \dim_C B(G)$, i.e., a is an invertible element..

Let now $m < n$. Then $\Phi_g^{(S)}$ is isomorphic to the left ideal of the algebra Z and is, hence, a cyclic Z-module $\Phi_g^{(S)} = Za$. The right annihilator of the element a in Z_1 is equal to the right annihilator of $\Phi_g^{(S)}$. The latter, however, is an ideal of Z_1 and is, therefore, equal to zero. Allowing for the fact that $as^g = sa$ for all $s \in S$, we see that $R_F a$ is an (R, S_1)-bimodule and, by lemma **3.9.3**, we have $R_F e \supseteq R_F \supseteq Ie$ for a suitable $e \in Z_1$. But, as $a(1-e) = 0$, we have $e = 1$, i.e., as above, a is an invertible element of $B(G)$. The lemma is proved.

3.11 Extension of Isomorphisms

A traditional problem of the Galois theory is that of finding criteria for an intermediate subring S to be a Galois extension over R^G. The conditions for a general case being quite complex, we are not going to discuss them here. The considerations of this problem are commonly based on the theorem on the extension of isomorphisms which will be given below with two applications concerning the sought criteria for outer groups and domains.

Let us begin with a simple example which shows the extension of isomorphisms over R^G between intermediate subrings not to always be possible for arbitrary M-groups.

3.11.1. Example (D.S.Passman). Let R be a ring of all matrices of the fourth order over a field $F \neq GF(2)$ and let G be a group of all (inner) automorphisms of this algebra. Let us set $S = \{diag(a, a, a, b) \mid a, b \in F\}$ and let $S_1 = \{diag(a, a, b, b) \mid a, b \in F\}$. Then $S \cong S_1$, in which case the corresponding isomorphism is identical on $R^G = \{diag(a, a, a, a)\}$. The both rings are Galois subrings, since their centralizers Z, Z_1 are generated by invertible elements. At the same time, the isomorphism between S and S_1 cannot be extended to the automorphisms of R, as Z is not isomorphic to Z_1.

The situation is better under the assumption that S, S_1 are Galois subrings of F-groups.

3.11.2. Theorem. *Let G be a regular group of automorphisms of a prime ring, S, S_1 be intermediate Galois subrings of quite regular groups. Then any automorphism $\varphi: S \to S_1$, which is identical on R^G, can be extended to the automorphism $\bar{\varphi} \in G$.*

This statement can be easily deduced from the correspondence theorem for a ring $R \oplus R$ with a group $G^2 \rtimes S_2$ (see theorem **5.9.2**). Here we shall give a direct proof of a somewhat more general statement which is going to be useful in a number of cases. Theorem **3.11.2** results from it by lemma **3.10.7**.

3.11.3. Proposition. *Let G be a regular group of automorphisms of a prime ring R, S, S_1 be intermediate subrings obeying condition* **BM**. *If $\varphi: S \to S_1$ is an isomorphism identical on R^G and if the ring S obeys condition* **SI** *for a set of isomorphisms $G \cap \varphi G$, then φ can be extended to the automorphism $\bar{\varphi} \in G$.*

Proof. Let us consider $\mathcal{B}(G)$ as a left Z-module, where $Z = Z(S)$. By lemma **3.9.4**, the algebra Z is semisimple and, hence, $\mathcal{B}(G)$ is a completely reducible Z-module. In particular, Z can be singled out as a direct addend from $\mathcal{B}(G)$: $\mathcal{B}(G) = Z \oplus V$, where V is a left Z-module. In Z let us choose a basis $z_1, \ldots, z_k$ over C, in such a way that $z_1 = 1$. Let us extend this basis with elements from V to the basis $b_1, \ldots, b_n$ of the algebra $\mathcal{B}$ over C. Let $b_1^*, \ldots, b_n^*$ be a dual basis.

Let then $H = A(S)$, $H_1 = A(S_1)$. Let us also set $G(S) = G_{in}H$, $G(S_1) = G_{in}H_1$, where G_{in} is a subgroup of inner automorphisms from G. Let us choose a system of representatives of the right cosets of the group $G(S)$ over the subgroup G_{in} which consist of elements $1 = h_1,\ldots, h_m$ lying in H, and a system of representatives of the right cosets of the group G over the subgroup $G(S)$ of the elements $1 = g_1,\ldots, g_n$. As G is a reduced-finite group, then both these systems are finite and, moreover, the set $\{h_i g_j\}$ forms a system of representatives of the right cosets of G over G_{in}. Performing the same operations for the group $G(S_1)$, we can find a system of representatives $\{h'_i g'_j\}$ for G_{in} in G, where $h'_i \in H_1$. For any x let us set $t_H(x) = \sum_i x^{h_i}$, $t_{H_1}(x) = \sum_i x^{h'_i}$. Now, by lemma **3.4.1**, we have two trace forms:

$$\tau(x) = \sum_j t_H(v(x))^{g_j}, \qquad \tau_1(x) = \sum_j t_{H_1}(v(x))^{g'_j},$$

where $v(x) = \sum_t b_t x b_t^*$. Let us note that these forms are equal. We have $h_i g_i = \tilde{a}\, h'_{\pi(i,\, j)} g'_{\varepsilon(i,\, j)}$, where $\tilde{a} \in G_{in}$. Hence, we have

$$\tau(x) = \sum_{i,\, j} v(x)^{h_i g_j} = \sum_{i,\, j} (v(x)^{\tilde{a}})^{h'_{\pi(i,\, j)} g'_{\varepsilon(i,\, j)}} = \sum_{i,\, j} v(x)^{h'_i g'_j} = \tau_1(x).$$

Let us write the equality $\tau(x)^{\varphi} = \tau_1(x)$ in detail:

$$[\sum_i t_H(v(x))^{g_i}]^{\varphi} = \sum_i t_{H_1}(v(x))^{g'_i}$$

Let $\varphi(\beta)$ be an arbitrary element from $\mathbb{L}(S_1) = \mathbb{L}(S^{\varphi})$. Then, applying $\varphi(\beta)$ to both parts of the latter equality, we get (assuming, as usual, $f(x)\cdot\beta = \sum s_j f(r_j x)$, $\varphi(\beta) = \sum r_j \otimes s_j^{\varphi}$, where $\beta = \sum r_j \otimes s_j \in \mathbb{L}(S)$):

$$[\sum_i t_H(v(x)\cdot g_i^{-1}(\beta))^{g_i}]^{\varphi} = \sum_i t_{H_1}(v(x)\cdot \varphi g_i'^{-1}(\beta))^{g'_i}. \qquad (18)$$

It should be remarked that $\Phi^{(S)}_{g_i^{-1}} = 0$ at $i \neq 1$. Indeed, since otherwise, by condition **SI**, one can find an invertible element $a \in B(G)$, such that $sa = as^{g_i^{-1}}$ for all $s \in S$. It means that the automorphisms $g_i^{-1}\tilde{a}^{-1}$ belongs to $A(S)$ and, hence, $g_i^{-1} = (\tilde{a}g_i)^{-1}\tilde{a} \in G(S)$, which affords $i = 1$.

Let us assume that $\Phi^{(S)}_{\varphi g'^{-1}_i} = 0$ for all i (including $i = 1$). Then we can write

$$1 \notin \sum_{m>k} Zb_m + \sum_{\substack{i \neq 1 \\ 1 \leq j \leq n}} \Phi^{(S)}_{g_i^{-1}} b_j + \sum_{i,\, j} \Phi^{(S)}_{\varphi g'^{-1}_i} b_j,$$

since the last two sums are zero. (It should be recalled that $b_m = z_m$ at $m \leq k$). By corollary **3.9.12**, we can find a $\beta \in L(S)$, such that $b_0 \stackrel{df}{=} 1 \cdot \beta \neq 0$ and, simultaneously, $b_m \cdot \beta = 0$ at $m > 1$; $b_j g_i^{-1}(\beta) = 0$ at all j, $1 \leq j \leq k$ and all $i \neq 1$; $b_j \cdot \varphi g'^{-1}_i(\beta) = 0$ at all i and all j. Now, according to equality (18), we get $t_H(v_1(x)) = 0$, where $v_1(x) = \sum_{m=1}^{k} b_m b_0 x b^*_m$. Or, in more detail,

$$\sum_{i,\, m} b_m^{h_i} b_0^{h_i} x^{h_i} (b^*_m)^{h_i} = 0 .$$

As at $i \neq 1$ the elements h_i do not lie in G_{in}, then this identity, by theorem **2.2.2**, yields equality $\sum b_m b_0 \otimes b^*_m = 0$ in the ring $Q \otimes Q$, which is impossible, as $\{b^*_m\}$ are linearly independent elements and $b_1 b_0 = b_0 \neq 0$.

Thus, we can choose an element $g = g'^{-1}_i \in G$, such that $\Phi^{(S)}_{\varphi g} \neq 0$. By condition **SI**, for φg one can find an invertible element $a \in B(G)$, such that $sa = as^{\varphi g}$. Hence, $s^{\varphi} = (a^{g^{-1}})^{-1} s^{g^{-1}} a^{g^{-1}}$, i.e., $g^{-1}\tilde{b}$, where $b = a^{g^{-1}}$ is the sought extension of φ. The proposition is proved.

3.11.4. Corollary. *Let G be a reduced-finite group of automorphisms of a domain. Then any isomorphism over R^G between intermediate subrings can be extended to an automorphism from $A(R)$.*

Indeed, it has already been noted that any intermediate subring of the domain obeys condition **BM**, and its centralizer in Q is a sfield. We are, therefore, to extend the group G to a quite regular one. In this case a fixed ring does not undergo changes and we can use proposition **3.11.3**.

3.11.5. Corollary. *Let G be a finite group of outer (for Q) automorphisms of a prime ring R. Then any isomorphism over R^G between intermediate subrings can be extended to an automorphism from G.*

3.11.6. Theorem. *Let G be a finite group of outer (for Q) automorphisms of a prime ring R. An intermediate ring S will be a Galois extension of R^G iff the group $A(S)$ is normal in G, in which case a Galois group of extension $S | R^G$ is equal to $G / A(S)$..*

Proof. Let $A(S)$ be normal in G. Let us first show that $Q(S)$ is a Galois extension of $Q(R^G)$. By theorem **3.8.2**, the subring $Q(S)$ is Galois and equal to $Q(R)^{A(S)}$ and, hence, it is invariant with respect to G: if $a \in Q(S)$, $g \in G$, $h \in A(S)$, then $(a^{ghg^{-1}})^g = a^g$, i.e., $a^g \in Q(R)^{A(S)} = Q(S)$. Therefore, the action of the group G is induced on $Q(S)$ and the induced group is isomorphic to $G / A(S)$. This group will be outer: if $\tilde{a} = g \in G$ on $Q(S)$, where $a \in Q(S)$, then for any $s \in R^G$ we have $s^{\tilde{a}} = s$, i.e., $a \in Z(R^G) = C$. It is also evident that a fixed ring of this group is equal to $Q(R^G)$. The task now is to show that the induced automorphisms belong to $A(S)$ (see **1.7**).

According to corollary **3.9.18** and proposition **1.4.13**, we have $A(S) = A(\bar{S})$, where $\bar{S} = IA(S) = Q(S) \cap R$ and, hence, it would be sufficient to find for every $g \in G$ ideals U_1 and U_2 from $F(\bar{S})$, such that $U_1 \subseteq U_2^g \subseteq \bar{S}$. By the definition of $A(R)$ there can be found ideals $I_1, I_2 \in F$, such that $I_1 \subseteq I_2^g \subseteq R$. Herefrom for $U_2 = I_2 \cap Q(S) = I_2 \cap \bar{S}$ we have $U_2^g = (I_2 \cap Q(S))^g = I_2^g \cap Q(S) \subseteq R \cap Q(S) = \bar{S}$, and for $U_1 = I_1 \cap \bar{S}$ we get $U_1 \subseteq I_2^g \cap R \cap Q(S) = I_2^g \cap Q(S) = (I_2 \cap Q(S))^g = U_2^g$, which is the required proof.

Let us, inversely, assume S to be a Galois extension of R^G with a group $H \subseteq A(S)$. Let us show that the automorphisms from H have an extension up to those from the group G. If $\varphi \in H$, then there are U_1, $U_2 \in F(S)$, such that $U_1 \subseteq U_2^{\varphi} \subseteq S$. According to corollary **3.11.5**, a restriction of φ on $U_2 + R^G$ can be extended to the automorphism $g \in G$. As the restriction of g on S belongs to $A(S)$ and $g = \varphi$ on $U_2 \in F(S)$, then $g = \varphi$ on S as well.

Let N be a group of all possible extensions of H. Then $N \subseteq G$, since G is a Galois group and $N \supseteq A(S)$. Moreover, $A(S)$ is a normal subgroup in N and $N / A(S) \cong H$. It should also be remarked that H is an outer group for S: if $\varphi(s) = a^{-1}sa$ for $a \in Q(S)$, $s \in S$, then at $s \in R^G$ we get $s = a^{-1}sa$, wherefrom we have $a \in Z(R^G) = C$.

Let us now calculate $Q(R)^N$. Since $A(S) \subseteq N$, then $Q(R)^N \subseteq Q(R)^{A(S)} = Q(S)$ and, hence, $Q(R)^N \subseteq Q(S)^H = Q(S^H) = Q(R^G) = Q(R)^G$. As N is also a Galois group, then $N = G$ and the theorem is proved.

3.11.7. Definition. Let G be a reduced-finite N-group of automorphisms of a domain R. Then a subgroup H of the group G is called *almost normal in* G if the least N-subgroup of the group G, which contains a normalizer $N_G(H)$, coincides with G.

3.11.8. Theorem. *Let G be a reduced-finite N-group of automorphisms of a domain R. The intermediate subring S is a Galois extension of R^G iff the group $A(S)$ is almost normal in G. In this case a Galois group of extension $S \subseteq R^G$ is equal to $N_G(A(S)) / A(S)$.*

Proof. Let $S^H = R^G$ for a group of automorphisms $H \subseteq A(S)$. For every $h \in H$ we can find nonzero ideals I, J of S, such that $I \subseteq J^h \subseteq S$. Now, by corollary **3.11.4**, the automorphism h can be extended from $J + R^G$ to an automorphism $g \in G$. One can easily check that g coincides with h on S, since $(s^h - s^g) J^h = 0$ for any $s \in S$. Let us denote by M a set of all extension of the elements of H. In order to show that $A(S)$ is almost normal in G we have to check the validity of the following two statements:

(a) the group $A(S)$ is normal in M; (b) $R^M = R^G$.

(a) Let $m \in M$, $a \in A(S)$. Let us choose a nonzero ideal I of S, such that $I^m \subseteq S$. Then a restriction mam^{-1} on I will be an identical automorphism. As an extension of automorphisms from the ideal on S (and even on $Q(S)$) is unique, then $mam^{-1} \in A(S)$.

(b) Let $r \in R^M$. Then $r \in R^{A(S)} = \bar{S}$, since $A(S) \subseteq M$. By corollary **3.9.18**, we can find a nonzero ideal I of S contained in S. As H is a reduced-finite group of automorphisms for the domain S, then the ideal has a nonzero intersection with $S^H = R^G$. Let t be a G-fixed element, such that $\bar{S}t \subseteq S$. Then $rt \in R^M \cap S = S^H = R^G$, i.e., $(r^g - r)t = (rt)^g - rt = 0$ for all $g \in G$. Hence, $r \in R^G$, and the proof is completed.

Inversely, let $A(S)$ be almost normal in G, and let M be its normalizer. Let us, first, show that a restriction of $m \in M$ on S is an automorphism of S, i.e., it belongs to $A(S)$.

Since S and $\bar{S} = R^{A(S)}$ have a shared nonzero ideal, then $A(S) = A(\bar{S})$. If now $m \in M$, $a \in A(S)$, then $ma = (mam^{-1})m$, in which case $mam^{-1} \in A(S)$. It means that $\bar{S}^m \subseteq Q^{A(S)}$. As $m \in A(R)$, then we can find nonzero ideals I, J of R, such that $I \subseteq J^m \subseteq R$. Hence, $(J \cap \bar{S})^m \subseteq Q^{A(S)} \cap R = \bar{S}$. Besides, $I \cap \bar{S} \subseteq J^m \cap \bar{S} = (J \cap \bar{S}^{m^{-1}})^m \subseteq (J \cap Q^{A(S)})^m = (J \cap R \cap Q^{A(S)})^m = (J \cap \bar{S})^m$.

Since $A(S)$ is a reduced-finite group, then $J \cap \bar{S}$ and $I \cap \bar{S}$ are nonzero ideals of $\bar{S}$ and, hence, a restriction of m on $\bar{S}$ belongs to $A(S)$.

The kernel of the mapping of the restriction $M \to A(S)$ is equal to $A(S)$ and, hence, the induced group is isomorphic to $M / A(S)$. The fixed ring of this group coincides with $R^M \cap S$. Now we have to remark that $R^M = R^G$, as when adding the inner automorphisms corresponding to the elements of the algebra of the group M, the fixed ring undergoes no changes. The theorem is proved.

REFERENCES

M.Artin [9];
L.N.Childs, F.R.De Meyer [27];
C.W.Curtis, I.Rainer [36];
J.Dieudonne [39];
J.M.Grousaud, J.L.Pascaud, J.Valette [50];
G.Hochschild [60,61];
N.Jacobson [62,64,65];

T.Kanzaky [66];
G.Kartan [67];
V.K.Kharchenko [68,69,76,79,80,81];
H.Kreimer [89];
Y.Miyashita [108];
S.Montgomery, D.S.Passman [118];
R.Moors [122];
T.Nakayama [124];
T.Nakayama, G.Azumaya [125];
E.Noether [126];
J.Osterburg [127,228,129];
A.Rosenberg, D.Zelinsky [141];
T.Sundström [147];
A.I.Shirshov [144];
H.Tominaga, T.Nakayama [150];
O.E.Villmayor, D.Zelinsky [152];

CHAPTER 4

THE GALOIS THEORY OF PRIME RINGS (THE CASE OF DERIVATIONS)

In this chapter we shall explore the problem of correlation between a given prime ring R and a subring of constants of a finite-dimensional Lie algebra L of its derivations. If the characteristic of a ring is equal to zero, then constants of finite-dimensional Lie algebras are related, but weakly, with the basic ring. Let, for instance, $R = F\langle x, y\rangle$ be a free algebra with no unit. Let us consider a derivations μ, such that $y^{\mu} = y$, $x^{\mu} = x$. Then its derivation has the only constant, i.e., zero, in which case the Lie algebra generated by μ is one-dimensional. A reasonable restriction which can be imposed in the case of a zero characteristic is that the associative algebra generated by the considered derivations in a ring of endomorphisms of an additive group of the ring R is finite-dimensional. In this case, however, theorem **2.7.8** yields that L consists only of the derivations which are inner for the ring of quotients $Q(R)$, and the problems are reduced to those of centralizers of finite-dimensional subalgebras, which have already been considered in **3.3**.

Therefore, of primary interest is the case of a positive characteristic p, in which case the p-th power of any derivation is again a derivation, and all the constants of the initial derivation are also those of its p-th power. Hence, a restricted Lie algebra has to be viewed as L.

It should be recalled that extend the situation a little under the assumption that derivations from L transfer R into $Q(R)$ but not necessarily to R, in which case, however, we set that for any $\mu \in L$ there is a nonzero ideal I of R, such that $I^{\mu} \subseteq R$. A set of all these derivations are designated by $D(R)$ (see **1.8**). This set forms a differential Lie C-algebra (see lemma **1.8.3**), i.e., it is also a space over a generalized centroid C. Since when multiplied by a nonzero central element its constants undergo no changes, then it would be natural to view L as a Lie ∂-subalgebra in $D(R)$ (see **1.2**).

Therefore, let L be a finite-dimensional restricted Lie ∂-algebra over C. If Λ is a subalgebra of derivations from L which are inner of Q, then Λ is an ideal in L and a subspace over C. Therefore, the action of a factor algebra L/Λ is induced on the ring of constants $R^{\Lambda} = \{x \in R \mid \forall\, \mu \in \Lambda (x^{\mu} = 0)\}$, in which case $R^{L} = (R^{\Lambda})^{L/\Lambda}$. This peculiarity, combined with the fact that R^{Λ} is a centralizer of a finite-dimensional C-subalgebra from Q, makes it possible to restrict ourselves to considering Lie ∂-algebras of outer derivations.

Henceforth in this chapter we shall assume that L is a *finite-dimensional restricted lie ∂-algebra of outer for Q derivations from $D(R)$ of a given prime ring R of a characteristic $p>0$.*

4.1 Duality for Derivations in Multiplication Algebra

Let S be a subring in a prime ring R. Let us denote by $L^{op}(S)$ a subring in the tensor product $Q\otimes Q^{op}$ (where, as usual, $Q^{op}$is a ring antiisomorphic to Q with the former additive group), generated by elements of type $s\otimes 1$, $1\otimes r$, where $s\in S$, $r\in R$. (It should be recalled that the ring $L(S)$ is generated by elements of the type $1\otimes s$, $r\otimes 1$ (see **3.9.8**)).

If $a\in R_F$, $\beta=\sum s_i\otimes v_i\in L^{op}(S)$, then, as before, we set $a\cdot\beta=\sum v_i a s_i$. If Δ is a mapping from S to R, then we set $\beta^{\Delta}=\sum s^{\Delta}\otimes v_i$. Let us denote by $a^{\perp S}$ a set of all $\beta\in L^{op}(S)$, such that $a\cdot\beta=0$, i.e., $a^{\perp S}=a^{\perp}\cap L^{op}(S)$. If $V\subseteq L^{op}(R)\overset{df}{=}=L(R)$, then we should recall that $V^{\perp}=\{a\in R_F\mid a\cdot V=0\}$.

4.1.1. Lemma. *Let S be a subring of R, such that every nonzero (R,S)-subbimodule of R contains a nonzero ideal of the ring R. Then the following formula is valid:*

$$\sum_{i=i}^{m} a_i Z+B^{\perp}=\Big(\bigcap_{i=1}^{m} a_i^{\perp S}\cap B\Big)^{\perp},$$

where a_i are arbitrary elements from R_F, B is an arbitrary right ideal from $L^{op}(S)$, Z is a centralizer of S in the ring of quotients R_F.

Proof. The inclusion of the left part in the right one is obvious. Let us carry out the proof of the inverse inclusion by induction over m. If $m=1$ and v is an element of the right part, then for any $\beta\in B$ the equality $a_1\cdot\beta=0$ implies $v\cdot\beta=0$. Hence, the mapping $\varphi\colon a_1\cdot\beta\to v\cdot\beta$, where β runs through B is correctly defined. We can choose a nonzero ideal I of R, such that $Iv \subseteq R$, $Ia_1 \subseteq R$. If now β runs through $\mathbb{B}(1 \otimes I)$, then the domain of φ values is contained in R, and the domain of the definition forms an (R, S)-subbimodule in R. If this subbimodule is zero, i.e., $a_1 \cdot \mathbb{B}(1 \otimes I) = 0$, then $a_1 \cdot \mathbb{B} = 0$ and, hence, $v \cdot \mathbb{B} = 0$, i.e., $v \in \mathbb{B}^{\perp}$. Therefore, the domain of φ definition can be assumed to contain a nonzero ideal of the ring R. It is also evident that φ is a homomorphism of the (R, S)-bimodules and, hence, there is an element $\xi \in R_F$, such that $(a_1 \cdot \beta)\xi = v \cdot \beta$. Substituting into this equality β with $\beta(s \otimes 1)$, we get $(a_1 \cdot \beta)\xi s = (a_1 \cdot \beta)s\xi$, i.e., $\xi \in Z$. Now the relation $(a_1 \xi - v) \cdot \beta = 0$ indicates that $a_1\xi - v \in (\mathbb{B}(1 \otimes I))^{\perp} = \mathbb{B}^{\perp}$. Therefore, $v \in \mathbb{B}^{\perp} + a_1 \xi$.

Let the lemma be proved for $m = k$ and let us set $\mathbb{B}_1 = \bigcap_{i=1}^{k} a_i^{\perp S} \cap \mathbb{B}$. Then, according to the case when $m = 1$, we have

$$a_{k+1}Z + \mathbb{B}_1^{\perp} = \Big(\bigcap_{i=1}^{k+1} a_i^{\perp S} \cap \mathbb{B}\Big)^{\perp}.$$

This is the required proof as, by the inductive supposition, we have

$$\mathbb{B}_1^{\perp} = \sum_{i=1}^{k} a_i Z + \mathbb{B}^{\perp}.$$

Lemma **4.1.1.** is proved.

4.1.2. Lemma. *Let S be a subring of the ring R, such that every nonzero (R, S)-subbimodule of R contains a nonzero ideal of the ring R. Let, then, Z be a centralizer of S in R_F. In this case, if $a_1, \ldots a_m \in R_F$ and $a_1 \notin a_2 Z + a_3 Z + \ldots + a_m Z$, then there is a $\beta \in \mathbb{L}^{op}(S)$, such that $a_1\beta \neq 0$, $a_2\beta = a_3\beta = \ldots = a_m\beta = 0$.*

Proof. If such a β does not exist, then $a_1 \in (a_2^{\perp S} \cap a_3^{\perp S} \ldots \cap a_m^{\perp S})^{\perp}$. By lemma **4.1.1**, the right-hand part of this inclusion is equal to $a_2 Z + \ldots + a_m Z$. The lemma is proved.

4.1.3. Lemma. *Let the rings R and S obey the conditions of* *lemma* **4.1.2,** $a_1, \dots, a_m \in R_F$ *and let us assume that* Z *is a sfield. Let* $\mu_1, \dots, \mu_v$ *be derivations determined on* S *and assuming their values in* R, *and let* $r_1 \neq 0$, $r_2, \dots, r_{v+1}$ *be elements from* R_F. *In this case one of the following statements is valid:*

(a) *there is an element* $\beta \in \bigcap_{i=1}^{m} a_i^{\perp S}$, *such that*

$$\sum_{i=1}^{v} r_i \beta^{\mu_i} + r_{v+1} \cdot \beta \neq 0 \tag{1}$$

(b) *there are elements* $z_1, \dots, z_m \in Z$ *and an element* $t \in R_F$, *such that for all* $s \in S$ *the following equality is valid:*

$$s^{\mu_1} + \sum_{i=2}^{v} z_i s^{\mu_i} = st - ts.$$

Proof. Let us consider a right Z-module $r_1 Z + r_2 Z + \dots + a_1 Z + \dots + a_m Z_v$. Let $d_1 = r_1, d_2, \dots, d_n$ be its basis. In this case we obviously have $\bigcap_{i=1}^{m} d_i^{\perp S} \subseteq \bigcap_{i=1}^{m} a_i^{\perp S}$. Let I be a nonzero ideal of R, such that $Ir_i, Id_j \subseteq R$ for all $i, j, 1 \leq i \leq v+1, 1 \leq j \leq n$. Let us consider a set $B = \bigcap_{j=2}^{n} d_j^{\perp S}$. By lemma **4.1.2**, the element r_1 does not belong to $B^{\perp}$ and, hence, $I(r_1 \cdot B)$ is a nonzero (R, S)-subbimodule in R, i.e., it contains a nonzero two-sided ideal of the ring R.

Let us assume that for any $\beta \in r_1^{\perp S} \cap B_1 \subseteq \bigcap_{j=1}^{n} d_i^{\perp S}$, where $B_1 = B(1 \otimes I) \subseteq B$, the left-hand part of the relation is zero. In this case the mapping

$$\xi : r_1 \beta \longrightarrow \sum_{i=1}^{v} r_i \cdot \beta^{\mu_i} + r_{v+1} \beta ,$$

where β runs through B_1, is well-defined. This mapping is a homomorphism of the left R-modules,

$$\xi[r(r_1\beta)] = \xi[r_1 \cdot \beta(1 \otimes r)] = \sum_{i=1}^{v} r_i \cdot \beta^{\mu_i}(1 \otimes r) +$$

$$+ r_{v+1}\beta(1 \otimes r) = \sum_{i=1}^{v} r(r_1 \cdot \beta^{\mu_i}) + r(r_{v+1} \cdot \beta) =$$

$$= r[\xi(r_1\beta)].$$

Since the domain of the ξ definition contains a nonzero ideal of the ring R, while the domain of its values lies in R, then there is an element $t \in R_F$, such that

$$(r_1\beta)t = \sum_{i=1}^{v} r_1\beta^{\mu_1} + r_{v+1}\beta. \tag{2}$$

Let $s \in S$, then

$$\begin{aligned}(r_1\beta)st &= [r_1 \cdot \beta(s \otimes 1)]t = \\ &= \sum_{i=1}^{v} r_i[\beta^{\mu_i}(s \otimes 1) + \beta(s^{\mu_i} \otimes 1)] + \\ &+ r_{v+1} \cdot \beta(s \otimes 1) = \\ &= \sum_{i=1}^{v} [(r_i\beta^{\mu_i})s + (r_i\beta)s^{\mu_i}] + (r_{v+1}\beta)s.\end{aligned} \tag{3}$$

If we multiply relation (2) from the right by s and subtract it from (3), we get

$$(r_1\beta)[s, t] = \sum_{i=1}^{v} (r_i \cdot \beta)s^{\mu_i}. \tag{4}$$

If $\beta \in \mathcal{B}_1$ and $r_i = r_1 z_i + \dots$, $z_i \in Z$ is a decomposition of r_i on the basis $\{d_j\}$, then $r_i \cdot \beta = (r_1\beta)z_i$ and equality (4) assumes the form

$$(r_1\beta)([s, t] - \sum z_i s^{\mu_i}) = 0.$$

By lemma **4.1.2**, the element r_1 does not belong to $\mathcal{B}_1^{\perp}$ and, hence, $r_1 B_1$ is a nonzero (R, S)-subbimodule in R, i.e., by the condition, it contains a nonzero two-sided ideal of the ring R. Therefore, the latter equality implies that statement (b) is fulfilled. The lemma is proved.

4.2 Transformation of Differential Forms

4.2.1. Definitions. A *differential form* will be an expression of the kind $\sum_i a_i x^{\Delta_i} b_i$, where $a_i, b_i \in R_F$ and Δ_i are words of derivations from $D(R)$. It should be recalled (corollary **2.3.2**) that if a form turns to zero on a nonzero ideal of the ring R, then it will also be zero on R, and even on R_F. In this respect the following agreement looks natural.

4.2.2. Agreement. If T is a subset of R_F, then for the form $f(x)$ we shall write $f(x) \subseteq T$ provided there is a nonzero ideal I of the ring R, such that $f(i) \in T$ for all $i \in I$. Analogously, if $f(x_1,..., x_n)$ is a D-polynomial (or a polylinear differential form), then the presentation $f(x_1,..., x_n) \subseteq T$ implies that there is a nonzero ideal I of R, such that $f(i_1,..., i_n) \in I$ for all $i_1,..., i_n \in I$.

4.2.3. Denotations. If $f(x)$ is a differential form, and the element $\beta = \sum s_i \otimes r_i$ belongs to $\mathbb{L}^{op}(S)$, then, as before, let us set $f(x) \cdot \beta = \sum r_i f(s_i x)$, and

$$f(x) * \beta = \sum f(xr_i) s_i .$$

4.2.4. Lemma. *If* $f(x) \subseteq T$, *then* $f(x) \cdot \beta \subseteq RT$, $f(x) * \beta \subseteq TS$ *for all* $\beta \in \mathbb{L}^{op}(S)$.
The proof is obvious.

4.2.5. Let $\mu_1,..., \mu_n$ be a set of derivations from $D(R)$. Let us fix an order on this set: $\mu_1 < \mu_2 < ... < \mu_n$ and extend it in a standard way onto a set of all the words from $\mu_1,..., \mu_n$, i.e., we assume (as before) that a longer word is bigger than a shorter one, while the words of the same length are compared from the right, as in a dictionary. It should be recalled that the word $\Delta = \delta_1 ... \delta_m$ is called *correct* if it contains no subwords of the type $\mu_i \mu_j, \overbrace{\mu_i ... \mu_i}^{p}$, where $\mu_i > \mu_j$ and p is a characteristic of the given ring. In other words, in a correct word we always have $\delta_1 \le \delta_2 \le ... \le \delta_n$, and this chain should not contain p signs of equality in a row.

For two correct words, Δ_1, Δ_2, let us denote by $\Delta_1 \circ \Delta_2$ a correct word obtained from a combination of all the derivations included in Δ_1 and Δ_2, i.e., if $\Delta_1 = \mu_1^{m_1} ... \mu_n^{m_n}$, $\Delta_2 = \mu_1^{t_1} ... \mu_n^{t_n}$ then $\Delta_1 \circ \Delta_2 = \mu_1^{m_1 + t_1} ... \mu_n^{m_n + t_n}$, in which case, if one of the sums $m_i + t_i$ proves to be greater then or equal to p, then we assume that $\Delta_1 \circ \Delta_2 = O$ is a zero word. A zero word should be told from an empty one. It is an empty word that the identical mapping $x^{\varnothing} = x$ corresponds to, which agrees with bringing a derivation to a zero power, $\mu^0 = \varnothing$. A zero word is corresponded to by a zero mapping O: $x \to 0$. An empty word is, hence, a subword of any other word, while a zero word cannot be a subword of any other word. As it is common practice in the theory of Lie algebras, a correct word $\Delta_1 \ne O$ will be called a *subword* of the correct word Δ, if $\Delta = \Delta_1 \circ \Delta_2$ for a correct word Δ_2. Herefrom, when considering words we shall assume, unless otherwise stated, that they are not zero. However, sometimes it would be more convenient to carry out induction over the length, assuming a zero word to be as long as -1.

4.2.6. Let $\Delta = \mu_k^{m_k} ... \mu_n^{m_n}$ be a correct word, $m_k > 0$. In this case well known is the fact (easily proved by induction over the length of the word Δ) that for any x, y the following equality is valid:

$$(xy)^{\Delta} = \sum_{\Delta' \circ \Delta'' = \Delta} C_{m_k}^{m'_k} C_{m_{k+1}}^{m'_{k+1}} \dots C_{m_n}^{m'_n} x^{\Delta'} y^{\Delta''} =$$

$$= xy^{\Delta} + m_k x^{\mu_k} y^{\tilde{\Delta}} + \dots \tag{5}$$

where $\Delta' = \mu_k^{m'_k} ... \mu_n^{m'_n}$, the length of the word $\tilde{\Delta} = \mu_k^{m_k - 1} \mu_{k+1}^{m_{k+1}} ... \mu_n^{m_n}$ is less than that Δ, and it is the least among the words different from Δ, with which the element y enters the right-hand part.

It affords that if $\beta \in L^{op}(R)$, then

$$(ax^{\Delta}b)\cdot\beta=(a\cdot\beta)x^{\Delta}b+m_k(a\cdot\beta^{\mu_k})x^{\tilde{\Delta}}b+\ldots,\tag{6}$$

where dots denote a sum of the terms of type $dx^{\bar{\Delta}}b$, where $\bar{\Delta}<\tilde{\Delta}$, in which case $\bar{\Delta}$ is a subword of the word Δ.

Now the following formula is valid:

$$(ax^{\Delta}b)=(axb)^{\Delta}+\sum_i(a_i xb_i)^{\Delta_i},\tag{7}$$

where Δ_i are correct subwords of the word Δ, with their length less than that of Δ. Indeed, if the length of the word Δ is one, then $ax^{\mu}b=(axb)^{\mu}-a^{\mu}xb-axb^{\mu}$. In a general case we get by induction

$$ax^{\Delta}b=ax^{\mu_k\tilde{\Delta}}b=(ax^{\mu_k}b)^{\tilde{\Delta}}+\ldots=[(axb)^{\mu_k}-a^{\mu_k}xb-$$
$$-axb^{\mu_k}]^{\tilde{\Delta}}+\ldots=(axb)^{\mu_1\tilde{\Delta}}+\ldots=(axb)^{\Delta}+\ldots\quad.$$

Formula (7) can be modified for two cofactors:

$$x^{\Delta}b=(xb)^{\Delta}-m_k(xb^{\mu_k})^{\tilde{\Delta}}+\ldots\quad,\tag{8}$$

where dots denote a sum of terms $(xd)^{\bar{\Delta}}$, where $\bar{\Delta}<\tilde{\Delta}$. Indeed, if the length of the word Δ is one, then $x^{\mu}b=(xb)^{\mu}-xb^{\mu}$. If $m_k>1$, then we get by induction

$$x^{\Delta}b=x^{\mu_k\tilde{\Delta}}b=(x^{\mu_k}b)^{\tilde{\Delta}}-(m_k-1)(x^{\mu_k}b^{\mu_k})^{\tilde{\tilde{\Delta}}}+\ldots=$$
$$=(xb)^{\mu_k\tilde{\Delta}}-(xb^{\mu_k})^{\tilde{\Delta}}-(m_k-1)(xb^{\mu_k})^{\mu_k\tilde{\tilde{\Delta}}}-$$
$$-(xb^{\mu_1^2})^{\tilde{\tilde{\Delta}}}+\ldots=(xb)^{\Delta}-m_k(xb^{\mu_k})^{\tilde{\Delta}}+\ldots.$$

If $m_k=1$, then we get in the same way:

$$x^{\Delta}b=x^{\mu_k\tilde{\Delta}}b=(x^{\mu_k}b)^{\tilde{\Delta}}+\ldots=(xb)^{\mu_k\tilde{\Delta}}-(xb^{\mu_k})^{\tilde{\Delta}}+\ldots.$$

Here we have made allowances for the fact that

$$\mu_k \mu_{k+1}^{m_{k+1}-1} \mu_{k+2}^{m_{k+2}} \cdots \mu_n^{m_n} < \tilde{\Delta} = \mu_{k+1}^{m_{k+1}} \cdots \mu_n^{m_n}.$$

By analogy with formula (6) we now see that

$$(axb)^{\Delta} * \beta = [ax(b \cdot \beta)]^{\Delta} - [ax(m_k b \cdot \beta^{\mu_k})]^{\tilde{\Delta}} + \ldots, \tag{9}$$

where dots denote a sum of the terms $(xad)^{\bar{\Delta}}$, where $\bar{\Delta} < \tilde{\Delta}$, and $\bar{\Delta}$ are subwords of the word Δ.

Let us now assume that $\{\mu_1, \ldots, \mu_n\}$ is a basis of a restricted Lie ∂-algebra over C. In this case the following equalities are valid: $\mu_i \mu_j - \mu_j \mu_i = \sum c_{ij}^{(k)} \mu_k$, $\mu_i^p = \sum c_i^{(k)} \mu_k$, i.e., using identities (1), (4) from **1.2**, the product of the words Δ_1, Δ_2 can be presented as a C-linear combination of some correct words, in which case we have

$$\Delta_1 \Delta_2 = \Delta_1 \circ \Delta_2 + \ldots, \tag{10}$$

where dots denote a linear combination of the words, the lengths of which are strictly less than the sum of the Δ_1 and Δ_2 lengths. In particular, if $\Delta_1 \circ \Delta_2 = 0$, then the word $\Delta_1 \Delta_2$ is presented as a linear combination of correct words of a strictly less length.

4.3 Universal Constants

Let L be a finite-dimensional restricted Lie ∂-algebra of outer derivations. Let us fix a basis $\mu_1, \ldots, \mu_n$ of the right vector space L over a generalized centroid C of a prime ring R. Let us assume that this basis is ordered: $\mu_1 < \mu_2 < \ldots < \mu_n$. Let $\Delta_1, \ldots, \Delta_m$ be a set of all correct words, Δ_1 be an empty word. It is evident that $m = p^n$.

Let us consider a set $D(L)$ of all D-polynomials with coefficients from Q which include derivations from L as operations.. Let us identify the elements from $D(L)$ provided their difference is an identity on R. It is obvious that $D(L)$ forms a ring, in which case L acts on it as derivations.

4.3.1. Definition. *A universal constant (for* L*)* is an arbitrary nonzero constant from $D(L)$.

4.3.2. Lemma. If $f(x_1,\ldots,x_n)$ is a universal constant, then

$$f(x_1,\ldots,x_n) \subseteq R^L. \tag{11}$$

Proof. With no restrictions on generality one can consider f to be reduced (see **2.1**), i.e., to be presented as a polynomial from $\{x_i^{\Delta_j} \mid 1 \le i \le k, 1 \le j \le m\}$. Let us choose a nonzero ideal I_1 of the ring R in such a way that $I_1^{\mu_i} \subseteq R, 1 \le i \le n$ and for all coefficients q occurring in the reduced presentation of $f(X)$, the inclusions $I_1 q \subseteq R$, $qI_1 \subseteq R$ are fulfilled. In this case for any integer $l > 1$ and any basic μ valid is the inclusion $(I_1^l)^{\mu} \subseteq I_1^{l-1}$. Therefore, $(I_1^{(p-1)n+1})^{\Delta_j} \subseteq I_1$ and, hence, setting $I = I_1^{(p-1)n+2}$, we see that the f values on I belong to R.

Now we are to remark that , according to the definition of a universal constant, $f^{\mu} = 0$ and, hence, $f(I) \subseteq Q^L \cap R = R^L$.

The lemma is proved.

4.3.3. Lemma-definition. *For any finite-dimensional restricted Lie ∂-algebra L there is a linear universal constant, $f(x) = \sum c_j x^{\Delta_j}$ with coefficients from C, which will be called a trace form.*

Proof. It would be sufficient to find an element $f \ne 0$ from the universal enveloping U for L (see **1.2.3**), such that $f \cdot \mu_i = 0$, $1 \le i \le n$, where $\{\mu_i\}$ is the basis of L over C. Indeed, presenting here f as a linear combination of correct words $\sum \Delta_j c_j$, we get $\sum x^{\Delta_j} c_j$ is the universal constant sought.

It should be remarked that U is a Frobenius algebra over the field of constants C^L. Indeed, let us consider an arbitrary C^L-linear epimorphism $\psi: C \to C^L$, and determine a linear mapping $\varphi: U \to C$ putting in correspondence to the linear combination $\sum \Delta_j c_j$ the coefficients at the word $\mu_1^{p-1} \ldots \mu_n^{p-1}$. In this case the linear function $\lambda = \varphi\psi$ transforms U in C^L and the kernel of this function contains no nonzero right and left ideals as for any $u \ne 0$ one can choose a $u_1 \in U$ in such a way that $c = \varphi(uu_1) = \varphi(u_1 u) \ne 0$. In this case we have $\lambda(uu_1c^{-1}) = \lambda(c^{-1}u_1u) = 1$, because in U the rule of commutation $c\Delta = \Delta c + \ldots$ acts, where dots denote a linear combination of words of a less length. Thus, U is a Frobenius algebra.

Therefore, valid is the equality $ann_r(ann_l U') = U'$, where U' is the right ideal in U generated by elements $\mu_1,..., \mu_n$. This ideal consists of linear combinations $\sum \Delta_j c_j$ of nonempty correct words. Hence, $1 \notin U'$ and, thus, $U \neq U'$, i.e., $ann_l U' \neq 0$, which is the required proof.

4.3.4. Let us now consider the case when R has a generalized identity (or, for short, R is a GI-ring).

A restriction η of derivations onto C is a homomorphism of the Lie ∂-algebra $Der\, Q$ in $Der\, C$. If R is a GI-ring, then the kernel of η consists of only inner derivations (lemma **2.4.2**) and it, therefore, does not intersect with L, i.e., $L \cong \eta(L) \subseteq Der\, C$.

4.3.5. Theorem. *If R is a prime GI-ring, then*

$$RC = (RC)^L \otimes_F C, \tag{12}$$

where $F = C^L$.

Proof. Let $f(x) = \sum x^{\Delta_j} c_j$ be a trace form for L. Let us consider a subring $R_1 = (RC)^L C$ and show that $R_1 \cong (RC)^L \otimes C$. For this purpose it suffices to show that if $q_1,..., q_n$ are elements from $(RC)^L$ which are linearly independent over F, then they are also linearly independent over C. Let, on the contrary, $\sum q_i \lambda_i = 0$. Then $\sum q_i \lambda_i c = 0$ for an arbitrary $c \in C$. Applying $\Delta_j c_j$ to both parts of the equality and carrying out summation over j, we find $\sum q_i f(\lambda_i c) = 0$. Since f is a trace form and q_i are linearly independent over F, then $f(\lambda_i c) = 0$ for all i. Allowing for the fact that c is an arbitrary element, we get an identity $f(x) = 0$ in the ring C, i.e., by theorem **2.2.2**, all coefficients c_j are equal to zero, which is a contradiction.

In order to show that $R_1 = R$, let us fix a $q \in RC$ and consider a relation valid for all $c \in C$:

$$f(cq) = \sum_j (cq)^{\Delta_j} c_j \in R_1 . \tag{13}$$

Through induction over the senior word let us show that a relation of the type (13), where $\varnothing = \Delta_1 < \dots < \Delta_k$ are correct words, $c_k \neq 0$, yields the following equality:

$$q = \sum_i c'_i f(c''_i q) \in R_1 , \tag{14}$$

where the central elements c'_i, c''_i are independent of the choice of q.

If the senior word in (13) is empty, then at $c = 1$ we get $q = c_1^{-1} f(q) \in R_1$.

Let now Δ_k be the senior word in (13). Let us substitute the product $c'\, c''$ instead of c and write (13) using formula (5), setting $x = c'$, $y = c''\, q$. In this case the senior term for $c''\, q$ will have the form $c_k c'(c''\, q)^{\Delta_k}$. Therefore, if we substitute $c = c''$ into relation (13), multiply it by c' and subtract $f(c'\, c''\, q)$, then we get a relation of the previous type, where c'' plays the part of c, while the senior word is less than Δ_k.

Let us choose an element c' in such a way that not all the terms in the obtained difference were cancelled. For this purpose let us calculate the coefficient at $(c''\, q)^{\tilde{\Delta}_k}$ (here $\Delta_k = \mu_s \tilde{\Delta}_k$, see formula (5)). The word Δ_k in decomposition (5) can be formed only for words Δ, such that $\Delta'' \circ \tilde{\Delta}_k = \Delta$ at a certain Δ''. As Δ occurs in presentation (13), then $\Delta \leq \Delta_k$, i.e., for Δ'' we have only two possibilities: $\Delta'' = \varnothing$; $\Delta'' = \mu_r < \mu_s$. Thus, let $s(r)$ be the number of the word $\mu_r \tilde{\Delta}_k$ in presentation (13), $s(0)$ be the number of the word $\tilde{\Delta}_k$. Now the coefficient at $(c''\, q)^{\tilde{\Delta}_k}$ is seen, by formula (5), to be equal to

$$c_k \cdot m_s(c')^{\mu_s} - \sum_{r \geq 0} c_{s(r)}(c')^{\mu_r}.$$

By theorem **2.2.2**, the element c' can be chosen in such a way that this coefficient be nonzero (in which case the choice of c' is q-independent).

Therefore, applying the inductive supposition to $c'\, f(c''\, q) - f(c'\, c''\, q)$, we get equality (14), which implies $RC = (RC)^L \otimes_F C$. The theorem is proved.

4.3.6. Remark. It should be noted that in the proof of the preceding theorem use has been made only of the fact that the homomorphism $\eta\colon L\to \mathrm{Der}\, C$ is an embedding, so that in the formulation of theorem **4.3.5** the condition "R is a GI-ring" can be replaced with the condition "$L\cong \eta(L)$".

4.3.7. Definition. A subring S of the ring R will be called *almost intermediate* if $f(x)\subseteq S$ for a certain trace form $f(x)$. The subring S is called *intermediate* if $R\supseteq S\supseteq R^L$.

4.3.8. Theorem. *If S is an almost intermediate subring, a, b are nonzero elements from Q, then $aSb\neq 0$. In particular, R^L is a prime ring.*

Proof. Let, on the contrary, $aSb=0$. Then for the trace form we have the relation $af(x)b=a\sum x^{\Delta_j}c_jb=0$, i.e., by theorem **2.2.2**, we have $a\otimes c_jb=0$, which yields $c_j=0$ at all j, which is a contradiction.

4.3.9. Theorem. *If a polynomial identity is fulfilled on R^L, then it is also fulfilled in R.*

Proof. Let $g(z_i)=0$ be an identity in R^L. Then, by lemma **4.3.2**, we can find a nonzero ideal on which a D-identity $g(\sum_j c_jx_i^{\Delta_j})=0$ holds. By theorem **2.4.1**, on the ring R_F the identity $g(\sum c_jz_{ij})=0$ can be found. If now $c_{j_0}\neq 0$, then we set $z_{ij}=c_j^{-1}z_i$ at $j=j_0$, and $z_{ij}=0$ at $j\neq j_0$. We thus get an identity $g(z_i)=0$ on R_F. The theorem is proved.

4.4 Shirshov Finiteness

If $R=C$ is a field, then one can show (see below corollary **4.4.4**) that it has a finite dimension over C^L. In a general case R is not a finitely generated right (or left) R^L-module.

4.4.1. Example. Let $C\langle x, y\rangle$ be a free associative algebra over a field of characteristic 2, and let μ be its derivation, such that $\mu(x)=0$, $\mu(y)=y$. Then $\mu^2=\mu$ and, hence, $L=\mu C$ is a one-dimensional restricted Lie ∂-algebra. The algebra R^L is generated by monomials which have an even degree over y. Any monomial not belonging to R^L can be presented as $x^k y \cdot w$, where $w \in R^L$. If now $\{x^{k_i} y w_i\}$ is a finite set of monomials, then $x^{\max(k_i)+1} y \notin \sum x^{k_i} y R^L + R^L$ and, hence, R cannot be generated as a module over R^L by a finite number of elements.

4.4.2. Theorem. *In a prime ring R there is a nonzero ideal locally-finite over R^L* (see definition **3.1.9**).

This theorem results from the following proposition.

4.4.3. Proposition. *For any nonzero element u of the ring R there are elements $a \neq 0$, $r_1, \dots, r_t \in R$, such that at all $x \in R$*

$$ax = \sum_{i=1}^{t} r_i u \tau_i(x), \tag{15}$$

where τ_i are mappings from R to R^L, which have the form $\tau_i = f(s_i x)$ for suitable $s_i \in R$ and a trace form f.

Let us show the way theorem **4.4.2** is deduced from this proposition. Let us arbitrarily choose an element u. Let us then consider a set I of elements a for which there exist elements $r_1, \dots, r_{m(a)} \in R$, such that at all $x \in R$ equality (15) holds. One can easily see that IR is a Shirshov locally-finite ideal over R^L. The fact that $I \neq 0$ follows directly from proposition **4.4.3**.

Proof of proposition 4.4.3. Let $f(x)$ be a trace form for L. Let us consider a relation

$$V_1 x + V_2 x^{\Delta_2} + \dots V_k x^{\Delta_k} = \sum r_i u f(t_i x), \tag{16}$$

where, as usual, $\varnothing = \Delta_1 < \Delta_2 < \dots < \Delta_k < \dots$ are correct words, $V_1, \dots V_k$, r_i, t_i are elements of the ring Q, in which case $V_k \neq 0$. It is obvious that in the ring R the relation of the type $\sum (c_j u) x^{\Delta_j} = uf(x)$ is fulfilled, where $f(x) = \sum c_j x^{\Delta_j}$.

Through induction over the senior word Δ_k let us show that if a relation of type (16) holds, then the proposition is valid. If $\Delta_k = \varnothing$, then relation (16) assumes the form

$$V_1 x = \sum_i \bar{r}_i u f(t_i x). \tag{17}$$

Let us choose a nonzero ideal I of the ring R in such a way that $f(I) \subseteq R$, $IV_1 \subseteq R$, $I\bar{r}_i \subseteq R$, $It_i \subseteq I$ for all i. Then, since R is prime, the product $IV_1 I^2$ is nonzero. Let us choose elements $a_0, a_1, a_2 \in I$, in such a way that $a = a_0 V_1 a_1 a_2 \neq 0$. Substituting into (17) the expression $a_1 a_2 x$ instead of x and multiplying by a_0 from the left, we see that relation (15) is valid for the elements a, $r_i = a_0 \bar{r}_i$ and mappings $\tau_i(x) = f((t_i a_1) a_2 x)$.

Let us carry out an induction step. If we apply an arbitrary operator $\cdot \beta$ to both parts of equality (16) and make use of equality (6), then, if $\beta \in V_k^{\perp}$, we again get a relation of type (16) with a smaller senior word. Now we are to show that one of the coefficients $V'_1, V'_2, \dots$ of the newly obtained relation does not turn into zero.

Let us calculate the coefficient at $x^{\tilde{\Delta}_k}$, where $\Delta_k = \mu_s \tilde{\Delta}_k$ (see formula (5)). The word $\tilde{\Delta}_k$ in decomposition (6) can be formed only for words Δ, such that $\Delta'' \circ \tilde{\Delta}_k = \Delta$ at a certain Δ''. Since Δ is encountered in presentation (16), then $\Delta \leq \Delta_k$, i.e., for Δ'' only the following relations are possible: $\Delta'' = \emptyset$; $\Delta'' = \mu_r < \mu_s$. Let $\delta(r)$ be the number of the word $\mu_r \tilde{\Delta}_k$, $\delta(0)$ be the number of the word $\tilde{\Delta}_k$. By formula (6) we can now find a new coefficient at $x^{\tilde{\Delta}_k}$:

$$V'_{\sigma(0)} = m_s(V_k \cdot \beta^{\mu_s}) + \sum_{r=0}^{k-1} V_{\sigma(r)} \cdot \beta^{\mu_r}. \tag{18}$$

Let us apply lemma **4.1.3** to the ring R and a sequence $\mu_s, \mu_{s-1}, \dots \mu_1$, assuming that $S = R$, $r_1 = m_s V_k$, $n = 1$, $a_1 = V_k$. Condition (b) cannot be fulfilled, as $C = Z$ and the linear combination of derivations μ_i over C lies in L. Therefore, one can choose an element $\beta \in V_k^{\perp}$ in such a way that $V'_{\sigma(0)} \neq 0$, which completes the proof of proposition **4.4.3**.

4.4.4. Corollary (**M. Weisfeld**). *Let* T *be a sfield of a* *characteristic* $p>0$, *and let* L *be a finite-dimensional restricted Lie* ∂-*algebra of derivations of the sfield* T *(possibly, the one containing inner derivations). Then the right and left dimensions of* T *over the sfield of constants,* T^L, *are finite.*

Since the conditions are symmetric, it suffices to show that the right dimension of T over T^L is finite. Let us carry out induction over the dimension of L. If $L=0$, then there is nothing to prove.

If Λ is an ideal of inner derivations, then $T^L=(T^{\Lambda})^{L/\Lambda}$, in which case the center Z of the sfield T^{Λ} can strictly contain C, but, nonetheless, $\dim_Z((L/\Lambda)\cdot Z)\leq \dim_C(L/\Lambda)$ and, hence, using inductive supposition, it is sufficient to consider the cases when $L=\Lambda$ or $\Lambda=0$.

In the former case the associative algebra $B(L)$ generated over C by the elements corresponding to derivations from L is finite-dimensional and is, hence, a sfield. Therefore, $|T:T^L|=|T:C(B)|\leq \dim_C B$, where $C(B)$ is a centralizer of $B(L)$ in the sfield T (and, certainly, $C(B)=T^L$).

If $\Lambda=0$, then, by theorem **4.4.2**, we immediately have $|T:T^L|<\infty$. The corollary is proved.

4.4.5. Corollary. *Let* S *be an almost intermediate subring. Then any nonzero* (R, S)-*subbimodule and any nonzero* (S, R)-*subbimodule of* R *contains a nonzero two-sided ideal of the ring* R.

Proof. Let U be an (R, S)-subbimodule and let u be its nonzero element. By proposition **4.4.3**, for suitable elements valid is equality (13), the right-hand part of which belongs to U at all x from an appropriate nonzero ideal I. Therefore, $aI\subseteq U$ and, taking into account the fact that U is the left ideal, we get $RaI\subseteq U$.

If U is an (S, R)-subbimodule, then, going over to a ring which is antiisomorphic to R, we immediately get the required statement. The corollary is proved.

4.5 The Correspondence Theorem.

4.5.1. Lemma. *The ring of constants* R^L *is rationally complete in* R.

Proof. If $Hr\subseteq R^L$, then for any $\mu\in L$ we have $Hr^{\mu}=0$ and, moreover, $RHR^Lr^{\mu}=0$, i.e., by theorem **4.3.8**, we get $r^{\mu}=0$, which is the required proof.

4.5.2. Theorem (Correspondence). *Let R be a prime ring of a characteristic $p>0$, and L be a finite-dimensional restricted Lie ∂-algebra of outer derivations from $D(R)$. Then the mappings $\Lambda \to R^{\Lambda}$; $S \to D_S(R)$ set a one-to-one correspondence between all restricted Lie ∂-subalgebras in L and all intermediate rationally-complete subrings in R.*

The proof will be divided into a number of statements.

4.5.3. Lemma. *The centralizer of an almost intermediate ring S in R_F is equal to C.*

Proof. Let $[S, a]=0$. If $f(x)=\sum c_j x^{\Delta_j}$ is a trace form, such that $f(x) \subseteq S$, then by formula (11) we have an identity $[f(x), a]=0$ on a nonzero ideal of the ring R. By theorem **2.2.2**, we get relations $c_j \otimes a - c_j a \otimes 1 = 0$. Since one of the coefficients is not equal to zero, then $1 \otimes a = a \otimes 1$, i.e., $a \in C$. The lemma is proved.

4.5.4. Theorem. *Let μ be a derivation from $D(R)$, such that $S^{\mu}=0$ for an almost intermediate ring S. Then $\mu \in L$.*.

Proof. Let $\mu_1, \ldots, \mu_n$ be a basis of L over C. Let us first suppose that $\mu, \mu_1, \ldots, \mu_n$ is a strongly independent set. Let $f(x)=\sum c_j x^{\Delta_j} \subseteq S$. In R the relation $f(x)^{\mu} \subseteq \{0\}$ holds, i.e., $\sum c_j x^{\Delta_j \mu} + \sum c_j^{\mu} x^{\Delta_j} \subseteq \{0\}$. By theorem **2.2.2**, we have $c_j \otimes 1 = 0$, which implies that all c_j are zero.

Therefore, the elements $\mu, \mu_1, \ldots, \mu_n$ cannot be strongly independent over C, i.e., there are elements $c, c_1, \ldots, c_n \in C$, such that $\mu c + \mu_1 c_1 + \ldots + \mu_n c_n = ad\xi$ is an inner derivation for Q. As L contains no inner for Q derivations, then $c \neq 0$ and, hence, $\mu + \sum \mu_i c_i c^{-1} = ad\xi c^{-1}$. By the condition $f(x)^{\mu_i} \subseteq \{0\}$, it means that $[\xi c^{-1}, f(x)] \subseteq \{0\}$, i.e., by lemma **4.5.3**, the element ξc^{-1} lies in C. Hence, $ad\xi c^{-1}=0$ and, therefore, $\mu = -\sum \mu_i c_i c^{-1} \in L$. The theorem is proved.

4.5.5. Definitions and notations. Let us fix an almost intermediate ring S and denote by $D_S(R)$ a set of all derivations from $D(R)$ which map S into zero. Then, by theorem **4.5.4**, we get that $D_S(R)$ is a Lie ∂-subalgebra in L. By $\bar{S}$ let us denote a subring $R^{D_S(R)}$. It is obvious that $\bar{S} \supseteq S$ and $\bar{S}$ is an intermediate ring.

4.5.6. Theorem. *The ring* S *contains a nonzero ideal of the ring* $\bar{S}$.

Let us choose a basis $\mu_1,\dots,\mu_k$ of the algebra $D_S(R)$ and extend it to the basis $\mu_1,\dots,\mu_n$ of the algebra L. Let us assume this basis to be ordered, $\mu_1 > \mu_2 > \dots > \mu_n$. Any correct word in this basis has the form $\Delta_j = \Delta_j^{(1)}\Delta_j^{(2)}$, where $\Delta_j^{(2)}$ is a correct word from $\mu_1,\dots,\mu_k$, and $\Delta_j^{(1)}$ is a correct word from $\mu_{k+1},\dots,\mu_n$.

Theorem **4.5.6** will now result from the following proposition.

4.5.7. Proposition. *There are elements* $a_1,\dots,a_k \in R$ *not all of which are zero, and there are elements* $s_1,\dots,s_t \in S$, *such that at all* $x \in R$ *the following equality is valid:*

$$\sum_{j=1}^{p^k} (xa_j)^{\Delta_j^{(2)}} = \sum_{i=1}^{t} \tau_i(x)s_i, \tag{19}$$

where τ_i *are mappings from* R *into* R^L, *which have the form* $\tau_i(x) = f(xr_i)$, *while* $\Delta_j^{(2)}$ *are correct words from* $\mu_1,\dots,\mu_k$.

4.5.8. Let us first show how theorem **4.5.6** results from it. The left hand- part, $V(x)$, of equality (19) does not turn identically to zero on R, since otherwise, using corollary **2.5.8,** we would immediately get all a_j equal to zero. Equality (19) shows that $V(I)$ is contained in S for a suitable nonzero ideal I of R. On the other hand, the words $\Delta_j^{(2)}$ act on $\bar{S}$ in a trivial way and, hence, $V(I) \supseteq V(\bar{S}I) = \bar{S}V(I)$, i.e., $V(I)$ generates a left ideal A of the ring $\bar{S}$ in S. Going over to a ring which is antiisomorphic to R, we can also find a right nonzero ideal of the ring $\bar{S}$, contained in S. Now ASB is a two-sided ideal of the ring $\bar{S}$, contained in S, where $ASB \neq 0$ by theorem **4.3.8**.

4.5.9. ***Lemma.*** *Let* S *be an almost intermediate ring. Then there is an almost intermediate ring* $S_1 \subseteq S$, *such that* $S_1^{\mu_i} \subseteq R$ *for all basic derivations* μ_i, *and* $\bar{S}_1 = \bar{S}$.

Proof. Let us consider a subring $S_1 = \{s \in S \mid s^{\mu_i} \in R,$

$1 \leq i \leq n\}$. As $S_1 \supseteq S \cap R^L$, then S_1 is an almost intermediate ring. If $D_{S_1}(R) = D_S(R)$, then there is nothing to prove. Let, therefore, there be elements $\mu \in L$, $s \in S$, such that $s^{\mu} \neq 0$, $S_1^{\mu} = 0$. Let us choose an ideal $I \in F$ in such a way that $s^{\mu_i} I \subseteq R$ for all i, $1 \leq i \leq n$. Let, then, J be a nonzero ideal, such that $f(J) \subseteq S$ and $f(J) \subseteq I$ (applying formula (11) to the ring I). In this case $(sf(J))^{\mu_i} \subseteq R$ and, hence, $sf(J) \subseteq S_1$. Therefore, $(sf(J))^{\mu} = 0$. This affords $s^{\mu} f(J) = 0$, i.e., by theorem **2.2.2**, $s^{\mu} = 0$ is a contradiction.

Proof of proposition 4.5.7. By lemma **4.5.9**, we can assume $s^{\mu_i} \subseteq R$.

Let $f(x) = \sum x^{\Delta_j} c_j \subseteq S$. Using formulas (7) and (8) we can rewrite the constant as

$$\sum_i \{ \sum_j (xr_{ij})^{\Delta_j^{(1)}} \}^{\Delta_i^{(2)}} = f(x) \qquad (= \sum_q \tau_q(x) s_q). \tag{20}$$

Let us replace the right-hand part of equality (20) with $\sum_q \tau_q(x) s_q$ and, starting from it, prove equality (19).

Let us choose among words $\Delta_i^{(2)}$, for which correct words can occur in braces, a senior word $\Delta_v^{(2)}$, and let $\Delta_w^{(1)}$ be a senior word in the sum $\sum_j (xr_{vj})^{\Delta_j}$, for which $r_{vw} \neq 0$. If such a choice is impossible to implement, then all words $\Delta_j^{(1)}$ are empty and equality (20) turns into (19). This peculiarity is a foundation for carrying out induction over $\Delta_v^{(2)}$.

An induction step will be carried out by induction over $\Delta_w^{(1)}$. In other words, we shall transform equality (20) to the same form but with a smaller word, $\Delta_w^{(1)}$ in braces at $\Delta_v^{(2)}$, and in such a way that for the words $\Delta_i^{(2)} > \Delta_v^{(2)}$ only empty words could be encountered in braces.

Let $\beta = \sum r_\lambda \otimes s_\lambda \in r_{vw}^{\perp S} \subseteq L^{op}(S)$. Let us apply the operator $*\,\beta$ to both parts of (20). Allowing for the fact that the words $\Delta_i^{(2)}$ act trivially on S, we see that only empty words will still be encountered in braces for the words $\Delta_i^{(2)} > \Delta_v^{(2)}$. Let us, using identities (9), transform the braces for the words $\Delta_i^{(2)} \leq \Delta_v^{(2)}$. Then the coefficient under $\Delta_w^{(1)}$ will be equal to $r_{vw} \cdot \beta = 0$, and we can make use of the induction supposition, having showed that not all coefficients r'_{vj} of the newly obtained expression are zero.

Let us consider a coefficient $r'_{vj(0)}$ for the word $\Delta_{j(0)}^{(1)} = \tilde{\Delta}_w^{(1)}$, where, as usual,

$$\Delta_w^{(1)} = \mu_1 \tilde{\Delta}_w^{(1)} = \mu_1^{m_1} \ldots \mu_1^{m_1}, 1 \leq m_1 < p.$$

Since all the words $\Delta_j^{(1)}$ do no exceed $\Delta_w^{(1)}$, then the word $\tilde{\Delta}_w^{(1)}$ in decomposition (9) can appear only for the words of type $\mu_t \tilde{\Delta}_w^{(1)}$, where either $\mu_t = \emptyset$, or $\mu_t \leq \mu_1$. Let $\mu_t \tilde{\Delta}_w^{(1)} = \Delta_{j(t)}^{(1)}$. Then, by formula (9), we have

$$r'_{vj(0)} = m_1(r_{vw} \cdot \beta^{\mu_1}) + \sum_t r_{vj(t)} \beta^{\mu_t}.$$

Allowing for corollary **4.4.5**, let us apply lemma **4.1.3** to the ring R and the subring S, assuming that $r_1 = m_1 r_{vw}$, $n = 1$, $a_1 = r_{vw}$. In this case statement (a) yields $r'_{vj(0)} \neq 0$, which is the sought result.

Let us show that statement (b) cannot be fulfilled. Indeed, it means that for a derivation $\mu = \sum \mu_{j(t)} c_t$ there can be found an element $\xi \in R_F$, such that $[s, \xi] = s^{\mu}$ for all $s \in S$. Setting $s = f(i)$, we get $[f(x), \xi] = 0$, i.e., by theorem **2.2.2**, the element ξ lies in C and $s^{\mu} = 0$. Therefore, $\mu \in D_S(r) = \mu_1 C + \ldots + \mu_k C$. On the other hand, however, μ belongs to the subspace $\mu_{k+1} C + \ldots + \mu_n C$ and, as all μ_i form the basis of L, then $\mu = 0$. This fact contradicts the linear independence of the elements $\{\mu_1, \mu_{j(t)}\}$. The proposition is proved.

4.5.10. From theorems **4.5.4**, **4.5.6** and lemma **4.5.1** one can now easily deduce the correspondence theorem formulated above.

Indeed, by lemma **4.5.1**, for the Lie ∂-subalgebra $\Lambda \subseteq L$ the ring of constants R^{Λ} is an intermediate rationally-complete subring in R. Inversely, if S is an intermediate rationally-complete subring in R, then, by theorem **4.5.6**, one cam find a nonzero ideal H of the ring $\bar{S}$, contained in S, and, hence, for any $\bar{s} \in \bar{S}$ there is an inclusion $H\bar{s} \subseteq H \subseteq S$, i.e., due to rational completeness, $\bar{s} \in S$. Therefore, $S = \bar{S}$ is a ring of constants for the Lie ∂-algebra $D_S(R)$. Finally, by theorem **4.5.4**, for any restricted Lie ∂-algebra Λ in L we have the equality $D_{R^{\Lambda}}(R) = \Lambda$, which fact completes the proof of the correspondence theorem.

4.6 Extension of Derivations

Let us now consider the problem in what cases the extension $S \supseteq R^L$ is a ∂-extension, where S is a rationally-complete subring.

4.6.1. Definitions and notations. Let Λ be a Lie ∂-subalgebra in L. Let us consider an idealizer $I(\Lambda) = = \{\mu \in L \mid [\Lambda, \mu] \subseteq \Lambda\}$ of this algebra. One can easily see that $I(\Lambda)$ is a differential Lie C^{Λ}-algebra, but can be not a space over C. Let us say that Λ is a *quasi-ideal* in L provided $I(\Lambda)\, C = L$. Obviously, if L acts trivially on C, then any quasi-ideal is an ideal and vice versa.

4.6.2. Theorem. *An intermediate ring* S, $R \supseteq S \supseteq R^L$ *will be a* ∂*-extension of the ring of constants* R^L *iff the corresponding to it Lie* ∂*-subalgebra* $D_S(R)$ *is a quasi-ideal of* L*. In this case*

$$D_{R^L}(S) \cong I(D_S(R)) \,/\, D_S(R).$$

This theorem will be deduced from the following one.

4.6.3. Theorem (Extension). *Any derivation of an almost intermediate ring* S *in* R*, mapping* $R^L \cap S$ *to zero, can be extended to a derivation from* L*.*

Proof Let us assume that $\mu\colon S \to R$ be a differential mapping, such that $(R^L \cap S)^{\mu} = 0$, but μ cannot be extended to a derivation from L.

Let us consider a trace form $f(x) = \sum c_j x^{\Delta_j} \subseteq S$.

By theorem **2.2.2**, this form does not turn to zero on R. For any

$s \in S$ the following equality can be written:

$$[\sum x^{\Delta_j} c_j s]^{\mu} - \{\sum x^{\Delta_j} c_j s^{\mu}\} = 0 \tag{21}$$

which is valid for all x running through a nonzero ideal of the ring R. One can assume that an element s can be chosen in such a way that the expression in braces does not turn to zero on R. Indeed, if such an element cannot be chosen, then we get $f(x)S^{\mu} = 0$, which yields $f(x) \cdot R^{L} \cdot S^{\mu} = 0$, which contradicts theorem **4.3.8.**

It would be more convenient to reduce to a contradiction a somewhat more general relation (which holds by formula (7))

$$\Psi(x) \overset{df}{=} [\sum_{j \in J} (xb_j)^{\Delta_j}]^{\mu} - \{\sum_{v \in V} (xr_v)^{\Delta_v}\} = 0. \tag{22}$$

Let us assume that in this relation the expression in square brackets assumes the values in S for any x from a suitable ideal of the ring R. Let us also assume that all the elements b_j, r_v are nonzero elements from RC, $V \neq \emptyset$.

Let us consider a set U of the words Δ_v occurring in presentation (22), such that $\Delta_v \neq \Delta_v^{(2)}$, i.e., with their first letters not lying in $D_S(R)$. By induction over a senior word in the set $U \cup \{\mu\Delta_j, j \in J\}$, and assuming that $\mu_1 > \mu_2 > ... > \mu_n > \mu$, let us reduce equality (22) to a contradiction.

The basis for the induction is the case when $U \cup \{\mu\Delta_j, j \in J\} = \emptyset$. In this case the expression in braces disappears and we see a contradiction to corollary **2.5.8.**

Let us carry out an induction step. Let us first assume that the senior word Δ is encountered in braces. Then all the words $\Delta_v, v \in V$ which are greater than Δ and encountered in braces will be the words from $\mu_1, ... \mu_k$, i.e., they will act trivially on S.

Therefore, for any $\beta \in \bigcap_{v \in V} r_v^{\perp} S \subseteq L^{op}(S)$, at $v \in V_0$ we get:

$$(xr_v)^{\Delta_v} * \beta = (x(r_v \cdot \beta))^{\Delta_v} = 0,$$

where V_0 is a set of those indices $v \in V$, for which the word Δ_v is greater than Δ.

Therefore, the application of operator $* \beta$ to the expression in braces results in the disappearance of all the terms containing in their presentation words greater than Δ, and new terms do not appear from them. The term containing the word $\Delta = \Delta_{v(0)}$ is also obvious to disappear, but new terms will arise from it.

Let us consider the terms arising when applying the operator $*\beta$ to relation (22) and containing in their presentation the word $\tilde{\Delta}$, where, as usual, $\Delta = \mu_s \tilde{\Delta} = \mu_s^{m_s} \ldots$ (see formulas (5), (9)). The word $\tilde{\Delta}$ can appear in decomposition (9) only from the words of type $\mu_t \tilde{\Delta}$, where $\mu_t \le \mu_s$. Let $\mu_t \tilde{\Delta} = \Delta_{v(t)}$; $\mu \tilde{\Delta} = \mu\Delta_{j(0)}$. Then, making allowances for the fact that

$$[A]^{\mu} * \beta = [A * \beta]^{\mu} - \{A * \beta^{\mu}\},$$

we can write the term of the expression $\Psi(x) * \beta$ containing in its presentation $\tilde{\Delta}$:

$$x(m_s r_{v(s)} \cdot \beta^{\mu_s} + \sum_t r_{v(t)} \cdot \beta^{\mu_t} - r_{j(0)} \cdot \beta^{\mu}),$$

in which case the last two addends may be absent (provided the words $\Delta_{v(t)}$ in braces and the word $\tilde{\Delta}$ in square brackets, respectively, do not participate in (22)).

Now we choose $\beta \in \bigcap_V r_V^{\perp} S$ in such a way that the written term does not turn to zero. Indeed, in the opposite case, by lemma **4.1.3**, the linear combination $\mu_s + \sum \mu_t c_t + \mu c_0$ acts trivially on S, in which case the corresponding element commutes with the elements of an almost intermediate ring $R^{L \cap S}$. By lemma **4.5.3**, we get

$$\mu_s + \sum_t \mu_t c_t + \mu c_0 = 0.$$

If in this relation $c_0 \ne 0$, then $-\mu_s c_0^{-1} - \sum \mu_t c_t c_0^{-1}$ is an extension of derivation μ on R. If $c_0 = 0$, then we have a contradiction with the linear independence of the derivations $\{\mu_s, \mu_t\}$.

Therefore, choosing β in the above mentioned way, we see that the inductive proposition can be applied to the expression $\Psi(x) * \beta$.

Let us now assume that the senior word has the form $\mu\Delta = \mu\Delta_{j(0)}$. Then all the words in braces with their length greater than Δ will be the words from $\mu_1,\dots \mu_k$ and, hence, choosing $\beta \in \bigcap_j b_j^{\perp S} \cap \bigcap_v r_v^{\perp S}$, we see that the senior word for $\Psi(x)*\beta$ gets smaller and, second, the term in braces containing the word Δ will have the form

$$x(b_{j(0)} \cdot \beta^{\mu}).$$

By lemma **4.1.3**, one can choose β in such a way that this term does not turn to zero. If we now use the inductive proposition, then the theorem is proved.

4.6.4. Theorem (Rigidity). *Any monomorphism of an almost intermediate ring S to R over R^L is identical.*

Proof. Let $\mu: S \to R$ be an monomorphism acting identically on $S \cap R^L$. Then for any trace form $f(x)$ we have $[f(x)]^{\mu} - \{f(x)\} = 0$, and we get a relation of type (22). Let us deduce this relation to a contradiction by induction over the length of the senior word occurring in braces, assuming that the expression in square brackets lies in S and does not turn to zero on R.

If the expression in braces is not present, then the contradiction results from the fact that the kernel of μ is zero.

As $S^{\mu} \supseteq R^L \cap S$, then S^{μ} is an almost intermediate ring.

If $\beta \in \bigcap_V r_V^{\perp S}$, then we can choose an element $\beta_0 \in \mathbb{L}^{op}(S)$, such that $\beta_0^{\mu} = \beta$. In this case we have

$$\Psi(x) * \beta = [T(x) * \beta_0]^{\mu} - \{T_1(x) * \beta\}.$$

It is obvious that the senior word in braces for $\Psi(x)*\beta$ has a less length and it would, therefore, suffice to choose β in such a way that the expression in square brackets does not turn to zero on R. Allowing for the fact that the senior term, $T(x)*\beta_0$, is equal to $(x(b_{j(0)} \cdot \beta_0))^{\Delta_{j(0)}}$, where $\Delta_{j(0)}$ is a senior word, we come to the conclusion that it suffices to choose β in such a way that $b \cdot \beta_0 \neq 0$ for a given nonzero $b = b_{j(0)}$.

Let us suppose that it is impossible, i.e., that for any $\beta_0^{\mu} \in \bigcap_{i=1}^{m} b_i^{\perp S^{\mu}}$ we have $b \cdot \beta_0 = 0$. Without any violation of generality one can assume that $\{b_i\}$ are linearly independent over C elements, while the number m is the least of all possible, i.e., $b \cdot \mathcal{B}_0 \neq 0$, where $\mathcal{B}_0^{\mu} \overset{df}{=} \bigcap_{i=2}^{m} b_i^{\perp} S^{\mu}$ and, by lemma **4.1.2**, $b_1 \cdot \mathcal{B}_0^{\mu} \neq 0$. In this case, if $\beta_0 \in \mathcal{B}_0$ and $b_1 \beta_0^{\mu} = 0$, then $b \cdot \beta_0 = 0$ and, hence, the mapping $\xi: b_1 \beta_0^{\mu} \to b \cdot \beta_0$, where β_0 runs through $\mathcal{B}_0$, is well-defined. One can easily see that ξ is a homomorphism of left R-modules, the domain of whose definition is an (R, S^{μ})-bimodule and, hence, by corollary **4.4.5**, it contains a nonzero ideal of the ring R. Therefore, by the definition of a ring of quotients, one can find a nonzero (as $b\mathcal{B}_0 \neq 0$) element $t \in R_F$, such that $b_1 \beta_0^{\mu} t = b \cdot \beta_0$ for all $\beta_0 \in \mathcal{B}_0$. If $s \in S$, then $\beta_1 = \beta_0(s \otimes 1) \in \mathcal{B}_0$ and, hence,

$$b_1 \cdot \beta_1^{\mu} t = b_1 \beta_0^{\mu} s^{\mu} t = b\beta_1 = b\beta_0 s = b_1 \beta_0^{\mu} ts.$$

Herefrom we have $b_1 \beta_0^{\mu}(s^{\mu} t - ts) = 0$ and, as $b_1 \mathcal{B}_0$ contains, by corollary **4.4.5**, a nonzero ideal of the ring R, then $s^{\mu} t = ts$ for all $s \in S$. In particular, choosing s from R^L and using lemma **4.5.3**, we see that t is a nonzero central element, i.e., the latter equality can be cancelled by t. Thus, $s^{\mu} = s$ for all $s \in S$. The theorem is proved.

Proof of theorem 4.6.2. Let us assume that a quasi-ideal $D_S(R)$ corresponds to the subring S. It means that in the idealizer $I(D_S(R))$ one can choose a basis of the space L over C. Let us, therefore, assume the basic derivations $\mu_1, \dots, \mu_n$ to lie in the idealizer.

Let us consider a restriction $\bar{\mu}$ of an arbitrary derivation $\mu \in I(D_S(R))$ on S and show this restriction to lie in $D(S)$, i.e., we shall find a nonzero ideal I of S, such that $I^{\mu} \subseteq S$. By lemma **4.5.9**, one can find a subring $S_1 \subseteq S$, such that $S_1^{\mu} \subseteq R$ and $\bar{S}_1 = \bar{S}$. By theorem **4.5.6**, one can find an ideal A of the ring $\bar{S}$ lying in S_1. If now $\mu_1 \in D_S(R)$, then $[\mu, \mu_1] \in D_S(R)$ and, hence, $(S_1^{\mu})^{\mu_1} \subseteq S_1^{[\mu, \mu_1]} + (S_1^{\mu_1})^{\mu} = 0$, and, hence, $S_1^{\mu} \subseteq \bar{S}$ and, thus, the ideal A^2 can be chosen as I.

Therefore, the mapping $\mu \rightarrow \bar{\mu}$ is a homomorphism from $I(D_S(R))$ on a restricted Lie subalgebra $\bar{L}$ of $D(S)$, in which case $S^{\bar{L}} = R^L$, as $I(D_S(R))$ contains the basis of L. It is also evident that the kernel of the latter homomorphism is equal to $D_S(R)$ and, hence,

$$\bar{L} \cong I(D_S(R)) / D_S(R).$$

As far as the homomorphism of the restriction is C^L-linear and L is finite-dimensional over C^L, then $\bar{L}$ is also finite-dimensional over C^L.

Let now, vice versa, $S^{\bar{L}} = R^L$ for a finite dimensional Lie ∂-algebra $\bar{L} \subseteq D(S)$.

Let us now show that $\bar{L}$ cannot contain inner derivations. Let $\bar{\mu} \in \bar{L}$. One can find a nonzero ideal I of the ring S, such that $I^{\bar{\mu}} \subseteq S \subseteq R$ and, hence, by the theorem of extension, $\bar{\mu}$ coincides with a certain $\mu \in L$ on $I + R^L$. It means that $\bar{\mu}$ coincides with μ on S as well. Indeed, if $s \in S$, $i \in I$, then $(si)^{\mu} = (si)^{\bar{\mu}}$ and $si^{\mu} = si^{\bar{\mu}}$, i.e., by theorem **4.3.8**, we get $\bar{\mu} = \mu$ on S. Let us assume that $\xi s - s\xi = s^{\mu}$ for a certain $\xi \in Q(S)$ and all $s \in S$. Having chosen an arbitrary element a from the ideal I of S, such that $I\xi \subseteq S$, $\xi I \subseteq S$, we get the equality $b_1 sa - asb_2 = as^{\mu}a$, where $b_1 = a\xi$, $b_2 = \xi a$. Let us assume that μ is a basic derivation, not lying in $D_S(R)$. Let a trace form $f(x) = \sum x^{\Delta_j} c_j$, where Δ_j are some words from basic derivations of $D_S(R)$, not turn to zero on R, and assume the values in $\bar{S}$ at $x \in J$, where J is a nonzero ideal of R. By theorem **4.4.6**, we can find a nonzero ideal A of the ring $\bar{S}$ lying in S. We have: $Af(x) \subseteq S$, and, hence, we get the following identity for all $d \in A$:

$$b_1(d \cdot f(x))a - a(d \cdot f(x))b_2 = a(df(x))^{\mu} a.$$

Using theorem **2.2.2**, we see that $a(d \otimes c_j)a = 0$, and since a is an arbitrary element from I, while the element d is also an arbitrary element from A, then $IA = 0$, which contradicts theorem **4.3.8.**

Let L_1 be a set of all derivations from L whose restrictions onto S belong to $\bar{L}$. Then the kernel of the natural homomorphism $L_1 \rightarrow \bar{L}$ coincides with $D_S(R)$, and the image is equal to $\bar{L}$, i.e.,

$D_{R^L}(S) = L_1 / D_S(R)$, as by theorem **4.5.4**, we have $\bar{L} = D_{R^L}(S)$.

Then, the Lie ∂-subalgebra L_1C of L has only elements from R^L as its constants and, hence, by the correspondence theorem, $L_1C = L$. The task now is to show that $L_1 = I(D_S(R))$. If $\mu \in D_S(R)$, $\mu_1 \in L_1$, then $\overline{[\mu_1, \mu]} = [\bar{\mu}_1, \bar{\mu}] = 0$, as $\bar{\mu} = 0$ and, hence, $[\mu, \mu_1] \in D_S(R)$. If $\mu_1 \in I(D_S(R))$ and $\mu \in D_S(R)$, then $\overline{[\mu_1, \mu]} = 0$, which affords $(S^{\mu_1})^{\mu} \subseteq S^{[\mu_1, \mu]} + (S^{\mu})^{\mu_1}$ and, hence, $S^{\mu_1} \subseteq R^{D_S(R)} = \bar{S}$, so that, using theorem **4.5.6**, one can find the ideal A of the ring S, such that $A^{\mu_1} \subseteq S$ or, in other words, $\bar{\mu}_1 \in D_{R^L}(S) = \bar{L}$, i.e., $\mu_1 \in L_1$. The theorem is proved.

REFERENCES

R.Baer [11];
A.Berkson [22];
N.Jacobson [63];
V.K.Kharchenko [75,77];
M.Weisfeld [153].

CHAPTER 5

THE GALOIS THEORY OF SEMIPRIME RINGS

In this chapter we shall, using the metatheorem, transfer the basic results of the Galois theory for automorphisms and derivations onto semiprime rings. The task is, in essence, purely syntaxical, i.e., to reformulate the results in the form of truthfulness of Horn formulas. One can easily see that the basic theorems cannot be formulated in an elementary language, which makes us analyze not the theorems proper, but the key statements in their proofs.

Let us begin with considering the simplest case of Lie ∂-algebras of outer derivations where the situation is much simpler since derivations (unlike automorphisms) act trivially on central idempotents. Considerations of the inner part will be postponed until automorphisms have been studied.

Let us fix notations for a semiprime ring R, its left and symmetrical Martindale rings of quotients R_F, Q and for a generalized centroid C. By E we shall denote a set of all idempotents of the ring C.

5.1 Essential Trace Forms

5.1.1. Definitions and notations. Let L be a finitely-generated C-submodule of derivations from $D(R)$, closed under the operations of a restricted Lie ∂-algebra, i.e., $[\mu, \mu_1] \in L$, $\mu^p \in L$, if $\mu, \mu_1 \in L$, where $0 < p$ is a characteristic of the ring R. Let us also assume that L contains no nonzero inner for Q derivations. By lemma **1.6.21**, we can expand L into a direct sum of cyclic submodules $L = \mu_1 C \oplus .. \oplus \mu_n C$. In this case $\mu_1,.. \mu_n$ will be strongly independent: if $\sum \mu_i c_i = ad\xi$, then $\sum \mu_i c_i = 0$, as L contains no other inner derivations and, hence, $\mu_i c_i = 0, 1 \le i \le n$. Let us also fix a Lie ∂-algebra L and its C-basis $\{\mu_1,.., \mu_n\}$.

The linear form $f = \sum_{j=1}^{m} x^{\Delta_j} c_j$, where $\Delta_1 < ... < \Delta_m$ are correct words from $\mu_1,.. \mu_n$; $c_j \in C$ will be called an *essential trace form* for L, if $f(x)^{\mu} = (\sum x^{\Delta_j} c_j)^{\mu} = 0$ and $\sup_j \{e(c_j)\, e(\Delta_j)\} = 1$. It should be recalled that $e(x)$ is a support of the element x, $e(\mu_1 .. \mu_k) = e(\mu_1) .. e(\mu_k)$.

5.1.2. Proposition. *The algebra* L *has an essential trace form.*

Proof. As the number of the correct words is finite (it equals $m = p^n$) and any word can be presented as their linear combination using the substitutions $\mu_1 \mu = \mu\mu_1 + [\mu_1, \mu]$, $\mu^p = \mu^{[p]}$, then we have $\Delta_j \mu_i = \sum_k \Delta_k c_{ij}^{(k)}$ and, hence,

$$(\sum_j \Delta_j c_j)\, \mu_i = \sum_j \Delta_j c_j^{\mu_i} + \sum_{k,\, j} \Delta_k c_j c_{ij}^{(k)},$$

i.e., the equality $f(x)^{\mu} = 0$ will result from the following system of relations for the coefficients c_j:

$$F_{k,\, i}(\vec{c}) \equiv c_k^{\mu_i} + \sum_j c_j c_{ij}^{(k)} = 0, \quad k = 1, \ldots, m, \ i = 1, \ldots, n.$$

Now the condition of the existence of the essential trace form is written as a Horn formula:

$$\exists\, c_1, \ldots, c_m \forall x \mathop{\&}_{j=1}^{m} P(c_j) \& \mathop{\&}_{k,\, i} F_{k,\, i}(\vec{c}) = 0 \&$$
$$\& [(\mathop{\&}_j x e_j c_j = 0) \to x = 0],$$

where $e_j = e(\Delta_j)$, while by P we denote a predicate which singles out a generalized centroid $P(x) = T \Leftrightarrow x \in C$. Since C is a closed subset in Q, then it is a strictly sheaf predicate (see **1.11.9**), in which case, by lemma **1.12.6**, $P(\bar{x}) = T$ on a stalk $\Gamma_{\mathfrak{p}}(Q)$ iff $\bar{x}$ belongs to the generalized centroid of the stalk.

As the relations $\bar{\Delta}_j \bar{\mu}_i = \sum \bar{\Delta}_k \bar{c}_{ij}^{(k)}$ are fulfilled for the derivations induced on the stalk, then, by lemma **4.3.3**, the above formula is also valid on all the stalks and, by the metatheorem, it is also valid on the ring Q of the global sections. The proposition is proved.

5.1.3. Corollary. *Let* $a, b \in R_F$, *and let* I *be an essential ideal of the ring* R. *If* $aI^L b = 0$, *then* $e(a)e(b) = 0$.

Proof. Let $f(x) = \sum x^{\Delta_j} c_j$ be an essential trace form. As $D(I) = D(R)$ (lemma **1.8.4**), then one can find an essential ideal J, such that $J^{\mu} \subseteq I$ for all basis derivations. Therefore, $(J^{p^n})^{\Delta_j} \subseteq I$ and, hence, the differential identity $af(x) \cdot b = 0$ holds on J^{p^n}. By theorem **2.2.2**, we get $e(c_j) \cdot e(\Delta_j) a \otimes b = 0$, which affords $e(a) \cdot e(b) =$

$e(a) \cdot e(b) \sup_j (e(c_j) e(\Delta_j)) = 0$, which is the required proof.

5.1.4. Theorem. *Let* μ *be a derivation from* $D(R)$, *such that* $(R^L)^{\mu} = 0$. *Then* $\mu \in L$.

Proof. Let us consider a C-module $L + \mu C$. By lemma **1.6.2**, the submodule L is obtained from it as a direct addend $L + \mu C = L \oplus \mu_0 C =$

$\mu_1 C \oplus ... \oplus \mu_n C \oplus \mu_0 C$. Let us bring in order the basis derivations $\mu_1 < \mu_2 < ... < \mu_n < \mu_0$.

If $\sum_{i=0}^{n} \mu_i c_i = ad\xi$, then the element ξ commutes with all constants for L. Let I be an essential ideal, such that for the essential trace form f the inclusion $f(I) \subseteq R$ holds. Then on I we have the identity $\xi f(x) - f(x)\xi = 0$, i.e., by theorem **2.2.2**, we get $(\xi x - x\xi) e(c_j)$

$e(\Delta_j) = 0$ for every j. Therefore, $\xi \in C$ and $ad\xi = 0$, i.e., $\mu_i c_i = 0$, $0 \le i \le n$, since the sum of the basic submodules is direct.

Thus, $\mu_0, \mu_1, ..., \mu_n$ is a strongly independent set of derivations and on I we have a reduced identity $0 = \sum_j x^{\Delta_j} c_j^{\mu_0} + \sum_j x^{\Delta_j \mu_0} c_j \equiv f(x)^{\mu_0}$. By theorem **2.2.2**, for every j we get $e(c_j) e(\Delta_j) e(\mu_0) = 0$, and, hence, $e(\mu_0) = 0$. The theorem is proved.

5.2 Intermediate Subrings

Our nearest aim is to transfer theorem **4.5.6** into a semiprime case. Preserving the notations of the preceding paragraph, let us fix an intermediate ring S, i.e., $R \supset S \supset R^L$. Let $\bar{S} = R^{\Lambda}$, where $\Lambda = D_S(R) = \{\mu \in D(R) \mid \mu(S) = 0\}$. By theorem, **5.1.4**, the inclusion $\Lambda \subseteq L$ is valid.

5.2.1. Lemma. *Λ is a finitely-generated submodule of the C-module L.*

Proof. Let us note that Λ is a closed submodule of the complete module L. If $\mu_0 = \lim \mu_\alpha$ and $\mu_\alpha \in \Lambda$, then $e_\alpha \mu_0(s) = e_\alpha \mu_\alpha(s) = 0$ at any α and any $s \in S$ and, hence, $\mu_0(s) = 0$ and $\mu_0 \in \Lambda$. By lemma **1.6.16**, the module Λ is injective and, therefore, it is obtained as a direct addent from the finitely-generated module L and is, hence, finitely-generated itself. The lemma is proved.

5.2.2. Theorem. *The ring S contains a two-sided ideal of the ring $\bar{S}$, the annihilators of which (both left and right) in R_F are zero.*

For the metatheorem to be used, it is necessary to single out elements from the set S. For this purpose it is natural to introduce a unary predicate determined by this set (see **1.11.9**). This predicate, however, will not be strictly sheaf if S is not orthogonally complete. Let us consider a closure $\widehat{SE}$ and a strictly sheaf predicate P_S determined by the set $\widehat{SE}$. In this case the value of P_S on the stalk Γ_p is determined by the set $\rho_p(\widehat{SE})$.

5.2.3. Lemma.

(a) *If $\mu \in D(R)$ and $\mu(S) = 0$, then $\mu(\widehat{SE}) = 0$.*

(b) *$\rho_p(\widehat{SE})$ is an almost intermediate subring on the stalk for the ∂-algebra of the induced derivations $\bar{L} = \rho_p(L)$.*

(c) *The set of derivations of the stalk Γ_p over $\rho_p(\widehat{SE})$ is equal to $\bar{\Lambda} = \rho_p(\Lambda)$.*

(d) *If $s \in \widehat{SE}$, then one can find an ideal I of R^L, such that $sI \cup Is \subseteq S$ and annihilators of I in R_F are zero.*

Proof.

(a) As all derivations act trivially on the central idempotents (**2.5.2**) and are continuous (**1.6.3**), then the equality $\mu(S) = 0$ implies $\mu(SE) = 0$

and $\mu(\widehat{SE}) = 0$.

(b) Let $f(x)$ be an essential trace form. Then $f(I) \subseteq S$ for a certain ideal $I \in F$. It affords $f(IE) \subseteq SE$ and, since f is continuous, we have $f(\widehat{IE}) \subseteq \widehat{SE}$. Now we are to recall that $\rho_p(\widehat{IE})$ is a two-sided ideal of the stalk $\rho_p(\widehat{RE})$ (see **1.12.5**), while the induced form $\bar{f}$ will be a trace form for $\bar{L}$ in the stalk.

(c) As $\bar{L}$ is a restricted Lie ∂-algebra acting on the stalk, while $\rho_p(\widehat{SE})$ is an almost intermediate ring, then any derivation $\bar{\mu}$ of the stalk which acts trivially on $\rho_p(\widehat{SE})$ belongs to $\bar{L}$, i.e., $\bar{\mu} = \rho_p(\mu)$, $\mu \in L$ (see **4.5.4**). Now the truthfulness of the formula $\forall\, x(P_S(x) \rightarrow \bar{\mu}(x) = 0)$ on the stalk implies that it is also true on the ring of sections over a certain neighborhood of the point p, i.e., there is a central idempotent $e \notin p$, such that $e\mu(\widehat{SE}) = 0$, i.e., $e\mu \in \Lambda$ and $\bar{\mu} = \rho_p(e\mu) \in \bar{\Lambda}$.

(d) Let U be a set of all elements s from $\widehat{SE}$ satisfying condition (d). Let us show that it is a closed set containing SE.

If $s = \sum s_i e_i$ and $Je_i \subseteq R$, then by corollary **5.1.3**, an ideal J^L has zero annihilators in R_F and $J^L e_i \subseteq S$. Therefore, if $s_i \in S$, then $sJ^L \cup Js^L \subseteq S$.

Let now $s = \lim_\alpha s$ and $s_\alpha I_\alpha \cup I_\alpha s_\alpha \subseteq S$. For every α let us choose an essential ideal J_α, such that $e_\alpha J_\alpha \subseteq R$. Then, by corollary **5.1.3**, the ideal J^L has zero annihilators and $e_\alpha J^L_\alpha \subseteq R^L$. Therefore, $s_\alpha e_\alpha J^L_\alpha I_\alpha \cup e_\alpha J^L_\alpha I_\alpha s_\alpha \subseteq S$. Let $I = \sum e_\alpha J^L_\alpha I_\alpha$. Then the annihilators of I in R_F are zero: if, for instance, $qI = 0$, then $(e_\alpha q) I^L_\alpha J_\alpha = 0$ for all α and, hence, $q = \sup_\alpha \{e_\alpha\}\, q = 0$. Moreover, $sI \subseteq \sum_\alpha s_\alpha e_\alpha J^L_\alpha I_\alpha \subseteq S$, and analogously, $Is \subseteq S$. The lemma is proved.

Proof of theorem 5.2.2. Let us expand the module Λ into a direct sum of cyclic submodules $\Lambda = \mu_1 C \oplus .. \oplus \mu_k C$. Let us recall that the proof of theorem **4.5.6** was based on proposition **4.5.7**, which can be written on the stalk $\Gamma_{\mathbb{P}}$ as the formula $\Phi(t)$ which depends on a natural parameter t:

$$\exists\, a_1, \ldots, a_{p^k}, s_1, \ldots, s_t, r_1, \ldots, r_t\, \forall x \mathop{\&}_{i=1}^{t} P_S(s_i) \,\&$$

$$\& \sum_{j=1}^{p^k} (xa_j)^{\Delta_j^{(2)}} =$$

$$= \sum_{i=1}^{t} f(xr_i)s_i \,\&\, \sup_j \{ e(a_j)e(\Delta_j^{(2)}) \} = 1,$$

where $\Delta_j^{(2)}$ are correct words from $\mu_1, \ldots, \mu_k$, while $f(x)$ is an essential trace form which can be considered as a strictly sheaf operation.

On every stalk Γ_p a formula $\Phi(t(p))$ is true, which results from proposition **4.5.7** and lemma **5.2.3.**

By corollary **1.11.21**, we see that formula $\Phi(t)$ is true for a certain t on the ring $\widehat{RE}$. Now the proof is completed, as was the case in theorem **4.5.6.** Let us set

$$F(x) = \sum_{j=1}^{p^k} (xa_j)^{\Delta_j^{(2)}}$$

and choose an essential ideal I in such a way that $f(Ir_i) \subseteq R$. Let, besides, A be an ideal of the ring R^L with zero annihilators in R_F, such that $s_i A \subseteq S$ (see lemma **5.2.3**). Then $V = F(I)A$ is a left ideal of the ring $\bar{S}$ contained in S. Its left annihilators is equal to zero: if $qF(I)A = 0$, then $qF(x)a = 0$ is a differential identity on I and, by theorem **2.2.2**, for all j valid is $e(q) \cdot e(a_j)\, e(\Delta_j) e(a) = 0$, which affords $e(q)a = 0$ and, since a is an arbitrary element from A, then $e(q)A = 0$, i.e., $q = 0$.

In an analogous way we shall find the right ideal W of the ring $\bar{S}$ contained in S, the right annihilator of which in R_F is zero. Now VR^LW is the sought ideal: if, for instance, $qVR^LW = 0$, then by lemma **5.1.3**, for any $v \in V$, $w \in W$ we have $e(qv)e(w) = 0$, which yields $we(qv) = 0$ or $We(qv) = 0$, i.e., $qv = 0$ and, hence, $qV = 0$ and, therefore, $q = 0$. The theorem is proved.

5.3 The Correspondence Theorem for Derivations

5.3.1. Theorem. *Let* L *be a restricted Lie* ∂*-algebra of outer derivations from* $D(R)$. *Let us assume that* L *is finitely-generated as a module over* C. *Then the mappings* $\Lambda \to R^{\Lambda}$, $S \to D_S(R)$ *set a one-to-one correspondence between restricted Lie* ∂*-subalgebras of* L, *which are finitely-generated over* C, *and intermediate rationally complete subrings. In this case an extension* $S \subseteq R^L$ *will be differential iff* $D_S(R)$ *is a quasi-ideal of* L.

It should be recalled that the subring S is called *rationally complete* if the inclusion $Hr \subseteq S$ implies $r \in S$ for an essential ideal H of S and an element $r \in R$. If Λ is a Lie ∂-subalgebra of L, then it is called *a quasi-ideal* if its idealizer $I(\Lambda) = \{l \in L | [l, \Lambda] \subseteq \Lambda\}$ contains a basis of L, i.e., $I(\Lambda)C = L$.

This theorem is deduced by standard considerations from theorems **5.1.4** and **5.2.2**, as well as from the following extension theorem.

5.3.2. Theorem. *Let* S *be an intermediate ring,* $\mu: S \to R$ *be a differential mapping acting trivially on* R^L. *Then* μ *is extended to a derivation from* L.

Proof. Let us first extend μ to a differential mapping from SE into RE. Let us set $\mu(\sigma) = \sum \mu(s_i)e$ for a linear combination $\sigma = \sum s_i e_i$, $s_i \in S$, $e_i \in E$. This is a correct definition: if $\sum s_i e_i = 0$, then for any constant $j \in R^L$, such that $e_i j \in R$, we get $e_i j \in R^L \subseteq S$ and, hence, $\sum \mu(s_i)e_i j = \sum \mu(s_i(e_i j)) = = \mu(\sum s_i e_i j) = 0$, i.e., by corollary **5.1.3**, we get $\sum \mu(s_i)e_i = 0$. Direct calculations show μ to be a differential E-linear mapping.

Let us extend μ to the differential E-linear mapping from $\hat{SE}$ into $\hat{RE}$. Let $s = \lim_{\alpha \in A} s_\alpha$ and let μ values on s_α be determined. Then, since the module $\hat{RE}$ is complete, there is a $\lim_A \mu(s_\alpha)$, which will be henceforth assumed to be $\mu(s)$. Let, simultaneously, $s = \lim_B t_\beta$ and we set $t = \lim' \mu(t_\beta)$. Then

$$(\mu(s) - t)e_\alpha e_\beta = \mu(s_\alpha)e_\alpha e_\beta - \mu(t_\beta)e_\alpha e_\beta =$$

$$\mu(s_\alpha e_\alpha e_\beta - t_\beta e_\alpha) = \mu(se_\alpha e_\beta - se_\beta e_\alpha) = 0$$

for all $\alpha \in A$, $\beta \in B$. As $\sup_{A\times B} e_\alpha e_\beta = 1$, then $\mu(s) = t$ and, hence, the definition of $\mu(s)$ is correct. The differentiality of μ is directly checked: if $s = \lim_A s_\alpha$, $t = \lim_B t_\beta$, then $[\mu(st) - - \mu(s)t - s\mu(t)]e_\alpha e_\beta = 0$, as $st = \lim_{A\times B} s_\alpha t_\beta$.

The next step will be an extension of μ to the E-linear mapping $\bar{\mu}: \hat{RE} \to \hat{RE}$. For this purpose let us consider $\hat{SE}$ as an E-module (see remark **1.6.9**). Since $\hat{SE}$ is complete, then it is an injective E-module (see **1.6.16**) and, by the definition of injectivity (**1.6.15**), the homomorphism of E-modules $\mu: \hat{SE} \to \hat{RE}$ is extended to a homomorphism $\hat{\mu}: \hat{RE} \to \hat{RE}$.

Let us now consider $\hat{\mu}$ as a unary operation on $\hat{RE}$. Due to E-linearity, this operation is strictly sheaf, its action on the stalks being of such a character that the restriction of $\hat{\mu}$ on $\rho_p(\hat{SE})$ (see **5.2.3**, (b)) will be a derivational mapping acting trivially on the constants of the almost intermediate ring $\rho_p(\hat{SE})$. By theorem **4.6.3**, the restriction of $\hat{\mu}$ on $\rho_p(\hat{SE})$ will be extended to a derivation from $\rho_p(L)$, i.e., on every stalk the following formula is valid:

$$\exists\, c_1,\dots, c_n \forall x \mathop{\&}_{i=1}^{n} P_C(c_i) \& (P_S(x) \to$$
$$\to \bar{\mu}(x) = c_1 x^{\mu_1} + \dots + c_n x^{\mu_n})$$

where $L = \sum \mu_i C$, $P_C(x) = T \Leftrightarrow x \in C$, $P_S(x) = I \Leftrightarrow x \in \hat{SE}$.

By the metatheorem, this formula is also true on $\hat{RE}$. The theorem is proved.

5.3.3. Theorem. *Any monomorphism* μ *of an intermediate ring* S *into* R *over* R^L *is identical.*

Proof. By analogy with the above theorem, let us extend μ to an E-linear monomorphism $\mu: \hat{SE} \to \hat{RE}$, and then to an E-linear mapping $\hat{\mu}: \hat{RE} \to \hat{RE}$. Then on every stalk a Horn formula $\forall x(P_S(x) \to \hat{\mu}(x) = x)$ will be true. The validity of this formula on $\hat{RE}$ will give the required statement.

5.4 Basic Notions of the Galois Theory of Semiprime Rings (the Case of Automorphisms)

5.4.1. Let us start with a simple example, vividly illustrating variations in some notions of the Galois theory when going over to a semiprime case.

Let $R = \prod_{\alpha \in A} F_\alpha$ be a direct product of isomorphic fields $F_\alpha \cong F$ and H be a finite group of automorphisms of the field F. Let us determine the action of H on the product R in a componentwise manner, $f^h(\alpha) = (f(\alpha))^h$, i.e., $(..., f_\alpha, ...)^h = = (..., f_\alpha^h, ...)$. On the other hand, on R one can naturally determine the action of the direct product $G = \prod_{\alpha \in A} H_\alpha$ of the groups H_α isomorphic to H: $f^g(\alpha) = f(\alpha)^{g(\alpha)}$, i.e., $(..., f_\alpha, ...)^{(..., g_\alpha, ...)} = = (..., f_\alpha^{g_\alpha}, ...)$. In this case the subrings of fixed elements for the group H and for the group G coincide and are isomorphic to the direct product of copies of F^H.

5.4.2. Reduced finiteness. The most natural and direct transfer of this notion is as follows: a module $\mathcal{B}(G)$ is finitely generated over C and a factor-group G / G_{in} is finite with respect to a subgroup of inner for Q automorphisms. Such a definition, however, excludes from consideration in the above example the group G, which is a Galois closure of a finite group H. By the same reason we have to transfer to a "local" variation of the same notion.

It should be recalled that for an automorphism $g \in A(R)$ of a semiprime ring R we have determined a conjugation module, $\Phi_g = \{a \in Q | \forall x \in R \;\; xa = ax^g\}$, which is a cyclic C-submodule $\Phi_g = \varphi_g C$, the support of which, $e(\varphi_g)$, has been denoted by $i(g)$. The action of the automorphism g on $i(g)Q$ coincides with the conjugation by the invertible (in $i(g)Q$) element φ_g. Moreover, for any $e \leq i(g)$ the automorphism g will be inner on eQ, which, in particular, implies that $G_e = \{g \in G | \; i(g) \geq e\}$ is a subgroup of the group G (for a fixed e).

5.4.3. Definition. A group of automorphisms $G \subseteq A(R)$ is called *reduced-finite* if its algebra $\mathcal{B}(G) = \sum_{g \in G} \Phi_g$ is finitely-generated as a C-module and $\sup\{e | | G: G_e| < \infty\} = 1$.

5.4.4. ***Closure of a group.*** Returning to example **5.4.1**, we should remark that the action of every $g \in G$ on cofactor F_α coincides with that of a certain $h \in H$, which is α-dependent. It is this peculiarity that results in a coincidence of the invariants of G and those of H. Under general conditions we, by analogy, come to the notion of a local belonginness to the group: the automorphism g *locally belongs* to the group H, provided there is a dense family of idempotents $\{e_\alpha \in C | \alpha \in A\}$, such that the action of g on $e_\alpha Q$ coincides with that of a certain $h_\alpha \in H$, i.e., $\sup\{e : g|_{eQ} \in H|_{eQ}\} = 1$.

5.4.5. Lemma. *If the automorphism g locally belongs to the group H, then $g \in Al(H)$, i.e., g acts identically on Q^H.*

Proof. Let g action on $e_\alpha Q$ coincide with that of $h_\alpha \in H$. For any $r \in Q^H$ we have $e_\alpha^g r^g = (e_\alpha r)^g = (e_\alpha r)^{h_\alpha} = e_\alpha^{h_\alpha} r$, but, by the condition we have $e_\alpha^{h_\alpha} = e_\alpha^g$ and, hence, $e_\alpha^g(r^g - r) = 0$, i.e., $e_\alpha(r - r^{g^{-1}}) = 0$. Therefore, $0 = \sup_\alpha e_\alpha \cdot (r - r^{g^{-1}}) = r - r^{g^{-1}}$, which is the required proof.

5.4.6. Definition. *A group Q is called closed if any automorphism locally belonging to the group G lies in G.*

We immediately get the following corollary.

5.4.7. Corollary. *Any Galois group is closed.*

5.4.8. Let us consider in detail the notion of closure for the case when R has a finite prime dimension, i.e., $Q = Q_1 \oplus .. \oplus Q_n$ is a direct sum of a finite number of prime rings (see **3.6.6**).

Let us start with an example somewhat more complex than that in **5.4.1**. Let $Q = \underbrace{Q_0 \oplus ... \oplus Q_0}_{n}$, where Q_0 is a prime ring. If a group H acts on Q_0, then on Q we have, first, a product $H^n = H \times .. \times H$ acting in a componentwise manner and, second a group of permutations S_n, rearranging the addends $(q_1 \oplus .. \oplus q_n)^\pi = q_{\pi^{-1}(1)} \oplus .. \oplus q_{\pi^{-1}(n)}$. Therefore, the action of a semidirect product $G = H^n \rtimes S_n$ is determined on Q. It is obvious that G is a closed group, the inverse statement being also valid.

5.4.9. Lemma. *Let a closed group G act transitively on the* *components of* Q *(which is equivalent to the fact that* Q *is G-prime). In this case rings* Q_i *are mutually isomorphic and they can be identified in such a way that* $G = H^n \rtimes S_n$.

Proof. Because of transitivity, we can find elements $g_i \in G$, such that $g_i(Q_i) = Q_{i+1}$, $1 \le i \le n-1$ (in particular, Q_i are mutually isomorphic). Let us set $\sigma_{ij} = g_i g_{i+1} \dots g_{j-1}$ at $j > i$ and $\sigma_{ij} = \sigma_{ji}^{-1}$ at $j < i$, $\sigma_{ii} = 1$. Let us identify addends Q_i with respect to the system of isomorphisms σ_{ij} (i.e., elements $q_i \in Q_i$ and $\sigma_{ij}(q_i) \in Q_j$ will be considered to be the same). In this case the transvection $(1, \dots, \sigma_{ij}, 1, \dots, 1, \sigma_{ji}, \dots, 1)$ permuting the the i-th and j-th components locally belongs to G and, hence, G contains the whole group of permutations S_n.

Let H_1 be a subgroup of all automorphisms from G acting trivially on the first component Q_1. To every automorphism $h_1 \in H_1$ let us put in correspondence an automorphism h, the action of which on Q_1 coincides with that of h_1, while on the other components it is identical. This automorphism belongs to G since the group is closed. Let H be a group of all such automorphisms. Components Q_i are acted upon by conjugated groups $\sigma_{1i}^{-1} H \sigma_{1i}$; and, due to the closure, the group G contains a direct product $H^n = \prod_{1 \le i \le n} \sigma_{1i}^{-1} H \sigma_{1i}$. Since we have identified Q_i with respect to the system of isomorphisms σ_{ij}, then the groups $\sigma_{1i}^{-1} H \sigma_{1i}$ are also identified. The groups H^n and S_n generate a semidirect product $H^n \rtimes S_n$.

Let now g be an arbitrary automorphism from G. This automorphism in some way permutes components of Q, i.e., there is a permutation π, such that $Q_i^g = Q_{\pi(i)}$. The automorphism $g\pi^{-1}$ acts invariantly on all the addends Q_i. Let g_i be an automorphism the action of which on Q_i coincides with that of $g\pi^{-1}$, while on the other components they are identical. Then $g_i \in \sigma_{1i}^{-1} H \sigma_{1i}$ and $g = g_1 \dots g_n \pi \in H^n \rtimes S_n$. The lemma is proved.

5.4.10. Note. It is obvious that in the preceding lemma the fixed ring Q^G is isomorphic to Q_1^H, where the isomorphism carries out a diagonal mapping $q \to q^{\sigma_{11}} + \dots + q^{\sigma_{1n}}$.

5.4.11. Let us now consider the structure of an arbitrary closed group of automorphisms of a ring R which has a finite prime dimension. Let $Q(R) = Q_1 \oplus \ldots \oplus Q_n$ and G be a closed group. Then every element $g \in G$ rearranges, in this or that way, components Q_i, i.e., G acts on a set of indices $\{1,\ldots, n\}$. This set falls into orbits $I_1 \cup I_2 \cup \ldots \cup I_k$. Let $\bar{Q}_\alpha = \sum_{i \in I_\alpha} Q_i$, $1 \le \alpha \le k$. Then we can apply lemma **5.4.9** to the ring $\bar{Q}_\alpha$ and to the group G, i.e., the restriction of G on $\bar{Q}_\alpha$ has the form $H_\alpha^{n_\alpha} \rtimes S_n$, where n_α is a number of elements of the orbit I_α. Now, since G is closed, we immediately get

$$G = (H_1^{n_1} \rtimes S_{n_1}) \times (H_2^{n_2} \rtimes S_{n_2}) \times \ldots \times (H_k^{n_k} \rtimes S_{n_k}).$$

Assuming, for simplicity, $\alpha \in I_\alpha$ at $1 \le \alpha \le k$, we also get

$$Q^G \cong Q_1^{H_1} \oplus Q_2^{H_2} \oplus \ldots \oplus Q_k^{H_k}.$$

5.4.12. Noether groups (N-groups). This notion remains unchanged: a group $G \subseteq A(R)$ is called an N-group provided every invertible element of its *algebra* $B(G) = \sum_{g \in G} \Phi_g$ determines an automorphism lying in G.

It is easy to see now that any Galois group is a closed N-group.

5.4.13. Maschke groups (M-groups). The definition is preserved: a reduced-finite group G is called an *M-group* provided its algebra is semiprime.

Let B be an arbitrary non-semiprime algebra with a unit over a regular self-injective commutative ring C. Let us assume that the module ${}_C B$ is projective and generated by a finite number of invertible elements. It is obvious that the algebra of any reduced-finite group of automorphisms obeys these conditions (non-semiprimeness, possibly, excluded).

5.4.14. Proposition. *There exists a semiprime ring R with a generalized centroid C and a reduced-finite group of automorphisms G, the algebra of which is isomorphic to B, while a fixed ring R^G is not semiprime.*

Proof. Since the module ${}_C B$ contains a free submodule, then it is a projective generating module in the category of modules over C. It means that the rings C and $R = End_C B$ are Morita-equivalent ([36], theorem **4.2.9**). In particular, R is a regular and self-injective ring. Therefore, it coincides with its complete left ring of quotients ([91], remark on p.152) and, moreover, $R = R_F$. The center of the ring R is isomorphic to C ([36], corollary **4.36**), so that C is also a generalized centroid of R.

Let us define the embedding of B into R using the operators of right multiplication $r_b(x) = xb$. For every invertible element $b \in B$ let us determine the inner automorphism of the ring R corresponding to the element r_b. Let G be a group of such automorphisms. As B is generated by invertible elements, then $\mathcal{B}(G) = r_B \cong B$, in which case the fixed ring of this group coincides with the centralizer of r_B in the ring R. Since B has a unit, then this centralizer coincides with the ring of left multiplications l_B. The latter ring is antiisomorphic to B and, hence, it cannot be semiprime. The proposition is proved.

This proposition shows that if, in the process of going over to fixed rings, we wish to remain within the class of semiprime rings with confidence, then we should consider the groups the algebras of which are semiprime. The condition of semiprimeness of an algebra does not greatly restrict the class of the groups considered. If, for instance, in R we have no nonzero nilpotent elements, then the ring of quotients Q does not have them either, i.e., the algebra of any reduced-finite group is semiprime.

Another important class of M-groups gives an analog of the Maschke theorem.

5.4.15. Theorem. *If G is a finite group of automorphisms of a semiprime ring R having no $|G|$-torsion, then the algebra of the group G is semiprime.*

Proof. Let us assume that in $\mathcal{B}(G)$ there exists a nonzero element $b = \sum_G \varphi_g c_g$, such that $b\mathcal{B}(G)b = 0$. Let us consider a finite Boolean algebra generated by idempotents $e(b)$, $e(\varphi_g c_g)$. If e is a minimal idempotent of this algebra, then for eb we have a presentation $\sum e\varphi_g c_g$, where g runs only through the automorphisms for which $e(\varphi_g c_g) \geq e$ and, moreover, $e(\varphi_g) \geq e$, i.e., $g \in G_e$. This implies that $eb \in \mathcal{B}(G_e)$ and the algebra of the group G_e is not semiprime.

By proposition **5.4.14** we can find a semiprime ring R_1 with a generalized centroid eC and a group of inner automorphisms G_1, such that

$\mathcal{B}(G_1) = e\mathcal{B}(G_e)$, in which case $R_1^{G_1}$ is not semiprime. This fact contradicts corollary **1.3.7** from the Bergman-Isaacs theorem. Indeed, G_e is a subgroup of the group G and, hence, eC and, likewise, R_1, have no additive $|G_e|$-torsion and, besides, conjugations on the elements $e\varphi_g$ determine the G_e action on R_1, in which case $R_1^{G_e} = R_1^{G_1}$. The theorem is proved.

5.4.16. ***Regular groups.*** If an M-group H is given, then we can extend it to an N-group by adding all the inner automorphisms corresponding to invertible elements from $\mathcal{B}(H)$. The obtained group will have the same algebra $\mathcal{B}(G) = \mathcal{B}(H)$ and will, therefore, be reduced-finite, the fixed ring $Q^G = Q^H$ remaining unchanged. We can now extend G to a closed group $\hat{G}$ by adding all the automorphisms locally belonging to G. In this case we also have $Q^{\hat{G}} = Q^H$ and $\mathcal{B}(\hat{G}) = \mathcal{B}(H)$. The latter equality needs proof and suggests the reader the idea of doing this easy task himself. Below **(5.5.6)** we shall see that the group $\hat{G}$ can be not reduced-finite (see corollary **5.5.6**).

5.4.17. Definition. An N-group G will be called ***regular*** if it is a closure of an M-group.

Below **(5.5.9)** we shall see that any closed N-subgroup of a regular group is a closure of a certain reduced-finite group with the same algebra.

5.5 Stalks of an Invariant Sheaf for a Regular Group. Homogenous Idempotents

Let R be a semiprime ring, G be a reduced-finite group of its automorphisms. The elements of the group G will be viewed as unary operations. In this case the invariant sheaf Γ is a correct stalk **(1.11.4)**. In this paragraph we shall see the way the stalks of an invariant sheaf are arranged in nearly all the orbits of the spectrum.

5.5.1. Definition. A nonzero central idempotent e is called *homogenous* if $G_e = G_f$ for any nonzero $f \le e$ and every orbit $\overline{p}$ has not more than one point from $U(e)$. Here $G_e = \{g \in G \mid i(g) \ge e\}$ (see **5.4.2**).

5.5.2. Lemma. *A set of all homogenous idempotents is dense in E.*

For **proof** it suffices to establish that any idempotent $f \in E$, such that $|G : G_f| < \infty$ (see **5.4.3**) has a homogenous subidempotent $e \le f$. We shall start from the following remark.

5.5.3. Remark. *If a point* $p \in U(f)$ *and* $g \in G_f$, *then* $p^g = p$.

Proof. Let $a \in p$. Then, in line with the von Neumann regularity of a centroid C, we have $e(a) \in p$. Allowing for the fact that g acts identically (by conjugation) on all the idempotents less than $i(g) \ge f$, we get $p \ni e(a)f = (e(a)f)^g = e(a)^g f = e(a^g)f$. Since $f \notin p$ and the ideal p is simple, then $e(a^g)$ and, hence, a^g lie in p, i.e., $p^g = p$, which is the required proof.

Therefore, a subgroup G_f of a finite index acts trivially on all the points from $U(f)$. This implies that every orbit generated by a point from $U(f)$ has a finite number of elements (not greater than the index). Let p be one of the points the orbit of which has the greatest possible number of elements $\overline{p} = \{p^{g_1} = p, p^{g_2}, \ldots, p^{g_k}\}$.

Using the fact that the spectrum is Hausdorff (see **1.9.13**), let us find a neighborhood $U(e_1)$ of the point p contained in $U(f)$ and containing no points $p^{g_2}, \ldots, p^{g_k}$ other than p. Let us consider a Boolean algebra $\mathcal{B}$ generated by the sets $U(e_1^g) = U(e_1)^g$, $g \in G$. Let $G = \bigcup_{i=1}^{n} G_f g_i$. Then, if $g = h \cdot g_i$ and $h \in G_f$, then $U(e_1^g) = U(e_1)^g$ and, hence, the Boolean algebra $\mathcal{B}$ is finitely generated and, therefore, finite.

Let $U(e)$ be the least set of this algebra containing p. Then $U(e^g)$ is the least subset of the algebra $\mathcal{B}$ containing the point p^g. In particular, different neighborhoods $U(e)^g$ do not intersect. If $p_1, p_1^g \in U(e)$ and $p_1 \ne p_1^g$, then $U(e)^g = U(e)$ and, hence, in every neighborhood $U(e^{g_i})$, $1 \le i \le k$ we can find two points of the orbit of p_1 which contradicts the maximality of k. Thus, the neighborhood $U(e)$ has not more than one point of every orbit.

Since $G_f \supseteq G_e$ at $f \le e$ and the index $|G: G_e|$ is finite, then, reducing, if necessary, e, we get that $G_f = G_e$ at $0 \ne f \le e$. The lemma is proved.

5.5.4. Proposition. *There exists a system of mutually orthogonal homogenous idempotents* $\{e_\alpha, \alpha \in A\}$, *such that*

$$\hat{RE} = \prod_{\alpha \in A} \bar{e}_\alpha \hat{RE}, \quad Q = \prod_{\alpha \in A} \bar{e}_\alpha Q,$$

where $\bar{e}_\alpha = \sup\{e_\alpha^g, g \in G\}$ *is a system of fixed mutually orthogonal idempotents.*

Proof. Let us denote through Σ a set of all sets of mutually non-intersecting domains of type $U(\bar{e})$, where e is a homogenous idempotent, $\bar{e} = \sup\{e^g, g \in G\}$. In this case Σ is a non-empty directed set. By the Zorn Lemma, in Σ we can find a maximal element $\sigma = \{U(\bar{e}_\alpha), \alpha \in A\}$. Let us show that $\bigcup_{\alpha \in A} U(\bar{e}_\alpha)$ is a dense set in the spectrum, i.e., $\sup_\alpha \bar{e}_\alpha = 1$. If this is not the case, then $e' \cdot e_\alpha = 0$ for a certain nonzero idempotent e'. Let us find a homogenous idempotent $e < e'$ (lemma 5.5.2). Then $\bar{e} \cdot \bar{e}_\alpha = 0$, since all $\bar{e}_\alpha$ are fixed and, hence, $\sigma \cup \{U(\bar{e})\} \in \Sigma$, which contradicts the maximality of σ.

Let us consider a homomorphism $\xi\colon r \to \prod_\alpha \bar{e}_\alpha r$ from Q into the direct product. As $\sup e_\alpha = 1$, then this homomorphism is an embedding. If $r_\alpha \in \bar{e}_\alpha Q$, then, by the definition, of a sheaf there is a global section $r \in Q$, such that $r\bar{e}_\alpha = r_\alpha$. This implies that ξ is an epimorphism. The proposition is proved.

Allowing for the fact that a closed group of automorphisms of a direct product of invariant components falls into a direct product of induced groups, let us concentrate our attention on homogenous idempotents and rings of sections over the domains determined by them.

5.5.5. Proposition. *Let* e *be a homogenous idempotent. Then there can be found automorphisms* $g_1 = 1, g_2, \ldots, g_k \in G$, *such that the ring of global sections* $\hat{RE}$ *is presented as a direct sum*

$$e\hat{RE} \oplus e^{g_2}\hat{RE} \oplus \ldots e^{g_k}\hat{RE} \oplus (1 - \bar{e})\hat{RE}, \tag{1}$$

where $\bar{e} = \sum_{i=1}^{k} e^{g_i} = \sup\{e^g, g \in G\}$. *If the components* $e^{g_i}Q$ *and* $e^{g_j}Q$ *are identified through the automorphisms* $g_i^{-1}g_j$, *then the following inclusions are valid:*

$$G \subseteq (H_1^k \rightthreetimes S_k) \times G' \subseteq \hat{G}, \tag{2}$$

where $\hat{G}$ *is a closure of the group* G*;* H_1 *is a projection of the stabilizer* $H = \{g \in G, e^g = e\}$ *on* eQ *which is identically extended onto* $(1-e)Q$*;* S_k *is a group of permutations rearranging the components* $e^{g_i}Q$*;* G' *is a group acting identically on* $\bar{e}Q$.

Proof. As $\bar{e}$ is a homogenous idempotent, then a subgroup G_e has a finite index in G, in which case $H = \{g \in G, e^g = e\} \supseteq G_e$, i.e., the stabilizer H of the idempotent e has a finite index in G. Let $G = \bigcup_{i=1}^{k} Hg_i$ be a decomposition of the group G into a union of cosets. Then $\{e^g | g \in G\} = \{e^{g_i}, 1 \le i \le k\}$. Besides, $\bar{e} = \sup_{g \in G} \{e^g\} = \sum_i e^{g_i}$ is a fixed idempotent, since, according to homogeneity, the idempotents e^{g_i} are mutually orthogonal. Now it is the decomposition of the unit, $1 = e + e^{g_2} + \ldots + e^{g_k} + (1 - \bar{e})$, that results in decomposition (1).

Let us connect to every automorphism $h \in H$ an automorphism h_1, the action of which on eQ coincides with that of h, while it is identical on the remaining addends of sum (1). Then we have $h_1 \in \hat{G}$, since $\hat{G}$ is closed. Now the group H_1 of all such automorphisms is easily identified with a projection of H onto eQ. On the addends $e^{g_i}Q$ the conjugated subgroups $g_i^{-1}H_1g_i$ are acting and, due to the closure, the group $\hat{G}$ contains a direct product $H_1 \times g_2^{-1}H_1g_2 \times \ldots \times g_k^{-1}H_1g_k$. If we identify the addends $e^{g_i}Q$ with respect to the system of isomorphisms $g_i^{-1}g_j$, then the groups $g_i^{-1}H_1g_i$ are also identified and the direct product will have the form H_1^k. Moreover, due to the closure, the group $\hat{G}$ will contain all the transvections $(1, \ldots, g_i^{-1}g_j, \ldots, g_j^{-1}g_i, \ldots, 1)$ and, hence, the group of permutations. Thus, $G \supseteq H_1^k \rightthreetimes S_k$.

So, if $g \in G$, then let us determine an automorphism g', the action of which on $(1-\bar{e})Q$ coincides with that of g^{-1}, while on $\bar{e}Q$ it is identical. Let G' be a subgroup of all such automorphisms. For every i, $1 \le i \le k$ we can find a j, $1 \le j \le k$, such that $e^{g_i g} = e^{g_j}$, i.e., g

rearranges, in some way, addends of $\bar{e}Q$ in decomposition (1). If π is a corresponding permutation component, then $(gg')\pi^{-1}$ acts invariantly on all the components $e^{g_i}Q$, i.e ., $(gg')\pi^{-1} \in H_1^k$ and, hence, $g \in (H_1^k \rightthreetimes S_k) \times G'$, which is the required proof.

5.5.6. Corollary. *If G is a closed and reduced-finite group, then the ring Q has a decomposition*

$$Q = Q_1 \oplus \prod_{\alpha} Q_\alpha^{n_\alpha}$$

into invariant components, which is corresponded to by a decomposition of the group

$$G = G_1 \times \prod (G_\alpha^{n_\alpha} \rightthreetimes S_{n_\alpha})$$

in which case the group G_1 is a group of inner automorphisms of the component Q_1, while the rings Q_α are prime.

Proof. By formula (2) and proposition **5.5.4**, it is sufficient to show that the homogenous idempotent e, for which $G_e \neq G$, is minimal. Let $g \notin G_e$ and $0 \neq f < e$. Let us denote by g_1 an automorphism coinciding with g on fQ and identical on $(1 - f)Q$; while by g_2 we shall denote an automorphism coinciding with g on $(1 - f)Q$ and identical on fQ. Then $g = g_1 g_2$, but $g_1 \in G_{e-f} = G_e = G_f \ni g_2$ is a contradiction.

This corollary shows that a closure of a reduced-finite group is, as a rule, not reduced-finite. If, for instance, the spectrum contains no isolated points, then only the groups of inner automorphisms will be reduced-finite and closed.

5.5.7. Theorem. *Let G be a reduced-finite group, e be a homogenous idempotent, $H = \{g \in G, e^g = e\}$. Then there are automorphisms $g_1 = 1, g_2, \ldots, g_k$ from the group G, such that for any point $p \in U(e)$ the following statements are valid:*

(a) *The orbit $\overline{p}$ is equal to $\{p^{g_1}, \ldots, p^{g_k}\}$.*

(b) *The stalk $\Gamma_{\overline{p}}(Q)$ is decomposed into a direct sum of mutually isomorphic stalks of a canonical sheaf*

$$\Gamma_{\overline{p}}(Q) = \Gamma_{p^{g_1}}(Q) \oplus \ldots \oplus \Gamma_{p^{g_k}}(Q). \tag{3}$$

(c) *If the stalks of* $\Gamma(Q)$ *at the points of the orbits are identified with the help of isomorphisms induced by automorphisms* $g_1,\ldots, g_k$, *then the closure of the group* $G_{\overline{p}}$ *will have the form* $H_p^k \leftthreetimes S_k$, *where* H_p *is a group induced by the group* H *on the stalk* Γ_p.

(d) *The group* H_p *is reduced-finite and its algebra equals the projections of the algebra of the group* G *on the stalk:* $\mathcal{B}(H_p) = \rho_p(\mathcal{B}(G))$.

(e) *If* G *is a Maschke group, then* H_p *is also a Maschke group.*

(f) *If* G *is a Noether group, then* H_p *is also a Noether group, in which case*

$$(\hat{G})_{\overline{p}} = H_p^k \leftthreetimes S_k = (\widehat{G_{\overline{p}}}).$$

Proof. Let $g_1,\ldots, g_k$ be automorphisms, the existence of which is claimed by proposition **5.5.5**. Then each of the neighborhoods $U(e^{g_i})$ contains exactly one point of the orbit $\overline{p}$, which proves (a).

(b) If $\bar{s}$ is an element of the stalk $\Gamma_{\overline{p}}$ determined by the section s over an invariant neighborhood W of the point p, then W will be a neighborhood of each of the points p^{g_i}, $1 \le i \le k$ and, hence, s will determine elements $\bar{s}_1,\ldots, \bar{s}_k$ of the stalks $\Gamma_{p^{g_1}},\ldots, \Gamma_{p^{g_k}}$, respectively. Let us show that the mapping ε: $\bar{s} \to \bar{s}_1 + \ldots + \bar{s}_k$ sets the isomorphism

$$\Gamma_{\overline{p}} \cong \Gamma_{p^{g_1}} \oplus \ldots \oplus \Gamma_{p^{g_k}}.$$

As the mappings $s \to \bar{s}_i$ are homomorphisms, then ε will also be a homomorphism.

Let all $\bar{s}_i$ be zero. It implies that there are neighborhoods $V_1,\ldots, V_k$ of the points $p, p^{g_2},\ldots, p^{g_k}$, such that the restrictions of s on each of them are zero. One can also assume that $V_1 \subseteq U(e)$. Let us set $V_0 = \bigcap_{i=1}^{k} V_i^{g_i^{-1}}$.

Then, using structure (2) of the group G, we get $W_1 = \bigcup_{g \in G} V_0^g =$

$= \bigcup_{i=1}^{k} V_0^{g_i} \subseteq \bigcup_{i=1}^{k} V_i$. As W_1 is an invariant neighborhood of the point p and the restriction of s on this neighborhood is zero, then $\bar{s} = 0$ in the stalk of an invariant sheaf.

Let now $\bar{s}_1,\dots, \bar{s}_k$ be arbitrary elements of stalks at the points of the orbit determined by sections $s_1,\dots s_k$ over the neighborhoods $V_1,\dots, V_k$ of the points of the orbit. Now we can assume that $V_i \subseteq U(e^{g_i})$, i.e., $s_i \in e^{g_i}Q$. In this case a section $s = s_1 + \dots + s_k$ over $U(\bar{e})$ determines the element $\bar{s}$ of the stalk $\Gamma_{\bar{p}}$, such that $\varepsilon(\bar{s}) = \bar{s}_1 + \dots + \bar{s}_k$.

(c) Now the isomorphisms $g_i \colon eQ \to e^{g_i}Q$ induce isomorphisms $\bar{g}_i \colon \Gamma_p \to \Gamma_p{}^{g_i}$. If we identify the stalks of the points of the orbit $\bar{p}$ with respect to the system of these automorphisms, then the groups S_k acting on $\bar{e}Q$ will be induced on the stalk $\Gamma_{\bar{p}}$. Besides, the group H is induced on the stalk Γ_p (and, by identification, $g_i^{-1} H g_i$ is induced on the stalk $\Gamma_p g_i$, since any neighborhood of the point p goes over, under the action of $h \in H$ to that of the point p). As the group $G_{\bar{p}}$ acts transitively on the components of decomposition (3), then, by **5.4.9**, its closure has the form $H_0^k \leftthreetimes S_k$, where H_0 is a group induced by a stabilizer $\{g \in G \mid p^g = p\}$, which, since the idempotent e is homogenous, is equal to H.

(d) Let us show that $\mathcal{B}(H_p) = \rho_p(\mathcal{B}(G))$. Let $g \in G$ and $\varphi \in \Phi_g$. If $\rho_p(\varphi) \neq 0$, then (lemma **1.9.18**), $e(\varphi) \notin p$, i.e., $p \in U(i(g))$. However, $g \in G_{i(g)}$, and, hence, g acts identically on all points of $U(i(G))$. In particular, $p^g = p$, i.e., $g \in H$. Let us apply a homomorphism ρ_p to the identity $x\varphi = \varphi x^g$. In this case we get $\bar{x}\bar{\varphi} = \bar{\varphi}\bar{x}^{\bar{g}}$, where $\bar{x}$ is any element of the stalk and g is an automorphism from H_p induced by g. Therefore, $\bar{\varphi} \in \Phi_{\bar{g}}$.

In order to prove the inverse inclusion, let us consider a ring R together with the action with the group H and an H-invariant sheaf Γ'. It should be remarked that the point p coincides with its orbit and the stalk of the sheaf Γ' on the orbit $\{p\}$ is equal to the stalk Γ_p (since any neighborhood V of the point p contains an H-invariant neighborhood $U(e) \cap V$).

Let $\varphi\bar{x}^h = \bar{x}\varphi$ for all $\bar{x} \in \Gamma_p$ and for a nonzero $\varphi \in Q(\Gamma_p)$. We can find elements $\bar{a}, \bar{b}$ of the stalk Γ_p, such that $0 \neq \bar{a}\varphi = \bar{d} \in \Gamma_p$, $0 \neq \varphi\bar{b} = \bar{v} \in \Gamma_p$ and come to the conclusion that the predicate

$$\forall\, x \quad \bar{d}x^h\bar{b} = \bar{a}x\bar{v}$$

is true on the stalk Γ_p. This is a strictly sheaf predicate for the sheaf Γ', i.e., there are preimages, such that $dx^h b = axv$ is an identity on Q, in which case $e(d), e(b) \notin p$. Let us transform this identity to a reduced form $d(1 - i(h))x^h b + d\varphi'_h x\varphi_h b = axb$. By theorem **2.2.2**, we get $d(1 - i(h)) \otimes b = 0$, i.e., $e(d) \cdot (1 - e(h)) \cdot e(b) = 0$, wherefrom we get $1 - e(h) \in p$ and, hence, $e(h) \notin \ker \rho_p$, i.e., $\rho_p(\varphi_h) \neq 0$. Since a conjugation by $\rho_p(\varphi_h)$ determines the same automorphism, h, on the stalk that a conjugation by φ does, then $\varphi = \bar{c}\bar{\varphi}_h$, , where $\bar{c}$ is an element of a generalized centroid of the stalk (equal to $\rho_p(C)$ (see **1.1.26**)). Thus, $\varphi \in \rho_p(B(G))$.

Let us show that H_p has a finite-reduced order. As $B(G)$ is a finitely-generated module over C, then $B(H_p) = \rho_p(B(G))$ is finite-dimensional over the field C_p. Then, all the automorphisms from G_e act by conjugations in the ring eQ and $G_e \subseteq H$, since $p \in U(e)$, and hence, the induced automorphisms $\rho_p(G_e)$ will be inner for the stalk: $\rho_p(G_e) \subseteq (H_p)_{in}$. Hence, $|H_p:(H_p)_{in}| \leq |H: G_e| < \infty$.

(e) Let us check if $B(H_p)$ is a semiprime algebra. Let us first remark that $B(G)$ is an injective C-module (since it is finitely-generated) and, hence, $B(G)$ is a closed set in the topology determined by idempotents. Therefore, a unary predicate, such that $P(s) = T \leftrightarrow s \in B(G)$ will be strictly sheaf, in which case on a stalk we shall have $P(\bar{s}) = T \leftrightarrow \bar{s} \in \rho_p(B(G)) = B(H_p)$. If now $\bar{s}B(H_p)\bar{s} = 0$, then the following sheaf predicate is true on the stalk:

$$\forall\, x \;\; P(x) \rightarrow \bar{s}x\bar{s} = 0.$$

Consequently, there is an idempotent $e_1 \notin p$ and a preimage $s \in e_1 \mathcal{B}(G)$, such that $s\mathcal{B}(G)s = 0$ and, hence, $s = 0$.

(f) Let $\bar{\varphi} \in H_p$ and $\bar{\varphi} \cdot \bar{\varphi}' = 1$. We can find an idempotent $e_1 \in p$ and elements $\varphi, \varphi' \in \mathcal{B}(G)$, such that $\varphi\varphi' = \varphi' \varphi = e_1$, $\rho_p(\varphi) = \bar{\varphi}$. Then $(1 - e_1 + \varphi)(1 - e_1 + \varphi') = = (1 - e_1 + \varphi')(1 - e_1 + \varphi) = 1$, i.e., an inner automorphism $\tilde{b}$, where $b = 1 - e_1 + \varphi$ belongs to H. Obviously, $\rho_p(\tilde{b})$ is an automorphism from H_p coinciding with the conjugation on $\bar{\varphi}$.

Let $v \in \hat{G}$. By statement (c), it suffices to show a projection $v_{\bar{p}}$ to belong to the closure of the group $G_{\bar{p}}$, i.e., to show that its action on every addend of (3) is induced by a certain automorphism from G. Let us, for instance, consider the first addend.

For every $g \in G$ let us denote by E_g a set of all central idempotents $\varepsilon \le e$, such that $v \equiv g(\text{mod}\, G_e)$ on εQ. Since G is a Noether group, and $\mathcal{B}(G_e) = e\mathcal{B}(G)$ is an orthogonally complete set, then E_g is also an orthogonally complete set. This implies that in E_g there is the biggest idempotent $\varepsilon(g)$. Let $G = \bigcup_{i=1}^{n} G_e g_i$. Since v locally belongs to the group G, then $\sup\{\varepsilon(g_i)\} = e$. Therefore, the point p belongs to one of the neighborhoods $U(\varepsilon(g_i))$. As in this neighborhood $v = \tilde{b}g_i \in G$, then $v_{\bar{p}} = (\tilde{b}g_i)_{\bar{p}}$ on the first addend of decomposition (3). The theorem is completely proved.

5.5.8. Corollary. *Let G be a reduced-finite Noether group, and e be a homogenous idempotent. If v locally belongs to G, then for a fixed idempotent $\bar{e}$ there is a decomposition into an orthogonal finite sum $\bar{e} = e_1 + ... + e_m$ in such a way that on $e_i Q$ the action of v coincides with that of a certain $g_i \in G$, $1 \le i < m$.*

Proof. It is the fact that we have established when proving point (f) of the preceding theorem.

5.5.9. Proposition. *Let G be a reduced-finite N-group. Then any closed N-subgroup F of the group $\hat{G}$ is a closure of a reduced-finite subgroup $H \subseteq F$, such that $\mathcal{B}(H) = \mathcal{B}(F)$.*

Proof. By proposition **5.5.4**, we have a decomposition $Q = \prod_{\alpha} \bar{e}_\alpha Q$.

Since F is a closed group, then the decomposition $F = \prod_{\alpha} F_\alpha$ corresponds to it.. If in each cofactor F_α we can find a dense reduced-finite subgroup H_α, such that $\mathcal{B}(H_\alpha) = \mathcal{B}(F_\alpha)$, then $H = \prod_{\alpha} H_\alpha$ will be the sought one. Therefore, with no violation of generality, one can assume that we have $1 = e^{g_1} + ... + e^{g_k}$ for a certain homogenous idempotent e and $g_i \in G$.

It is sufficient for us to find a reduced-finite group $H \subseteq F$ containing $\tilde{\mathcal{B}}(F) = F_{in}$, such that $H_{\overline{p}} = F_{\overline{p}}$ for a dense set of points $\overline{p}$ of the space of orbits.

Indeed, in this case for every $v \in F$ there is an $h \in H$, such that on the stalk $\Gamma_{\overline{p}}$ the formula $\forall\, x \;\; x^v = x^h$ is valid. Viewing v and h as strictly sheaf operations for a G-invariant sheaf (see corollary **5.5.8**) and applying the metatheorem, we see that v coincides with h in a certain neighborhood of the point $\overline{p}$. As $\overline{p}$ is arbitrary from a dense set, then $v \in \hat{H}$.

Let us fix a nonzero idempotent $e_1 \le e$ and show that there exists a nonempty neighborhood $\overline{U(f)}$, $f \le e_1$ and a reduced-finite group $H(f) \subseteq F$, such that $H(f)_{\overline{p}} = F_{\overline{p}}$ for all $\overline{p} \in \overline{U(f)}$. Let us choose a point $\overline{p}$ in such a way that the index $|F_{\overline{p}} : (F_{\overline{p}})_{in}|$ had the biggest value (by theorem **5.5.7**, point (f), it is not greater than $k! \cdot |G : G_e|$). Let $\bar{\varphi}_1, ..., \bar{\varphi}_n$ be representatives of cosets of $(F_{\overline{p}})_{in}$ in $F_{\overline{p}}$. Let us fix preimages of these automorphisms, $\varphi_1, ..., \varphi_n \in F$ and consider them to be strictly sheaf operations for a G-invariant sheaf (see corollary **5.5.8**). On the stalk the following Horn formulas hold:

$$\exists\, b_{ij} \forall\, x \quad b_{ij} (x^{\varphi_i})^{\varphi_j} = x^{\varphi_m} b_{ij} \,\&\, e(b_{ij}) = 1$$

$$\forall\, b [(\forall\, x \qquad b x^{\varphi_i} = x^{\varphi_j} b) \longrightarrow b = 0],$$

where in the former series of formulas i, j run through all the values from 1 to n, while in the latter one $i \neq j$. By the metatheorem, these formulas set sheaf predicates, i.e., by definition **1.11.11**, one can find a neighborhood $\overline{U(f)}$ of the point $\overline{p}$, over which these formulas are valid. As F is a closed group, then, violating no generality, one can assume that outside $\overline{U(f)}$ the operations $\varphi_1, ..., \varphi_n$ act identically. Let $H(f)$ be a group generated by $\varphi_1,\dots,\varphi_n$ and by inner automorphisms corresponding to the elements from $\bar{f}B(F)$ (extended onto $(1-\bar{f})Q$ by an identical action). In this case the validity of the above mentioned formulas indicates that $|H(f):H(f)_{in}|=n$, i.e., $H(f)$ is reduced-finite. By corollary **1.11.18**, these formulas are also true on all the stalks $\Gamma_{\bar{q}}$ under $\bar{q}\in U(f)$, i.e., the index $|H(f)_{\bar{q}}:(F_{\bar{q}})_{in}|=n$ has the biggest possible value, which fact implies that $H(f)_{\bar{q}}=F_{\bar{q}}$.

Now, as was the case in proposition **5.5.4**, from a dense set of fixed idempotents $\{\bar{f}\}$ one can choose a dense class $\{\bar{f}_\alpha\}$ of mutually orthogonal idempotents. In this case $Q=\prod\limits_\alpha \bar{f}_\alpha Q$ and the group $H=\prod\limits_\alpha H(f_\alpha)$ is the sought one. The proposition is proved.

By way of concluding this paragraph, let us consider Maschke groups for rings of a finite prime dimension. In this case the set E of central idempotents of Q is finite and, hence, the topology determined by them is discrete. By lemma **3.6.6**, the ring Q is decomposable into a direct sum of ideals which are prime rings

$$Q=Q_1\oplus\ldots\oplus Q_m$$

where $Q_i=Qe_i$, e_i are minimal idempotents. The neighborhood $U(e_i)$ contains only a point $p_i=(1-e_i)C$ and the stalk $\Gamma_{\mathbf{p}_i}(Q)$ is equal to Q_i and, hence, $\Gamma_{\mathbf{p}}(R)=Re_i$. Each of the idempotents e_i is homogenous and the ring Q can be presented as a direct sum of stalks of the invariant sheaf

$$Q=\bar{e}_1Q\oplus\bar{e}_{k_2}Q\oplus\ldots\oplus\bar{e}_{k_t}Q. \tag{4}$$

Let us consider these stalks in detail. By theorem **5.5.7**, we have a decomposition of, for instance, the first addend

$$\bar{e}_1Q=e_1Q\oplus e_1^{g_2}Q\oplus\ldots\oplus e_1^{g_n}Q,$$

in which case on the prime ring e_1Q an M-group H_1 acts, so that the closure of a restriction G on $\bar{e}Q$ has the form $H_1^n\leftthreetimes S_n$. Now the mappings

$$\begin{aligned} &a\longrightarrow e_1a, && a\in\bar{e}_1Q\\ &\alpha\longrightarrow\alpha+\alpha^{g_2}+\ldots+\alpha^{g_n}, && \alpha\in e_1Q \end{aligned} \tag{5}$$

set an isomorphism

$$\bar{e}_1 Q^G \cong Q_1^{H_1}.$$

Returning to the initial ring, we get the following result.

5.5.10. Theorem. *Let a ring R have a finite prime dimension, G be an arbitrary M-group of its automorphisms. Then a certain subring S containing an essential ideal K of R is decomposed into a direct sum of prime rings*

$$S = S_1 \oplus \ldots \oplus S_m$$

on each of which a Maschke group H_i acts, in which case

$$S^G \cong S_{k_1}^{H_1} \oplus \ldots \oplus S_{k_t}^{H_t},$$

where all components $S_1, \ldots, S_n$ occur (to the accuracy of an isomorphism) among the components $S_{k_1}, \ldots, S_{k_t}$, while the number t is equal to the invariant prime dimension of the ring R.

Proof. Let I be an essential ideal of R, such that $IE \subseteq R$. Let us choose an essential ideal J in such a way that $J^{g_i} \subseteq I$ for all automorphisms $g_i \in G$ participating in the construction of all isomorphisms (5).

Let

$$S = Je_1 + (Je_1)^{g_2} + \ldots + Je_{k_2} + (Je_{k_2})^{g_2'} + \ldots$$

where the idempotents $e_1, e_{k_2}, \ldots$ are determined by decomposition (4), while the automorphisms $g_2, \ldots, g_2', \ldots$ carry out the identification of stalks in theorem **5.5.7** (point (c)), thus participating in the definition of isomorphisms (5). We come to the conclusion that $S \subseteq R$ and mappings (5) set the isomorphism

$$S^G \cong (Je_1)^{H_1} \oplus \ldots \oplus (Je_{k_t})^{H_t}.$$

The task now is to prove if S contains an essential ideal. By the definition of the group $A(R)$ and using the equality $A(R) = A(J)$, one can find an essential ideal K, such that $K \subseteq J^{g_i}$ for all g_i. Then we have

$$S \supseteq Ke_1 + Ke_1^{g_2} + \ldots + Ke_{k_2} + Ke_{k_2}^{g_2'} + \ldots = K.$$

The theorem is proved.

5.6 Principal Trace Forms

In this paragraph we shall denote through e a homogenous idempotent, while for groups of automorphisms, idempotents, etc. we preserve the notations from proposition **5.5.5**. Using the metatheorem, we shall construct the "principal" trace forms $\tau_e(x)$ and $\bar{\tau}_e(x)$ for the groups H and G, respectively. For this purpose it is sufficient to formulate our problems as Horn formulas, the validity of which for prime rings has been established.

Let us start enumerating strictly sheaf and sheaf predicates and functions which will be used (not only in the present paragraph).

5.6.1. Let us put the unary operations of projections π_C, π_B, π_g into correspondence with each of the sets C, $B(G)$, Φ_g.

In more general terms, let S be a closed E-linear set in Q. Then, considering S as a module over E (see **3.6.19**), we see that it is an injective module. Consequently, it is singled out as a direct addend. In particular, the E-linear projection $\pi_S\colon Q \to S$ is determined. This projection is a strictly sheaf operation. But, it goes without saying, that the projection π_S is not uniquely determined by the set S. In particular, if S is a C-submodule, then π_S can be chosen as a homomorphism of C-modules. For us of major importance is the very existence of this projection, but it is going to be used for eliminating excessive implications: for instance, $\forall\, b\, P_S(b) \to \psi(b)$ can be replaced with $\forall\, x \psi(\pi_S(x))$.

5.6.2. The unary operation of support $e(x)$, as we know, is strictly sheaf (see **1.1.10**).

5.6.3. Let us fix a system of coset representatives $h_1 = 1, h_2, \ldots, h_n$ for a subgroup H_{in} of inner for eQ automorphisms in H (see **5.5.5**). This system is finite, since $H_{in} \supseteq G_e$. These automorphisms can be viewed as unary strictly sheaf operations for H-invariant and G-invariant sheaves. In this case one has to remark that the restriction of an H-invariant sheaf on a closed domain $U(e)$ does not differ from that of a canonical sheaf, as the automorphisms from H do not replace the points of $U(e)$ and do not change the idempotents less than e .

5.6.4.. Let S_k be a subgroup determined by formula (2). Permutations from S_k (as well as all automorphisms from $\hat{G}$) can be viewed as strictly sheaf operations for a G-invariant sheaf, transvections $(1i)$ being, in particular, such operations.

5.6.5. Let us get down to constructing principal trace forms. Let us seek the form τ_e as a sum

$$\tau_e(x) = \sum_{j=1}^{n} \sum_{i=1}^{m} a_{ij} x^{h_j} b_{ij} \tag{6}$$

Let us write down the required properties of the form as formulas.

(1) *The invariance:*

$$I \Leftrightarrow \forall x \forall y \mathop{\&}_{i=1}^{n} \tau_e(x)\pi_B(y) =$$

$$= \pi_B(y)\tau_e(x) \mathop{\&} \mathop{\&}_{j=1}^{m} \tau_e(x)^{h_j} = \tau_e(x),$$

where, for simplicity, instead of the right part of (6), where a_{ij} and b_{ij} are replaced with $\pi_B(a_{ij})$ and $\pi_B(b_{ij})$, respectively, we have written the left one.

(2) *The non-degeneracy:*

$$\Gamma \Leftrightarrow \forall r \exists x \quad e(\tau_e(rx)) = e(r) \cdot e$$

$$\Gamma' \Leftrightarrow \forall r \exists x \quad e(\tau_e(xr)) = e(r) \cdot e.$$

Now the formula $\exists a_{ij} \exists b_{ij} \; I \& \Gamma \& \Gamma'$ expresses all the properties required. The validity of this formula for prime rings has been established (see lemma **3.4.3** and remark **3.4.5**) and, by the metatheorem, we can assume the existence of the required form τ_e to have been proved, provided

G is a Maschke group.

The form τ_e can now be considered as a strictly sheaf operation for a canonical sheaf over the domain $U(e)$. In this case its values τ_p at the points $p \in U(e)$ will be principal trace forms on stalks Γ_p.

Using decomposition (1) and identification, one can present the form $\bar{\tau}_e$ as a sum of copies $\tau_e \oplus \ldots \oplus \tau_e$ Or, in more exact terms,

$$\bar{\tau}_e = \tau_e + \tau_e^{g_2} + \tau_e^{g_3} + \ldots + \tau_e^{g_k}.$$

Therefore, $\bar{\tau}_e$ can be also viewed as a strictly sheaf operation for a G-invariant stalk.

As an illustration of this strictly stalk operation, let us prove the following useful statement.

5.6.6. Lemma. *Let G be a Maschke group of automorphisms of a semiprime ring R. If I is an essential ideal of the ring R, then the left (right) annihilator of I^G in R_F is zero and, in particular, I^G is an essential ideal of the ring R^G.*

Proof. Let $a \cdot I^G = 0$. Then for the principal trace form $\bar{\tau}_e$ one can find an essential ideal J, such that $\bar{\tau}_e(J) \subseteq I$ and, hence, $a\bar{\tau}_e(x) = 0$ at $x \in J$. By corollary **2.3.2**, this identity is fulfilled on Q as well. Let us project this identity on the stalk Γ_p, $p \in U(e)$, thus getting an identity $\bar{a}\tau_H(x) = 0$ in the prime ring Γ_p. By theorem **2.2.2**, we have $\bar{a}B(H_p) = 0$ and, hence, $\bar{a} = 0$. As the point p has been chosen arbitrarily, then $ea = 0$, and we are now to make use of the fact that a set of homogenous idempotents is dense. The lemma is proved.

5.6.7. Theorem. *If a semiprime ring R has an infinite prime dimension and G is an M-group of its automorphisms, then R^G also has an infinite prime dimension.*

Proof. Let us assume that R^G has a finite prime dimension. If e is a homogenous idempotent and $e_1, \ldots, e_n, \ldots$ are pairwise orthogonal nonzero idempotents less than e, then the fixed idempotents $\bar{e}_1, \bar{e}_2, \ldots, \bar{e}_n, \ldots$ are also pairwise orthogonal. Let us consider the principal trace forms $\bar{\tau}_{e_1}, \ldots, \bar{\tau}_{e_n}, \ldots$. For each of them let us find an ideal $I_n \in F$, such that $\bar{\tau}_{e_n}(I_n) \subseteq R$. Then we have an infinite direct sum of ideals in R^G

$$\bar{\tau}_{e_1}(I_1) + \ldots + \bar{\tau}_{e_1}(I_n) + \ldots$$

All the ideal in this chain are nonzero: if, for instance, $\tau_{e_1}(I_1)=0$, then, by corollary **2.3.2**, we have $\bar{\tau}_{e_1}(Q)=0$ and, projecting this equality on any stalk Γ_p, $p \in U(e_1)$, we get a contradiction.

Therefore, the idempotent e has only a finite number of less idempotents.Analogously, in Q there are no infinite sets of pairwise orthogonal idempotents $\bar{e}_\alpha$, where e_α is homogenous, and we are to use now proposition **5.5.4**. The theorem is proved.

Let us formulate the principal property of trace forms in a somewhat different way.

5.6.8. Proposition. *Let* L *be a nonzero one-sided ideal of a ring* R. *Then if* $\bar{\tau}_e(L)=0$, *then* $eL=0$.

Proof. Let L be a left ideal, $v \in L$. Then, by corollary **2.3.2**, the identity $e \cdot \bar{\tau}_e(xv)=0$ holds on the ring R_F. The formula Γ which defines the form τ_e shows that $e(v)e=0$, i.e., $ve=0$ and $Le=0$, which is the required proof.

By way of concluding this paragraph, let us consider the problem of uniqueness of principal trace forms.

5.6.9. Theorem. *The principal trace form* $\bar{\tau}_e$ *is uniquely determined to the accuracy of the replacing variable* $x_1=bx$, b *being an invertible element from* $e\mathcal{B}(G)$.

Proof. By lemma **3.4.3** and remark **3.4.5**, the projection $(\tau_e)_p$ will be a principal trace form on the stalk. By theorem **3.4.9**, this projection is uniquely determined to the accuracy of the substitution $x_1=bx$. If now $\tau_e^{(1)}$ is another trace form, then, viewing it as a strictly sheaf operation, we see the validity of the following formula on the stalk

$$\exists\, b_1,\, b\ \forall\, x \quad \tau_e^{(1)}(x)=\tau_e(bx)\,\&\, b_1b= \\ = bb_1=1\ \&\ P_{\mathcal{B}}(b)\&\ P_{\mathcal{B}}(b_1).$$

By the metatheorem, this formula is true on $e\mathcal{B}$ as well, i.e., $\tau_e^{(1)}(x)=\tau_e(bx)$ and, hence, $\bar{\tau}_e^{(1)}(x)=\bar{\tau}_e(bx)$, which is the required proof.

5.7 Galois Groups

5.7.1. Theorem. *Let G be an M-group of automorphisms of a semiprime ring R. Then the centralizer of a fixed subring R^G in a ring of quotients R_F is equal to the algebra $\mathcal{B}(G)$ of the group G.*

Proof. Let us fix a homogenous idempotent e and let a be an element which commutes with fixed elements of the ring R. It suffices to show that $\rho_p(a) \in \mathcal{B}(H_p)$ for all points $p \in U(e)$, since in this case the sheaf predicate $x = \pi_B(x)$ is true at $x = \rho_p(a)$ on the stalks at nearly all points of the spectrum and, by the metatheorem, $a = \pi_B(a)$, i.e., $a \in \mathcal{B}(G)$.

Let I be an essential ideal of the ring R, such that $\tau_e(I) \subseteq R$. Then, making allowances for decomposition (1), we get $a\tau_e(x) - \tau_e(x)a = 0$ for all $x \in I$. By corollary **2.3.2**, this identity is also valid on R_F. Going over to a stalk, we find on it an identity $\bar{a}\tau_p(x) = \tau_p(x)\bar{a}$. As, by the construction, the form τ_p is principal, then $\tau_p(\Gamma_p)$ is an almost intermediate ring of the stalk, i.e., by theorem **3.5.1**, $\bar{a} \in \mathcal{B}(H_p)$, which is the required proof.

5.7.2. Theorem. *An automorphism h belongs to the Galois closure of an MN-group G iff h locally belongs to G, i.e., $A(R^G) = \hat{G}$.*

Proof. Let $h \in A(R^G)$. Let us first show that h affects identically the G-fixed idempotents (which may not belong to R). Let, on the contrary, $f^h \neq f$ for such an idempotent. Let us choose an essential ideal I in such a way that $fI \subseteq R$. Then $fI^G \subseteq R^G$ and, hence, $(f - f^h)I^G = 0$, i.e., lemma **5.6.6** yields the equality $f = f^h$.

Now we can consider h as a strictly sheaf unary operation for a G-invariant stalk. It is sufficient to show that for every point $p \in U(e)$, where e is a homogenous idempotent, there can be found an automorphism $g \in (\hat{G})_{\bar{p}}$, the action of which on the stalk $\Gamma_{\bar{p}}$ coincides with that of h on this stalk. Indeed, in this case the formula $\forall x \; x^h = x^g$ which is true on the stalk by the metatheorem, will also be true in a certain invariant neighborhood of the point p, i.e., h will locally belong to the group G.

It should be remarked that the idempotents $\rho_{\bar{p}}(e^{g_1})$ in decomposition (3) are minimal central idempotents of the stalk. Therefore, the automorphism h rearranges them in some way, i.e., one can find a permutation $\delta \in S_k$

(see **5.5.7**), such that $h\delta$ affects all the components of decomposition (3) of the stalk $\Gamma_{\bar{p}}$ in a fixed way. Now we are to show that the restrictions of $h\delta$ on this components lie in H_p.

We have $\bar{\tau}_e(x)^h = \bar{\tau}_e(x)$ for all x from a suitable essential ideal I. This equality holds on R_F as well (see **2.3.2**). If we project it on the first component of the stalk, then we get an identity $\tau_p(x)^{h\delta} = \tau_p(x)$ on the stalk $\Gamma_p(R_F)$. Since the form τ_p is principal, then $\tau_p(R_F)$ is an almost intermediate ring and, hence, by theorem **3.5.2**, the restriction $h\delta$ on the first component belongs to H_p. Analogously, the restriction $h\delta$ on the other (identified) components also belongs to H_p. Therefore, $h\delta \in H_p^k$ and, hence, $h \in H_p^k \leftthreetimes S_k$. The theorem is proved.

5.8 Galois Subrings for Regular Groups

Let G be a regular group of automorphisms of a semiprime ring R, S be an intermediate ring , $R \supseteq S \supseteq R^G$. By theorem **5.7.1**, a centralizer of S in the ring R_F is contained in the algebra $\mathcal{B}(G)$ of the group G. *This centralizer will be denoted by* Z. The set Z coincides with the intersection of the kernels of all mappings $x \to sx - xs$, $s \in S$ and, hence, Z is a closed set in the topology determined by the central idempotents. Hence, we have a strictly sheaf unary operation of projection π_Z (see **5.6.1**).

Let us recall the conditions on an intermediate ring arising when considering the prime case.

BM (*Bimodule condition*). Let e be an idempotent from $\mathcal{B}(G)$, such that $se = ese$ for any $s \in S$. Then there is an (idempotent) $f \in Z$, such that $ef = f$, $fe = e$.

SI (*Sufficiency of invertible elements*). A C-algebra Z is generated by its invertible elements and if for an automorphism $g \in G$ there is an element $b \in \mathcal{B}(G)$, such that $sb = bs^g$ for all $s \in S$, then there is an invertible element in $e(b)Q$ with the same property.

RC (*Rational completeness*). If A is an essential ideal of S, and $Ar \subseteq S$ for a certain $r \in R$, then $r \in S$.

When going over to the semiprime case the formulation of condition

SI has been somewhat changed.

5.8.1. Theorem. *Any intermediate Galois subring of an* M*-subgroup of the group* G *obeys conditions* **BM, SI** *and* **RC.**

This theorem will result from the following somewhat more general statement.

5.8.2. Proposition.. *Let* G *be an* M*-group of automorphisms of a semiprime ring* R. *Then*

(a) *if* a *is an element from* R_F, *such that* $sa = asa$ *for all* $s \in R^G$ *then* $a\rho = \rho$, $\rho a = a$ *for an idempotent* $\rho \in B(G)$;

(b) *if* $g \in A(R)$ *and* b *is an element, such that* $sb = bs^g$ *for all* $s \in R^G$, *then there can be found an invertible element in* $e(b)Q$ *with the same property;*

(c) *the fixed ring* R^G *is rationally complete in* R.

In order to deduce the theorem from this proposition, it is sufficient to remark that, by theorem **5.7.1**, a centralizer Z of the fixed ring R^H of an M-subgroup $H \leq G$ coincides with the algebra $B(H)$ of this subgroup (and, in particular, is generated by invertible elements).

Proof of proposition **5.8.2.**

(a) Let e be a homogenous idempotent. Then by the condition, $\bar{t}_e(x)a = a\bar{t}_e(x)a$ for all x of an essential ideal of the ring R. By corollary **2.3.2**, this equality is also valid for all $x \in R_F$. If $p \in U(e)$, then when projecting this equality on the stalk Γ_p, we get $\tau_p(x)\bar{a} = \bar{a}\tau_p(x)\bar{a}$. As $\tau_p(\hat{RE})$ is an almost intermediate ring of the stalk, then, b y **3.7.6.**, we see that the following formula is true on the stalk (it should be recalled that the algebra of the group H_p equals $\rho_p(B)$ (see **5.5.7**)):

$$\exists f \;\; \bar{a}f = f \;\&\; f\bar{a} = a \;\&\; \pi_B(f) = f.$$

Since the homogenous idempotent e and the point $p \in U(e)$ have been arbitrarily chosen, then, by the metatheorem, the formula under consideration (with $\bar{a}$ replaced by a) is true on the ring of global sections, which is the required proof.

(b) For simplicity, let call an element φ *almost invertible* if it is invertible in the ring $e(\varphi)Q$. Let T be a set of all almost invertible elements for which the condition of statement (b) is fulfilled. It is obvious that T is a a closed set and, hence, the set $E(T)$ of the supports of elements from T will be also closed. Our task is to show that $e(b) \in E(T)$, for which purpose it is sufficient to find a dense family $\{f\}$ of the idempotents for which $e(b)f \in E(T)$.

Let e be an arbitrary homogenous idempotent , $e \leq e(b)$. Then, as was the case in the preceding point of discussion, we can find an identity $\bar{\tau}_e(x)b = b\bar{\tau}_e(x)^g$ on the ring R_F. If this is a reduced identity, then, by theorem **2.2.2**, we have $\bar{\tau}_e(x)b = b\tau_e(x) = 0$, which is impossible. Non-reducibility of the identity implies that $e(b)\Phi_{hg} \neq 0$ for a certain $h \in G$ (naturally, under the condition that $\bar{\tau}_e(x)$ is reduced). Therefore, we can find an almost invertible nonzero element φ, such that $x\varphi = \varphi x^{hg}$ for all $x \in R_F$, in which case $e(\varphi) \leq e(b)$. If we assume $x = s \in R^G$, then we shall get $\varphi \in T$. As the set of all homogenous idempotents is dense and e is arbitrary among them, then we have $e(b) \in E(T)$, which is the required proof.

(c) Let A be an essential ideal of the ring R^G. Let us show that its left annihilator in Q is zero. If $Ar = 0$, then for any principal trace form $\bar{\tau}_e$ and any ideal I we have $A\bar{\tau}_e(rI) = \bar{\tau}(ArI) = 0$. Choosing I in such a way that $\bar{\tau}_e$ values lie in R, we get $\bar{\tau}_e(rI) = 0$, which implies that $\tau_e(rx)$ is a zero form. However, τ_e is a principal trace form and, hence, (see **5.6.7**), $e(r) \cdot e = 0$ and, consequently, $e(r) = 0$, which is the required proof.

If now $Ar \subseteq R^G$, then $A(r - r^g) = 0$ at all $g \in G$. Hence, $r = r^g$, and the proposition is proved.

Let us now get down to proving the inverse statement.

5.8.3. Theorem. *Let G be an MN-group. Then any intermediate ring obeying conditions* **BM**, **SI** *and* **RC** *is a Galois subring of a regular subgroup of the group G.*

This theorem will obviously result from the following theorem.

5.8.4. Theorem. *Let G be an MN-group, S be an intermediate ring obeying conditions* **BM** *and* **SI**. *Then $A(S)$ is a regular group and the ring S contains an ideal W of the ring $\bar{S} = \mathbb{I}\,A(S)$, the annihilators of which in Q are zero.*

5.8.5. Let us start with describing the situation arising on nearly all stalks of canonical and invariant sheaves.

By proposition **5.5.9**, the group $A(S)$ is a closure of a reduced-finite N-group V, the algebra of which is equal to that of the group $A(S)$. Since Z is generated by its own invertible elements, then $B(A(S)) = Z$ and $B(V) = Z$.

It is obvious that the set of idempotents homogenous for both G and V is dense. Let us fix one of such idempotents, e, and let $p \in U(e)$. By theorem **5.5.7**, the projection $(\hat{G})_{\bar{p}}$ has the form $H_p^k \leftthreetimes S_k$, where H_p is a regular (i.e., an MN-) group of automorphisms of the stalk Γ_p. Since $A(S)_{\bar{p}}$ is a closed subgroup in $(\hat{G})_{\bar{p}}$, then, by **5.4.11**, we get $A(S)_{\bar{p}} = (F^m \leftthreetimes S_m) \times G''$. Carrying out, if necessary, a rearrangement, one can assume that S_m acts on the first m components of decomposition (3), while G'' acts on them identically.

5.8.6. Lemma. *Let us denote by* $\widehat{S \cdot E^G}$ *a closure of* $S \cdot E^G$ *in the topology determined by G-fixed idempotents, and let* $S_1 = \rho_{\bar{p}}(\widehat{S \cdot E^G})$. *Then* S_1 *is an almost intermediate ring of the stalk* $\Gamma_{\bar{p}}$, *obeying conditions* **BM** *and* **SI**, *in which case the group* $A(S_1)$ *coincides with the group* $A(S)_{\bar{p}}$.

Proof. If I is an essential ideal of the ring R, such that $\bar{\tau}_e(I) \subseteq R^G \subseteq S$, then $\bar{\tau}_e(\widehat{I \cdot E^G}) \subseteq \widehat{S \cdot E^G}$ and, hence, $\bar{\tau}_e(\rho_p(\widehat{I \cdot E^G})) \subseteq S_1$, i.e., S_1 is an almost intermediate subring.

Let us denote by π_S a strictly sheaf projection $\pi_S : Q \to \widehat{S \cdot E^G}$ (for an invariant sheaf). Then conditions **BM** and **SI** will be presented as implications:

$$[\forall x \quad \pi_S(x)e = e\pi_S(x)e \,\&\, e^2 = e \,\&\, \pi_B(e) = e] \longrightarrow$$

$$\longrightarrow \exists f \quad ef = f \,\&\, fe = e \,\&\, f = \pi_Z(f),$$

$$[\forall x \ \ \pi_S(x)b = b\pi_S(x)^h] \longrightarrow \exists b', b'' \ \forall y$$

$$\pi_S(y)b' = b'\pi_S(y)^h \,\&\, b'b'' = b''b' = e(b),$$

where π_B, π_Z are strictly sheaf projections (it should be recalled that B, Z are closed). As $x = \pi_Z(x)$ iff $\forall y \, \pi_S(y) x = x\pi_S(y)$, then a centralizer Z_1 of the subring S_1 in $Q(\Gamma_{\bar{p}})$ coincides with the image of π_Z, i.e., $Z_1 = \rho_{\bar{p}}(Z)$. Allowing for the fact that for any h, b, e the above mentioned implications are valid on all rings of sections, we come to the conclusion (proposition **1.11.18**) that they are also valid on $\rho_{\bar{p}}(Q)$, i.e., S_1 obeys conditions **BM** and **SI**.

Since $\hat{G}_{\bar{p}}$ is a regular group, then it is a Galois group and, hence, $A(S_1) \subseteq \hat{G}_{\bar{p}}$. Let $g \in G$, and the formula $\forall x \, \pi_S(x)^g = \pi_S(x)$ be valid on the stalk $\Gamma_{\bar{p}}$. We can find a fixed idempotent $f \notin p$, such that $\pi_S(xf)^g = \pi_S(xf)$ for all $x \in R \cdot \hat{E}^G$. Let us determine $x^h = x^g f + (1 - f)x$. Then $s^h = s$ for all $s \in S \cdot \hat{E}^G$, i.e., $h \in A(S)$ and, hence, $\bar{g} = \bar{h} \in A(S)_{\bar{p}}$, i.e., $A(S_1) \subseteq A(S)_{\bar{p}}$. The inverse inclusion is obvious as $A(S \cdot \hat{E}^G) = A(S)$. The lemma is proved.

5.8.7. Lemma. *The algebra of the group* $A(S)$ *is semiprime.*

Proof. Since, by the condition, Z is generated by its invertible elements, then the algebra of the group $A(S)$ coincides with Z. Let us consider the formula

$$\forall x[\forall y \;\; \pi_Z(x)\pi_Z(y)\pi_Z(x) = 0 \longrightarrow \pi_Z(x) = 0].$$

By lemma **3.9.4**, this formula is true on the stalk Γ_p. As this is a Horn formula, and the point p is arbitrary, then this formula is also true on the ring of global sections. The lemma is proved.

5.8.8. Proposition. *In a ring of global sections* $\hat{RE}$ *there are elements* $a, r_1, \ldots, r_k, v_1, \ldots, v_k$ *and elements* $s_1, \ldots, s_k \in S \cdot \hat{E}^G$, *such that* $e(a) = e$, *and for any* x, y *the following equality holds*

$$\tau_{A(S)}(yax) = \sum_{i=1}^{k} \tau_e(yr_i) s_i \tau_e(v_i x), \tag{7}$$

where $\tau_{A(S)}$ *is a principal* V*-trace form determined by the homogenous idempotent* e*, and* V *is a reduced-finite group, such that* $A(S) = \hat{V}$, $B(A(S)) = B(V)$.

Proof. Let us consider the forms $\tau_{A(S)}$ and $\bar{\tau}_e$ as strictly sheaf operations for the invariant sheaf. The predicate $P_S(P_S(x) = T \Leftrightarrow x \in S \cdot \hat{E}^G)$ will also be strictly sheaf. Since the statement of the theorem is set by a Horn formula ψ_k, then, by corollary **1.11.20**, it is sufficient to establish that it is true on the stalk $\Gamma_{\bar{p}}$.

It should be recalled that we are under the conditions of lemma **3.9.22**, and it is only necessary to check if the projection of $\Phi_f^{(S)} \leftthreetimes (1\ j)$ on the first component of the decomposition of $\Gamma_{\bar{p}}$ equals zero at any $f \in H_p^k$, $m+1 \le j \le k$.

Let us assume that this is not the case. Then, by condition **SI**, we can find an invertible in Γ_p element b, such that

$$sb = bs^{h_1 \times \ldots \times h_k \leftthreetimes (1\ j)}$$

for all $s \in S_1$, where $S_1 = \rho_{\bar{p}}(S \cdot \hat{E}^G)$. Let us consider an automorphism

$$g = (\tilde{b}\, h_j^{-1} \times 1 \times \ldots \times 1 \times h_j \tilde{b}^{-1} \times 1 \times \ldots \times 1) \leftthreetimes (1\ j).$$

Let $s = s_1 \oplus \ldots \oplus s_n \in S_1$. In this case

$$s_1 b = sb = bs^{h_1 \times \ldots \times h_n \leftthreetimes (1\ j)} = bs_j^{h_j}.$$

From here we get $s_1^{\tilde{b}} = s_j^{h_j}$ and, hence, $s^g = s$. Therefore, $g \in A(S_1) = A(S)_{\bar{p}}$, but $A(S)_{\bar{p}} = (F^m \leftthreetimes S_m) \times G''$ (see **5.8.5**). However, $g \notin (F^m \leftthreetimes S_m) \times G''$, as $j > m$. This is a contradiction and, hence, the proposition is proved.

5.8.9. Lemma. *For any finite set* $\{\hat{s}_1, \dots, \hat{s}_k\}$ *of elements from* $\widehat{S \cdot E^G}$ *there is an ideal* $I \in F(R)$, *such that* $\hat{s}_i \bar{\tau}_e(I) \subseteq S$ *for all* i, $1 \le i \le k$.

Proof. It suffices to show that the set $V = \{v \in Q \mid \exists\, I_v \in F(R), v\bar{\tau}_e(I_v) \subseteq S\}$ is closed in the topology determined by fixed central idempotents, and contains $S \cdot E^G$.

Let $v = s_1 e + \dots + s_n e_n$, where $s_i \in S$, $e_i \in E^G$. Let us find an ideal $J \in F$, such that $e_i J \subseteq R$. Then let us choose an $I \in F$ in such a way that $\bar{\tau}_e(I) \subseteq J$. We get $e_i \bar{\tau}_e(I) \subseteq Q^G \cap R = R^G \subseteq S$ and, hence, $v\bar{\tau}_e(I) \subseteq S$.

Let, then, $v = \lim v_\alpha$, i.e., $ve_\alpha = v_\alpha e_\alpha$, where $v_\alpha \in V$, $e_\alpha \in E^G$, $\sup\{e_\alpha\} = 1$. Let I_α, J_α be ideals from F, such that $v_\alpha \bar{\tau}_e(I_\alpha) \subseteq S$, $e_\alpha J_\alpha \subseteq R$. Then $I = \sum e_\alpha I_\alpha J_\alpha$ is an essential ideal of the ring R. In this case we get

$$v\bar{\tau}_e(I) \subseteq \sum_\alpha v\bar{\tau}_e(e_\alpha I_\alpha J_\alpha) = \sum_\alpha ve_\alpha \bar{\tau}_e(I_\alpha J_\alpha) =$$

$$= \sum_\alpha v_\alpha \bar{\tau}_e(I_\alpha(e_\alpha J_\alpha)) \subseteq \sum_\alpha v_\alpha \bar{\tau}_e(I_\alpha) \subseteq S.$$

The lemma is proved.

5.8.10. The **proof** of lemma **5.8.4** now becomes evident. Let W be a sum of all ideals of the ring $\bar{S} = \mathbb{I}\,A(S)$ contained in S, and let us assume that $1W = 0$. Let us choose an idempotent $e \le e(1)$, which is homogenous for both groups, G and V, and make use of proposition **5.8.8**. By lemma **5.8.9**, we can find an ideal $I \in F$, such that at $y, x \in I$ the right-hand part of relation (7) will lie in S. Therefore, $\tau_{A(S)}(IaI) \subseteq W$ and, hence, $1\tau_{A(S)}(IaI) = 0$. Let us recall that $\tau_{A(S)}(x) = \sum_1^m \tau_e^{(1j)}(x)$ and, hence, $\tau_{A(S)}(x) = \tau_{A(S)}(ex)$. Therefore, the form $1\tau_{A(S)}(x)$ turns to zero on $IaI + (1 - e)I$. Reducing, if necessary, I, we can assume the last sum to lie in R and to form, therefore, an ideal from F. Thus (see theorem **2.2.2**), $1\tau_{A(S)}(x)$ is a zero form. Consequently, $1\tau_e(x)$ will also be a zero form. Going now over to the stalk Γ_p, we get $\rho_p(1) = 0$, as τ_e is a principal trace form. Allowing for the fact that p is an arbitrary point from $U(e)$, we get $e \cdot e(1) = 0$, while our choice was $e \leq e(1)$. Thus, this is a contradiction and the theorem is proved.

5.9 Correspondence and Extension Theorems

Now we can summarize the results obtained above as a correspondence theorem. We will also show that a certain form of the extension theorem can be easily deduced from it.

5.9.1. Correspondence theorem. *Let G be a regular group of automorphisms of a semiprime ring R. In this case mappings $H \to I(H)$, $S \to A(S)$ set a one-to-one correspondence among all regular subgroups of the group G and all intermediate subrings obeying conditions* **BM**, **RC** *and* **SI**.

Proof. It immediately results from theorems **5.7.2, 5.8.1, 5.8.3** and corollary **5.4.7.**

5.9.2. Extension theorem. *Let G be a regular group of automorphisms of a semiprime ring R and S', S'' be intermediate Galois subrings of M-subgroups. If $\varphi: S' \to S''$ is an identical on R^G isomorphism, and the ring S' obeys condition* **SI** *for mappings from φG, then φ is extended to an isomorphism from G.*

Proof. Let us consider a ring $\bar{R} = R \oplus R$ with a group $\bar{G} = G^2 \leftthreetimes S_2$. Then $S = \{ s \oplus s^{\varphi}, s \in S' \}$ will be an intermediate subring. Its centralizer in $Q(\bar{R}) = Q(R) \oplus Q(R)$ is equal to a direct sum of centralizers of the rings S' and S''. Therefore, S obeys condition **BM** and the first part of **SI**. The second part of **SI** results from the fact that S' obeys this condition for the mapping from φG. Finally, condition **RC** can also be easily verified: if I is an essential ideal of S' and $r_1 \oplus r_2 \in \bar{R}$, in which case $[i \oplus \varphi(i)] \cdot \cdot (r_1 \oplus r_2) \in S$ for all $i \in I$, then $Ir_1 \subseteq S'$, $\varphi(I) r_2 \subseteq S''$ and, hence, $r_1 \in S'$, $r_2 \in S''$ and, in particular, $r_2 = \varphi(s_1)$ for a certain $s_1 \in S'$; as $\varphi(ir_1) = \varphi(i)r_2$, then $\varphi(ir_1) = \varphi(is_1)$, i.e., $I(r_1 - s_1) = 0$ and, hence, $r_1 = s_1$. Therefore, $r_1 \oplus r_2 = s_1 \oplus \varphi(s_1) \in S$.

Let us now remark that $A(S)$ is not contained in G^2. Indeed, since in the opposite case, due to closure, $A(S) = H_1 \times H_2$ and, hence, $S = I\, A(S) = R^{H_1} \oplus R^{H_2}$, which is impossible, since φ is an isomorphism.

Let $(g_1 \times g_2) \rtimes (1, 2) \in A(S)$. Then for any $s_1 \in S'$ we have $s_1 \oplus \varphi(s_1) = \varphi(s_1)^{g_2} \oplus s_1^{g_1}$, which shows that g_1 is the sought extension. The theorem is proved.

5.10 Shirshov Finiteness. Structure of Bimodules

The use of an invariant sheaf when transferring the results of the Galois theory from prime rings onto semiprime ones, enables one to ignore some facts on invariants of M-groups. Nonetheless, these facts are of their own interest and can also be obtained using sheaves. In this paragraph we prove the theorem on finiteness in the Shirshov sense (see **3.1.9**) and describe the structure of (R, R^G)-subbimodules of R_F.

5.10.1. Theorem. *Let G be a Maschke group of automorphisms of a semiprime ring R. Then R has an essential ideal which is locally-finite (in the Shirshov sense) over a fixed ring R^G.*

Proof. For simplicity, let us call an element $a \in R$ *finite* if the right ideal aR is contained in a certain finitely-generated R^G-submodule of the ring R:

$$aR \subseteq \sum_{i=1}^{n} a_i R^G$$

It is obvious that a set of all finite elements forms a two-sided ideal W of the ring R. Our task is to prove that its annihilator in R is zero. For this purpose it is sufficient to show that $We \neq 0$ for any homogenous idempotent e.

Let e be a homogenous idempotent, and let us consider a sequence of Horn formulas:

$$\Phi_n \Leftrightarrow \exists\ a, t_1, \ldots, t_n, r_1, \ldots, r_n \quad ea \neq 0\ \& \tag{8}$$

$$\&\ \forall\ x \quad ax = \sum_{i=1}^{n} et_i \bar{t}_e(r_i x),$$

where $\bar{\tau}_e$ is a principal trace form viewed as a strictly stalk operation (see **5.6**). Since by the construction $e\bar{\tau}_e = \tau_e$, then, by proposition **3.7.2** (where $V = \{1\}$), on every stalk $\Gamma_{\bar{p}}$ of the invariant sheaf one of formulas Φ_n holds at $p \in U(e)$. According to metatheorem **1.11.13** and corollary **1.11.20**, one of formulas Φ_n holds on the ring of sections $\bar{e}\,\widehat{RE}$. Let us choose an essential ideal I in such a way that $\bar{\tau}_e(r_i I) \subseteq R$, $Iet_i \subseteq R$, $Ia \subseteq R$, and then we get $IaI \subseteq W$, as if $u, v \in I$, then

$$(uav)x = ua(vx) = \sum_{i=1}^{n} (uet_i)\bar{\tau}_e(r_i vx).$$

We have $eW \supseteq eIaI = IeaI \neq 0$. The theorem is proved.

5.10.2. Theorem. *Let G be a Maschke group of automorphisms of a semiprime ring R. Then for any (R, R^G)-submodule V of R_F there is an idempotent $e \in \mathbf{B}(G)$ and an essential ideal I of R, such that $Ie \subseteq V = Ve$.*

Proof. Let us show that the right annihilator of V in the algebra $\mathbf{B}(G)$ is generated by an idempotent. Let us write this statement as a Horn formula

$$\exists f \quad f^2 = f \;\&\; \pi_B(f) = f \;\&\; \forall t \quad \pi_V(t)f = 0 \;\&$$
$$\&\; \forall d (\forall w \quad \pi_V(w)\pi_B(d) = 0 \longrightarrow f\pi_B(d) = \pi_B(d)).$$

Here π_B, π_V are strictly sheaf projections on $\mathbf{B}$ and $\widehat{VE}$, respectively (see **1.11.9**). As the annihilators of V and $\widehat{VE}$ coincide, then this formula really gives the required statement. If e is a homogenous idempotent and $p \in U(e)$, then the stalk $\Gamma_{\bar{p}}$ has a finite prime dimension and , hence, the algebra $\mathbf{B}(G)_{\bar{p}}$ is semisimple, i.e., its any one-sided ideal is generated by its idempotent. Now we are to use the metatheorem.

Let $e = 1 - f$, where $ann_{\mathbf{B}}^{(r)} V = f\,\mathbf{B}(G)$. Let us consider an element $a \in R$ to be *finite with respect to* V, if there is a finite set $v_1, \ldots, v_n \in V$, such that

$$aRe \subseteq \sum_{i=1}^{n} Rv_i R^G.$$

It is obvious that the set W of all finite with respect to V elements forms a two-sided ideal of the ring R, and the task now is to show that the annihilator of this set is zero.

Let ε be an arbitrary homogenous idempotent. Let us consider a sequence of Horn formulas:

$$\Phi_n \Leftrightarrow \exists\ a, v_1, \ldots v_n, r_1, \ldots, r_n, t_1, \ldots, t_n \forall\ x$$

$$x \ \underset{j=1}{\overset{n}{\&}}\ \pi_V(v_j) = v_j \,\&\, axe =$$

$$= \sum_{i=1}^{n} \varepsilon t_i v_i \bar{\tau}_\varepsilon(r_i x) \ \&\ \varepsilon a \neq 0.$$

As $\varepsilon\bar{\tau}_\varepsilon = \tau_\varepsilon$, then by proposition **3.7.2**, on every stalk $\Gamma_{\bar{p}}$, $p \in U(e)$, one of formulas Φ_n holds. By the metatheorem and corollary **1.11.20**, we see that one of these formulas holds on $\bar{\varepsilon}\ \widehat{RE}$.

Let us show that for any $w \in \widehat{VE}$ there is an ideal $I \in F$, such that $Iw \subset V$. For this purpose it is sufficient to show that the set $\{w \in R_F \mid \exists\ I \in F,\ Iw \subseteq V\}$ contains VE and is closed. If $w = \sum v_i e_i$, where $v_i \in V$, $e_i \in C$, then we can find an ideal $I \in F$, such that $Ie_i \subset R$, and we get $Iw \subseteq \sum (Ie_i) v_i \subseteq V$. Finally, if $w = \lim_{\alpha \in A} w_\alpha$, $w_\alpha I_\alpha \subseteq V$ for suitable $I_\alpha \in F$, then we set $I = \sum I_\alpha J_\alpha e_\alpha$, where $J_\alpha \in F$ and $J_\alpha e_\alpha \subseteq R$. In this case $I \in F$ (as $\sup e_\alpha = 1$) and $Iw \subseteq \sum I_\alpha J_\alpha w \subseteq \sum I_\alpha J_\alpha w_\alpha \subseteq V$.

Therefore, we can find an ideal $I \in F$, such that $I\varepsilon t_i v_i \subseteq V$, $Ia \subseteq R$, $\bar{\tau}_\varepsilon(r_i I) \subseteq R$. Now $IaI \subseteq$ W and, hence, εW $\supseteq IaεI \neq 0$. Since a set of one-sided homogenous idempotents is dense, then W is an essential ideal. The theorem is proved.

In a symmetrical way we get the following theorem.

5.10.3. Theorem. *Let G be an M-group of automorphisms of a semiprime ring R. Then for any (R^G, R)-subbimodule W of R_F there can be found an idempotent $e \in B(G)$ and an essential ideal I of R, such that*

$eI \subseteq W = eW$.

Theorem **5.10.1** has a left analog as well. Let us formulate it in a somewhat different way, taking into account the fact that the coefficients $\bar{\tau}_e(r_i x)$ in formula (8) are linear from the right with respect to x over R^G.

5.10.4. Theorem. *Let G be a Maschke group of automorphisms of a semiprime ring R. Then a set W of elements a of the ring R, such that the left R^G-module Ra is embeddible into a finite direct sum of copies of a regular module R^G, contains the essential ideal of the ring R.*

Proof. Let e be a homogenous idempotent. Let us denote by W_e a set of all elements $a \in R$ for which the left analog of formula (8) holds. In this case W_e is a nonzero ideal of the ring R and $eW_e \neq 0$. Let $a \in W_e$ and I be an essential ideal of the ring R, such that $\bar{\tau}_e(Ir_k) \subseteq R$, $1 \leq k \leq n$. Then formula (8) shows that at $i \in I$ the mapping

$$xia \longrightarrow \bar{\tau}_e(xir_1) \dot{+} \ldots \dot{+} \bar{\tau}_e(xir_n)$$

carries out an embedding Ria into $\underbrace{R^G + \ldots + R^G}_{n}$. Therefore, $IW_e \subseteq W$ and $\sum_e IW_e$ is the sought ideal. The theorem is proved.

Theorems **5.10.2** and **5.10.3** make it possible to obtain a statement which is, in a certain sense, inverse to lemma **5.6.6**.

5.10.5. Corollary. *Let G be an M-group of automorphisms of a semiprime ring R, ρ be an essential ideal of a fixed ring. Then the left ideal $R\rho$ (and the right ideal ρR) contains the essential ideal of the ring R.*

Proof. It suffices to show that the right (left) annihilator ρ in $B(G)$ is zero. If $\rho a = 0$, then for any principal trace form $\bar{\tau}$ one can find an essential ideal I of R, such that $\bar{\tau}(aI) \subseteq R$. We have $\rho\bar{\tau}(aI) = \bar{\tau}(\rho aI) = 0$, i.e., $\bar{\tau}(aI) = 0$ and, according to the basic property of principal trace forms, we get $a = 0$ (see **5.6.8**).

REFERENCES

J.M.Goursaud, J.-L.Pascaud, J.Valette [50];
J.M.Coursaud, J.Osterburg, J.-L.Pascaud, J.Valette [51];
V.K.Kharchenko [70,80];
J.-L.Pascaud [132,333];
A.I.Shirshov [144];
A.V.Yakovlev [155].

CHAPTER 6

APPLICATIONS

This concluding chapter is of a somewhat mosaic character and is mainly intended for the advanced reader. Here some problems are considered, which are not interrelated but they are in this or that way associated with the same basic topics: free algebras, non-commutative invariants, radicals, Goldie rings, rings of quotients, Noetherian rings, etc. It is clearly impossible to try to give a closed presentation of such a wide range of material within a monograph (let alone a single chapter). Therefore, use without proof published elsewhere will be made of classical theorems of the ring theory, such as the Cohn criteria on freedom of subalgebras, Goldie theorems, the lemma on composition (the Gröbner basis), etc.. The last section explores the general concept of the action of Hopf algebras embracing both the cases of automorphisms and derivations, and presents certain approaches to studying skew derivations from this point of view.

6.1 Free Algebras

Let $F\langle X\rangle$ be a free associative (but non-commutative) algebra over a field F with a set of generators X. On $F\langle X\rangle$ one can define a positive-degree function v, by setting arbitrarily the value of v on the generators and expanding v naturally, first on the monomials and then on the whole of the algebra. In this case we assume the degree of a polynomial to be equal to the maximum of the degrees of the monomials occurring in it with nonzero coefficients. In a free algebra the weak algorithm with respect to the degree function introduced in such a way ([34], proposition **4.2**,) holds. Let us fix a function v throughout the section.

An automorphism of a free algebra will be called *homogenous* if it transforms homogenous elements into homogenous ones with their degree preserved. For instance, if v is a usual X-degree, i.e., $v(x_i)=1$, where $\{x_i\}=X$, then the homogeneity of an automorphism g implies that it acts on the generators linearly, i.e., $x_i^g=\sum\alpha_{ij}x_j$, $\alpha_{ij}\in F$.

6.1.1. Theorem. *Let G be a finite group of homogenous automorphisms of a free algebra. Then the mappings*

$$S \to A(S) = \{g \in \mathrm{Aut}F\langle X\rangle \mid \forall\, s \in S(s^g = s)\},$$
$$H \to I(H) = \{s \in F\langle X\rangle \mid \forall\, h \in H(s^h = s)\}$$

set up a one-to-one correspondence between all subgroups of the group G *and all intermediate free algebras. A subalgebra* $S \supseteq I(G)$ *will be a Galois extension of* $I(G)$ *iff it is invariant relative* G. *In this case a subgroup* $A(S)$ *is normal in* G *and the Galois group of the extension* $I(G)$ *of* S *is isomorphic to* $G/A(S)$.

To prove this theorem, it will be sufficient to check the validity of the following statements:

(1) The algebra of invariants of any finite group of homogenous automorphisms is free.

(2) Any finite group of homogenous automorphisms is a Galois group, i.e., a group of all automorphisms of the algebra $F\langle X\rangle$ over its algebra of invariants.

(3) Any free subalgebra which contains the algebra of invariants of a finite group of homogenous automorphisms is an algebra of invariants of a certain group.

(4) Any automorphism of an intermediate algebra which leaves the elements of the algebra of invariants of a given group of homogenous automorphisms fixed, can be extended to an automorphism of the algebra $F\langle X\rangle$.

We shall show that statement (1) is valid for any (not necessarily finite) group of homogenous automorphisms, while statements (2) and (3), as well as (4) hold for a finite group of any (not necessarily homogenous) automorphisms. In this case it should be noted that any embedding of an intermediate subalgebra into $F\langle X\rangle$ which leaves the elements of the algebra of invariants of a finite group fixed, can be extended to an automorphism of the algebra $F\langle X\rangle$.

6.1.2. Proposition. *The algebra of invariants of any group of homogenous automorphisms of a free associative algebra is free.*

Proof. Using Cohn's criterion on freedom of a subalgebra of a free algebra ([50], proposition **7.2**), it is sufficient to show that the subalgebra $I(G)$ satisfies the weak algorithm relative the degree function v.

Let $a_1, a_2, \dots, a_n$ be a right v-dependent set of nonzero elements from $I(G)$, and let $v(a_1) \le v(a_2) \le \dots \le v(a_n)$. Since the elements $a_1, \dots, a_n$ are v-dependent in the algebra $I(G)$, they will also be v-dependent in $F\langle X\rangle$. Let us choose the least number k, such that $a_1, a_2, \dots, a_k$ are still v-dependent in $F\langle X\rangle$. Since in a free algebra the weak algorithm holds, one of the elements (it will be a_k, since k is minimal) will be right v-dependent on the preceding ones in $F\langle X\rangle$, i.e.,

$$v(a_k - \sum_{i<k} a_i b_i) < v(a_k). \tag{1}$$

In this formula under the sum sign we can eliminate the terms with their degree strictly less than $v(a_k)$. Therefore, one can assume that either $v(a_i b_i) = v(a_i) + v(b_i) = v(a_k)$ or $b_i = 0$. Additionally, inequality (1) is not violated if every b_i is replaced with its highest degree homogenous component, i.e., one can assume that b_i are homogenous elements. Now let g be an arbitrary automorphism from the group G. As g is homogenous, we have

$$v(a_k - \sum_{i<k} a_i b_i) = v([a_k - \sum a_i b_i]^g) = v(a_k - \sum_{i<k} a_i b_i^g).$$

Allowing for the fact that the degree of the difference of any two elements is not greater than the maximum of the degrees of these elements, we get

$$\begin{aligned} v(\sum a_i(b_i - b_i^g)) &= v([a_k - \sum a_i b_i^g] - [a_k - \sum a_i b_i]) < \\ &< v(a_k) = \max\{v(a_i) + v(b_i)\} \end{aligned} \tag{2}$$

If now for a certain i we have $b_i - b_i^g \neq 0$, then $v(b_i - b_i^g) = v(b_i)$ because the elements b_i, b_i^g are homogenous and, hence, relation (2) implies the right v-dependence of the set $a_1, \dots a_{k-1}$, which contradicts the choice of k.

Therefore, $b_i \in I(G)$, and relations (1) show a_k to be right v-dependent on $a_1, a_2, \dots, a_{k-1}$ in the algebra $I(G)$. This proves the proposition.

6.1.3. Lemma. *Let R satisfy a two-term weak algorithm. Then, either $Q(R) = R$ or R is not a ring of skew polynomials, where $Q(R)$ is a two-sided Martindale ring of quotients.*

Proof. It should be remarked that in $Q(R)$ there are no zero divisors. If $\varphi\psi = 0$, $\varphi \neq 0$, $\psi \neq 0$, then for some $p, q \in R$ we have $0 \neq p\varphi \in R$, $0 \neq \psi q \in R$ and we come to a contradiction with the fact that $(p\varphi)(\psi q) = p(\varphi\psi)q = 0$.

Let us assume that $Q(R) \neq R$. Among the elements $\varphi \in Q(R)$ not lying in R, let us choose one in such a way that the set $R\varphi \cap R$ has a nonzero element $p\varphi$ of the least possible degree. Then for a certain nonzero $q \in R$ we have $\varphi q \in R$. In the ring R let us consider the equality $(p\varphi)q = p(\varphi q)$. Since the two-term weak algorithm holds in this ring, we have two options [34]:

(1) $p\varphi = p\lambda + s$, where $\lambda, s \in R$, $v(s) < v(p)$ (if $v(p\varphi) \geq v(p)$);

(2) $p = (p\varphi)\lambda + s$, where $\lambda, s \in R$, $v(s) < v(p\varphi)$ (if $v(p\varphi) < v(p)$).

In the former case we have $p(\varphi - \lambda) = s$, i.e., $v(p(\varphi - \lambda)) = v(s) < v(p\varphi)$. Given the choice of the element φ we get $\varphi - \lambda \in R$, i.e., $\varphi \in R$, which is a contradiction.

In the second case we get $p(\varphi\lambda - 1) = - s$, i.e., as above, $\varphi\lambda - 1 \in R$. It implies $\varphi\lambda \in R$. The equality $p(\varphi\lambda) = p - s$, however, gives $v(\varphi\lambda) = 0$. Since in a ring with a two-term weak algorithm a set of elements of a zero degree forms a sfield, then $\varphi r = 1$ for a certain $r \in R$ and, as $Q(R)$ has no zero divisors, $r\varphi = 1$. Let us now show that if the degree of r is positive, then R is a ring of skew polynomials.

Let x be an element of the least positive degree from R. Let $0 \neq I$ be an ideal of the ring R such that $\varphi I \subseteq R$. Let us arbitrarily choose a nonzero element $i \in I$. Then we have $r(\varphi x i) = xi$. Therefore, due to the two-term weak algorithm, $r = x\mu + \delta$, $\mu, \delta \in R$, $v(\delta) = 0$. If now y is another element of the least positive degree, then $r = y\mu' + \delta'$, or $x\mu - y\mu' = \delta' - \delta$, in which case $\max\{v(x\mu), v(y\mu')\} > 0 = v(\delta' - \delta)$. Using the two-term weak algorithm, we get $y = x\delta_1 + \delta_2$, where $v(\delta_1) = v(\delta_2) = 0$. As $v(\delta x) = v(\delta) + v(x) = v(x)$ for all $\delta \in \Delta = \{a| v(a)\} = 0$, then $\delta x = x\delta^h + \delta^\sigma$. It is clear that h is a monomorphism of the sfield Δ, while σ is its h-derivation. The task now is to show that x and Δ generate the ring R. Let us carry out induction over the degrees.

If $s \in R$ and $v(s) > v(r)$, then the equality $r(\varphi s i) = si$ yields $s = r\mu' + s'$, in which case $v(s') < v(s)$, $v(\mu'') = v(s) - v(r) < v(s)$ and one can make use of the supposition of induction.

If $v(s) \leq v(r)$, then the equality $r(\varphi s i) = si$ affords $r = s\mu' + s' = x\mu + \delta$, where $v(s') < v(s)$. We, therefore, get $s\mu' - x\mu = \delta - s'$ and, hence, since the algorithm is two-term and weak, $s = x\mu'' + s''$, in which case $v(\mu'') = v(s) - v(x) < v(s)$, $v(s'') < v(s)$, i.e., one again make use of the induction proposition.

As the condition of the existence of an n-term weak algorithm is left-right symmetric [34], then h is an automorphism. This is the proof of the lemma.

6.1.4. Proposition. *Any finite group of automorphisms of a free associative algebra is a Galois group.*

Proof. Since the algebra of a group is contained in $Q(F\langle X\rangle)$, then, by the previous lemma, it is also contained in $F\langle X\rangle$. All the invertible elements of the algebra $F\langle X\rangle$, however, are exhausted by the elements of the field F. Therefore, the algebra of the group coincides with F and the only inner automorphism, which is identical, belongs to any group. Now if we, for instance, make use of theorem **3.2.2**, the proposition will be proved.

6.1.5. Proposition. *Any free subalgebra of a free associative algebra which contains the algebra of invariants of a finite group, is itself an algebra of invariants of a certain group.*

Proof. Let $F\langle Y\rangle$ be a free subalgebra, G be a finite group of automorphisms and $I(G)\subseteq F\langle Y\rangle$. Let, then, $IA(F\langle X\rangle)=S$. While proving the previous proposition we have seen the algebra of the group G to coincide with F. Therefore, for the group G and ring $F\langle Y\rangle$ the conditions of corollary **3.9.20** hold (as, incidentally, for any finite group of automorphisms of a ring with no zero divisors). Thus, by this theorem, the ring $F\langle Y\rangle$ contains a nonzero two-sided ideal A of the algebra S. Then it is evident that A is also an ideal in $F\langle Y\rangle$. Therefore, $Q(F\langle Y\rangle)=Q(A)=Q(S)$. As $F\langle Y\rangle$ is a free algebra, it is a ring of skew polynomials only when its rank is equal to one, i.e., $|Y|=1$. In particular, in this case the algebra is commutative and, hence, the following identity with automorphisms holds:

$$\Big(\sum_{g\in G}x^g\Big)\Big(\sum_{g\in G}y^g\Big)=\Big(\sum_{g\in G}y^g\Big)\Big(\sum_{g\in G}x^g\Big).$$

As the group G contains no non-trivial inner automorphisms, this is a reduced identity. By theorem **2.3.1**, in $F\langle X\rangle$ we find an identity $xy=yx$, but from the very beginning $F\langle X\rangle$ has been assumed to be non-commutative (i.e., $|X|>1$).

Therefore, one can use lemma **6.1.3**. We have $F\langle Y\rangle=Q(F\langle Y\rangle)\subseteq S$, i.e., $F\langle Y\rangle=S$, which is the required proof.

6.1.6. Proposition. *Let G be a finite group of automorphisms of the algebra $F\langle X\rangle$ and S be a subalgebra containing $I(G)$. Then any embedding of S into the algebra $F\langle X\rangle$, which is identical on $I(G)$, can be extended to an automorphism of $F\langle X\rangle$.*

This proposition results immediately from corollary **3.11.4**. One has, however, to remark that, by lemma **6.1.3**, the group $A(F\langle X\rangle)$ coincides with the group of automorphisms of $F\langle X\rangle$.

Theorem **6.1.1** is deduced from the propositions proved above using standard considerations. It should be remarked that in this theorem there is no need to assume that $|X|>1$, as it is evident that the theorem is also valid when $|X|=1$.

Let us now go over to a free associative algebra over a field of a positive characteristic, p.

6.1.7. Definition. A derivation μ of the algebra $F\langle X\rangle$ is called *homogenous* if it transforms homogenous elements into either homogenous ones with the their degree preserved, or to zero. For instance, if v is a usual degree, i.e., $v(x_i)=1$, $X=\{x_i\}$, then homogeneity implies that μ acts linearly on the generators, i.e., $x_i^{\mu}=\sum x_j\alpha_{ij}$.

6.1.8. Theorem. *Let* L *be a finite-dimensional restricted Lie algebra of homogenous derivations of the free algebra* $F\langle X\rangle$. *In this case the mappings* $\Lambda\rightarrow F\langle X\rangle^{\Lambda}$, $S\rightarrow \mathrm{Der}_S F\langle X\rangle$ *sets up a one-to-one correspondence between the restricted Lie algebras of* L *and intermediate free subalgebras. Then the differential extensions of* $F\langle X\rangle^{L}$ *correspond to the ideals of* L *and vice versa.*

Proof. By theorem **4.5.2** (on correspondence) and by theorem **4.6.2** (on extension), it is sufficient to show that any free subalgebra is rationally-complete in $F\langle X\rangle$, that a set of the constants of a finite-dimensional restricted Lie algebra of homogenous derivations forms a free subalgebra, and also that a homogenous nonzero derivation cannot be inner (for Q).

The rational completeness of free subalgebras results from lemma **6.1.3.** The same lemma also affords that the derivations which are inner for $Q(F\langle X\rangle)$ will also be inner for $F\langle X\rangle$, however, the equality $v(x)=v(xf-fx)$ for $x\in X$, $f\in F\langle X\rangle\setminus F$ results immediately in a contradiction and, hence, L consists of outer derivations. The following proposition remains also valid for infinite-dimensional Lie algebras over a field of an arbitrary characteristic.

6.1.9. Proposition. *A set of constants of any Lie algebra of homogenous derivations forms a free subalgebra.*

Proof. It coincides with the proof of proposition **6.1.2** nearly to the word. One only has to replace the word "automorphism" with the word "derivation" and to consider a "derivative" b_i^g instead of the difference $b_i-b_i^g$.

6.1.10. Well known is the fact that any finite group is presented by linear transformations of a certain finite-dimensional space V. If we consider an arbitrary basis of V as X, then we see that any finite group is presented by homogenous automorphisms of a free algebra $F\langle X\rangle$. Therefore, theorem **6.1.1** shows, in particular, that the lattice of free subalgebras which contain a fixed free subalgebra can be antiisomorphic to the lattice of subgroups of an arbitrary finite group.

An analogous problem of presentation arises for an arbitrary restricted Lie algebra as well. It is still more interesting , since, according to the Baer theorem, a restricted Lie ∂-algebra of derivations of a finite extension of fields $K \supseteq k$ is generated over K by one element and, in particular, can be arbitrary in no way. The following proposition shows that any restricted differential Lie algebra over a field C is presented by homogenous derivations of a certain ring $C\langle X\rangle$. In this case if the initial algebra acts trivially on C (i.e., it is a usual Lie p-algebra), then we have it presented by derivations of a free algebra.

6.1.11. Proposition. *Any restricted differential Lie C-algebra can be presented as an algebra of outer derivations of a prime ring with a generalized centroid C.*

Proof. Let us first show that any restricted Lie ∂-algebra D is embedded into its universal enveloping UD.

Let M be a certain quite ordered basis of the space D over a field C, and let $\{c_i\}$ be the basis of C over a field of constants, P. Let us consider a free associative algebra $F_0 = P < \sigma_\mu, 1_a >$, which is freely generated by the symbols $\sigma_\mu, 1_a$, where $\mu \in M$, $a \in \{c_i\}$. Let U be the factor-algebra of F_0 determined by the relations

$$\sigma_\mu \sigma_\rho = \sigma_\rho \sigma_\mu + \sum \alpha_i \sigma_{\mu_i} 1_{c_i},$$

where $\mu > \rho \in M, [\mu\rho] = \sum \alpha_i \mu_i c_i$, $\alpha_i \in P$;

$$\sigma_\mu^p = \sum \alpha_i \sigma_{\mu_i} 1_{c_i},$$

where $\mu^{[p]} = \sum \alpha_i \mu_i c_i$, $\mu, \mu_i \in M$, $\alpha_i \in P$;

$$1_c \sigma_\mu = \sigma_\mu 1_c + \sum \alpha_i 1_{c_i},$$

where $c^{\mu}=\sum\alpha_i c_i$, c, $c_i\in\{c_i\}$, $\mu\in M$, $\alpha_i\in P$;

$$l_{c_1}l_{c_2}=\sum\alpha_i l_{c_i},$$

where $c_1c_2=\sum\alpha_i c_i$, c_1, c_2, $c_i\in\{c_i\}$, $\alpha_i\in P$.

In the set of symbols $\{\sigma_\mu, l_a\}$ let us introduce a certain complete order in such a way that $\sigma_\mu>\sigma_\rho\Leftrightarrow\mu>\rho$ and $\sigma_\mu<l_a$ for all μ, a. On the set of words from $\{\sigma_\mu, l_a\}$ the order $>>$ will be determined in the following way. Let us assume that the word f is bigger than the word g ($f>>g$), if the word f_M obtained from f by crossing out all the symbols from $\{l_a\}$ is bigger than the word g_M under a standard order of words. If $f_M=g_M$, then bigger is the word which is bigger under a standard order (with respect to the order $>$).

Now we can easily check that the set of the written relations of the algebra F_0 is closed relative the compositions (see [23] and [24]). This implies that the images of the words not containing subwords of the type $\sigma_\mu\sigma_\rho$ at $\mu>\rho$; σ_μ^p; $l_c\sigma_\mu$; $l_{c_1}l_{c_2}$, form a basis of the algebra F_0 (the Gröbner basis).

Let us now denote by $\hat{\mu}$, L_a the images of the symbols σ_μ, l_a in U, respectively. In this case the mapping $\xi:\sum\alpha_i\mu_i c_i\to\sum\alpha_i\hat{\mu}_i L_{c_i}$ determines a certain embedding of the space D into U. It is evident that $\xi(D)$ generates U as an associative P-algebra with a unit. Then, U turns into a C-module when identifying L_c with an element $c\in C$, since $\{\sum\alpha_i L_{c_i}\}\cong C$. In this case $c\hat{\mu}=L_c\hat{\mu}=\hat{\mu}L_c+\sum\alpha_i L_{c_i}=\hat{\mu}c+c^{\mu}$, where $c^{\mu}=\sum\alpha_i c_i$. Therefore, the task now is to check if ξ is a homomorphism of restricted Lie rings. Using identities (4) (**1.1.2**), we get

$$\begin{aligned}
&\xi([\mu_1c_1,\mu_2c_2])=\xi([\mu_1,\mu_2]c_1c_2+\mu_1c_2c_1^{\mu_2}+\mu_2c_1c_2^{\mu_1})=\\
&=[\hat{\mu}_1,\hat{\mu}_2]c_1c_2+\hat{\mu}_1c_2c_1^{\mu_2}+\hat{\mu}_2c_1c_2^{\mu_1}=[\hat{\mu}_1c_1,\hat{\mu}_2c_2],\\
&\xi((\mu_1+\mu_2)^{[p]})=\xi(\mu_1^{[p]}+\mu_2^{[p]}+W(\mu_1,\mu_2))=\\
&=\hat{\mu}_1^{\,p}+\hat{\mu}_2^{\,p}+W(\hat{\mu}_1,\hat{\mu}_2)=(\hat{\mu}_1+\hat{\mu}_2)^p,\\
&\xi((\mu c)^{[p]})=\xi(\mu^{[p]}c^p+\mu T(c))=\hat{\mu}^{\,p}c^p+\hat{\mu}T(c)=(\hat{\mu}c)^p.
\end{aligned}$$

Therefore, U is a universal enveloping and ξ_U is an embedding.

Let us add to the set of generators of the algebra F_0 one more symbol, y, which is less then the former ones, and let us add to the set of relations relations of the type $1_c y = y 1_c$, $c \in \{c_i\}$. In this case the obtained set of relations is again closed with respect to the compositions. Hence, D does not intersect with the center of $U\langle \hat{y} \rangle$, where $\hat{y}$ is the image of y under a natural homomorphism. In particular, D is a restricted Lie ∂-subalgebra of the factor-algebra $U\langle \hat{y} \rangle / Z$, where Z is the center of $U\langle \hat{y} \rangle$.

Thus, in order to prove the proposition, it is sufficient to show that the factor-algebra over the center D/Z of any restricted ∂-algebra D is presentable in an appropriate way.

Let $C\langle M \rangle$ be a free associative C-algebra generated by the set σ_μ, $\mu \in M$. On the generating elements $C\langle M \rangle$ let us determine the action of the elements from D by the formula $\sigma_\mu^\nu = \sum \sigma_{\mu_i} c_i$, where $[\mu, \nu] = \sum \mu_i c_i$. Now the formula of derivation $(xy)^\mu = x^\mu y + xy^\mu$ makes it possible to propagate the action of D onto $C\langle M \rangle$. One can easily check that

$$\sigma_\mu^{[\nu_1, \nu_2]} = ((\sigma_\mu)^{\nu_1})^{\nu_2} - ((\sigma_\mu)^{\nu_2})^{\nu_1},$$

$$\sigma_\mu^{\nu^{[p]}} = ((\sigma_\mu^\nu)\ldots)^\nu.$$

Thus, we get a homomorphism of the restricted Lie ∂-algebras, $\varphi\colon D \to \mathrm{Der}C\langle M \rangle$, the kernel of which is equal to Z. The proposition is proved.

6.1.12. By way of concluding the section, let us remark that it is still unknown if we can be free from the condition of homogeneity in theorems **6.1.1** and **6.1.8**. We have seen that to this end one should show that the algebras of invariants of a finite group of automorphisms is free (and the algebra of constants of a finite-dimensional restricted Lie algebra is also free). These problems have been considered elsewhere. Thus, for instance, P. Cohn ("On the Automorphism Group of the Free Algebra of Rank 2", Preprint, 1978) has showed any finite group of automorphisms of an invertible order of a free algebra of rank 2 to be conjugated with a certain linear (i.e., affecting the generators linearly) group. Hence, the algebra of its invariants is free.

The Cohn's approach is based on the Cherniakievich-Makar-Limanov theorem stating that every automorphism of a free algebra of rank 2 is tame.

Since so far no examples of not tame automorphisms of free algebras have been known, our problem is not less interesting for a group of tame automorphisms as well. It should also be added that so far no examples of even infinite groups with non-free algebras of invariants have been known.

For derivations the situation is somewhat different. G.Bergman gives an example with a non-free kernel.

In a free algebra of rank 3 let us consider the derivation μ, determined by the formulas

$$x^{\mu} = xyx + x\ ,\ y^{\mu} = -\ yxy\ ,\ z^{\mu} = -\ x.$$

One can show that the kernel of this derivation is generated by the elements

$$\varphi = xyz + x + z \qquad q = yx + 1$$
$$r = xy + 1 \qquad s = zyx + x + z$$

These elements are related by $\varphi q = rs$, with the help of which, using the Cohn's theory, one can easily show that the subalgebra generated by them is not free.

If the basic field has the characteristic $p > 0$, then the Lie p-algebra generated by μ is infinite-dimensional.

By way of conclusion let us formulate the problems arising in their explicit form.

6.1.13. Will a free algebra of invariants of an arbitrary group of automorphisms of a free algebra be free? The same problem is also open for discussion for tame, finite, and finite groups of an invertible order.

6.1.14. Will an algebra of constants of a finite-dimensional restricted Lie algebra of derivations of a free algebra over a field of positive characteristic be free?

REFERENCES

G.Bergman [19];
G.Bergman, P.Cohn [20];
Bokut' [23, 24];
P.Cohn [34,35];
A.Czerniakiewich [37];
V.K.Kharchenko [83];
A.T.Kolotov [87];
D.R.Lane [92];
L.G.Makar-Limanov [101];
W.S..Martindale, S.Montgomery [107];
A.I.Shirshov [143].

6.2 Noncommutative Invariants

In this section we shall dwell on the problem of finite generality of algebras of noncommutative invariants of linear groups. The Koryukin's theorems presented here show finite generality not to exist, as a rule, and explain the reason of this phenomenon, which can be attributed to the fact that symmetrical groups act on homogenous components. This peculiarity taken into account, we come to an analog of the Hilbert-Nagata theorem.

6.2.1. Definitions. Let V be a finite-dimensional space over a field F, $\dim V > 1$, G be a certain group of its linear transformations. Let us denote by $F\langle V\rangle$ the tensor algebra of this space.

$$F< V> = F \dot{+} V \dot{+} V\otimes V \dot{+} V\otimes V\otimes V \dot{+} \dots$$

If we fix a basis $X = \{x_1,\dots, x_n\}$, then the algebra $F\langle V\rangle$ can be viewed as a free associative algebra $F\langle X\rangle$.

The action of the group G is uniquely extended onto $F\langle V\rangle$. The *noncommutative invariants* or simply *invariants* are fixed elements of the algebra $F\langle V\rangle$. A set of all invariants will be denoted by $Inv G$.

6.2.2. Definition. The *support space* of a subset $A\subseteq F\langle V\rangle$ is the least, by inclusion, subspace $W\subseteq V$, such that $A\subseteq F\langle W\rangle$.

6.2.3. Lemma. *Any subset has a support space. If a subset A is stable relative G, i.e., $A^g = A$, $g\in G$, then its support space is also stable relative the action of G.*

Proof. It is obvious that for subspaces U and W the equality $F\langle U\rangle \cap F\langle W\rangle = F\langle U\cap W\rangle$ is valid. Therefore, if $A\subseteq F\langle U\rangle$, $A\subseteq F\langle W\rangle$, then $A\subseteq F\langle U\cap W\rangle$, which implies the existence of a support space. If $A^g = A$, $g\in G$ and $A\subseteq F\langle W\rangle$, then $A = A^g \subseteq F\langle W^g\rangle \cap F\langle W\rangle = F\langle W^g\cap W\rangle$ and, hence, if W is a support space, then $W^g \supseteq W$ and, since the space W is finite-dimensional, $W^g = W$. The lemma is proved.

6.2.4. Theorem. *The algebra of noncommutative invariants $Inv G$ is finitely generated iff the group G acts on the support space W of the algebra $Inv G$ as a finite cyclic group of scalar transformations.*

In one respect the theorem is obvious: if the restriction of G on W is generated by the transformation $g: W \to W$, such that $gw = \alpha w$, $\alpha^m = 1$, then the algebra $InvG$ is equal to $F\langle W^{\otimes m}\rangle$ and, hence, it is finitely generated (here m is the order of g, i.e., $\alpha^k \neq 1$ at $0 < k < m$).

The proof of the inverse statement will be divided into a number of steps.

6.2.5. It should be recalled that on the n-th tensor degree of the space V, besides the linear group $GL(V)$, a group of permutations S_n acts, rearranging (in a similar way) the factors in all tensor sums:

$$(v_1 \otimes v_2 \otimes \ldots \otimes v_n)^{\pi} = v_{\pi(1)} \otimes v_{\pi(2)} \otimes \ldots \otimes v_{\pi(n)}.$$

In this case the action of $GL(V)$ and S_n commute and, hence, homogenous component of the degree n of the algebra $InvG$ is invariant relative the action of S_n, i.e., for every invariant r of the degree n we have $n!$ new invariants (not necessarily different), r^{π}, $\pi \in S_n$.

6.2.6. Let $X = \{x_1, \ldots, x_n\}$ be a basis of the space V, with its elements called letters. Every element f of the tensor algebra is uniquely presented as a linear combination $\sum_w \alpha_w w$ of different words from X. If $\alpha_w \neq 0$, then we shall say that the word w *is used* in the presentation of f. If A is a certain subset of a tensor algebra, then we shall say that the letter x *occurs* in A if x occurs in at least one of the words used in the presentation of elements of A. And, finally, a sequence of letters $y_1, y_2, \ldots, y_m, \ldots$ will be considered *agreed with* A, if at least one of the words of type $y_1 y_2 \ldots y_m$ is used in the presentation of a certain element $f \in A$.

6.2.7. Lemma. *If a multiplicative closure of a finite nonempty set of words M is closed relative the action of symmetric groups, then any infinite sequence of letters $y_1, y_2, \ldots, y_m, \ldots$ occurring in M agrees with M.*

Proof. Let m be the maximum of the lengths of the words from M. Let us consider a word $w = w_1 \ldots w_m$, where w_i is a word from M with the letter y_i included. Let us rearrange the letters in the word w in such a way that the new word had the form $w^{\pi} = y_1 y_2 \ldots y_m z$. By the condition, this word occurs in the multiplicative closure of the set M, i.e., $w^{\pi} = v_1 \ldots v_s$, where v_i lies in M. Since the length of v_i is not greater than m, then, comparing both presentations of the word w^{π} we come to the conclusion that $v_1 = y_1 .. y_k \in M$. The lemma is proved.

6.2.8. Lemma. *If the algebra* (A) *generated by a finite set* A *of the elements of a tensor algebra is homogenous and closed relative the action of symmetric groups, then any infinite sequence of the letters occurring in* A *agrees with* A.

In order to prove the lemma, it is sufficient to choose as M a set of all the words used in the presentation of A and to make use of the previous lemma, taking into account that the fact that the word m_1 occurs in the presentation of $a_1 \in (A)$, and the word m_2 occurs in the presentation of $a_2 \in (A)$ implies, due to homogeneity, that the word $m_1 m_2$ occurs in the presentation of $a_1 a_2$.

6.2.9. Lemma. *Let* $g \in GL(V)$ *and a field* F *be algebraically closed. Then, if* g *is not a scalar transformation, then there is a basis* X *of the space* V *and an infinite sequence of letters (from* X*) which does not agree with* $Inv(g)$.

Proof. Let us first assume that in a certain basis X a transformation g is set by a diagonal matrix $x_i^g = \alpha_i x_i$, $\alpha_i \in F$. Let also, for definiteness, $\alpha_1 \neq \alpha_2$. Let us construct, by induction, a sequence of letters $y_1,..., y_i,...$. If $\alpha_1 \neq 1$, then we set $y_1 = x_1$; if $\alpha_1 = 1$, then we assume $y_1 = x_2$. If $y_1,..., y_{i-1}$ are constructed and the element $y_1 .. y_{i-1} x_1$ is not fixed relative g, then we set $y_i = x_1$; in the opposite case we set $y_i = x_2$. By the construction, all the words $y_1 .. y_m$ are not fixed relative g. However, in the case under discussion the algebra $Inv\langle g \rangle$ is a linear enveloping of fixed words. This fact implies that none of the words $y_1 .. y_m$ is used in the presentation of the elements $f \in Inv\langle g \rangle$, i.e., the sequence constructed does not agree with $Inv\langle g \rangle$.

In a general case let us choose a basis X where the transformation g has a Jordan normal form: $x_1^g = \alpha x_1 + x_2$ and $x_2^g = \alpha x_2 + x_3$, or $x_2^g = \alpha x_2$, in which case the subspace $x_3 F + .. + x_n F$ is invariant relative g. Let us consider a homogenous invariant f of the automorphism g and present it as

$$f = \beta\, x_1^m + \gamma\, x_1^{m-1} x_2 + f_1\,,$$

where $\beta, \gamma \in F$, and the presentation of f_1 does not include the words

$x_1^m, x_1^{m-1}x_2$. Then we have

$$f^g = \beta\alpha^m x_1^m + (\gamma\alpha^m + \beta\alpha^{m-1}) x_1^{m-1} x_2 + f_2 ,$$

in which case in the presentation of f_2 no use is made of the monomials $x_1^m, x_1^{m-1}x_2$. As $f = f^g$, we get

$$\beta\alpha^m = \beta , \qquad \gamma\alpha^m + \beta\alpha^{m-1} = \gamma.$$

Allowing for the fact that the eigenvalue α of the non-degenerate transformation g is other than zero, we get $\beta = 0$. This implies that the sequence of letters $x_1, x_1, \ldots, x_1, \ldots$ does not agree with $Inv\langle g \rangle$. This is the proof of the lemma.

6.2.10. Lemma. *Let* K *be an extension of a field* F; G *be a group of linear transformations of the space* V *over* F. *In this case the algebra of invariants of the group* G *in* $F\langle V \rangle$ *is finitely generated iff the algebra of invariants of the group* G *in* $F\langle V \otimes K \rangle$ *is finitely generated over* K.

The proof is obvious.

6.2.11. Let us go over to proving theorem **6.2.4**. Let $InvG$ be finitely generated. By lemma **6.2.10**, the field F can be considered algebraically closed. Let W be a support space of the algebra of invariants. By lemma **6.2.3**, this space is invariant and, hence, by restricting the action of G on W one can assume, with no generality violated, that $V = W$.

If g is not a scalar automorphism from G, then, by lemma **6.2.9**, there is a basis of the space $V = W$ and a sequence of letters from this basis which does not agree with $Inv\langle g \rangle$. However, $InvG \subseteq Inv(g)$, and, hence, such a sequence of letters does not agree with $InvG$ either and, by lemma **6.2.8**, the algebra $InvG$ cannot be finitely generated.

Thus, G is a group of scalar transformations. If it is infinite, then it has no nonzero invariants at all and, hence, $0 = V = W$ is a contradiction. Therefore, G is finite. Since G is embeddible into F (to every scalar transformation we put into correspondence its eigenvalue), then it is, as a finite subgroup of the multiplicative group of the field, cyclic. This is the proof of the theorem.

6.2.12. Corollary. *Let* G *be an almost special group of matrices (i.e., the kernel of a mapping* $g \to \det g$ *has a finite index in* G*). If the algebra* $InvG$ *is finitely generated, then* G *is a finite cyclic group of scalar matrices.*

Proof. It is sufficient to show that the support space of the algebra of invariants coincides with the whole space V. Let us construct a standard polynomial

$$S = \sum(-1)^{\pi} x_{\pi(1)} \cdots x_{\pi(n)},$$

where summation is carried out by all the permutations π. One then can easily see that $S^g = \det g \cdot G$. Since $\det G$ is a finite group in the field, we can find a number m, such that $(\det g)^m = 1$ for all $g \in G$. Therefore, we have $S^m \in Inv\, G$. Let W be a support space of the element S^m. Let us choose its basis $v_1,\dots, v_k$ and extend it to the basis $v_1,\dots, v_n$ of the space V. Let H be a linear transformation $v_i \to x_i$. In this case $S(x_1,\dots, x_n) = S(v_1,\dots, v_n)^h = \det h \cdot S(v_1,\dots, v_n)$ and, hence, $S^m(v_1,\dots, v_n) \in F(v_1,\dots v_k)$, which is impossible as $k < n$. Therefore, $r = n$, $V = W$, and the corollary is proved.

6.2.13. Corollary. *The algebra of noncommutative invariants of an irreducible group is either trivial or finitely generated.*

The proof is obvious.

6.2.14. The proof of theorem **6.2.4** shows that the main obstacle for finite-generatedness of an algebra of invariants is the action of symmetric groups on homogenous components. Allowing for the fact that the construction of the invariant f^{π} by an invariant f and a permutation π is of no calculation difficulty, it is natural to study the algebra of invariants as an associative algebra, on the homogenous components of which symmetrical groups are acting. Now the question arises: if there exists a finite number of invariants, such that all the rest are expressed through them by using the operations of algebra and the actions of symmetrical groups on homogenous components? The answer to this question is positive for reductive groups (it should be recalled that a reductive group is a linear group with all its rational representations completely reducible).

6.2.15. Definition. The subalgebra $A \subseteq F(V)$ is called an *S-subalgebra* if it is homogenous and closed relative the actions of symmetric groups on homogenous components. The ideal I of A is called an *S-ideal* if it is an *S-subalgebra.*

6.2.16. Theorem. *The algebra of noncommutative invariants of a reductive group is finitely generated as an S-algebra.*

A general scheme of this theorem is the same as that in the Nagata-Hilbert theorem.. Let us start with an analog of the Hilbert theorem on the basis.

6.2.17. Theorem. *Any strictly ascending chain of S-ideals in* $F(V)$ *is finite.*

The proof of this theorem is based on the Higman lemma.

6.2.18. Lemma. *For any infinite sequence of words of a finite alphabet one can choose an infinite subsequence, every word of which can be obtained from any subsequent word of the subsequence by crossing out the letters.*

Proof. On the set of all the words let us introduce a relation of a partial order $u \leq v$ iff u can be obtained from v by crossing out letters. Let us call the set of words P *Higman* if for any infinite subsequence of words from P we can choose an increasing subsequence. Our task is to show a set of all the words from $X = \{x_1, \dots, x_n\}$ is Higman.

For the word w let us set $w^{\perp} = \{v | w \not\leq v\}$. Since $x_i^{\perp}$ is a set of all the words not including the letter x_i, we can assume that all the sets $x_i^{\perp}$ are Higman ones. By induction over the length of the word w let us show that all the sets $w^{\perp}$ are also Higman.

Let $w = xv$, where x is a letter, v is a word for which $v^{\perp}$ is a Higman set. An arbitrary word $u \in w^{\perp}$ which does not belong to $x^{\perp}$ can be presented as axb, where $a \in x^{\perp}$, $b \in v^{\perp}$. To this end it is sufficient to find the first letter x occurring in the presentation (since, if $b > v$, then $axb > xv$).

Let us consider an arbitrary sequence of words from $w^{\perp}$:

$$a_1xb_1, a_2xb_2, a_3xb_3, \dots, a_nxb_n, \dots$$

where $a_i \in x^{\perp}$, $b_i \in v^{\perp}$, in which the cofactors xb_i could be absent. Since $x^{\perp}$ is a Higman set, one can eliminate a part of the terms from this sequence in such a way that in the remaining infinite sequence the cofactors a_i will be increasing. As $v^{\perp}$ is also a Higman set, from the sequence obtained one can eliminate a part of the terms as well, so that in the remaining infinite subsequence the right cofactors b_i will be also increasing, which is the required proof.

Therefore, all the sets $w^{\perp}$ are Higman. Let $w_1, w_2, \dots w_n, \dots$ be a sequence of words from which one cannot choose an increasing subsequence. In this case not all the words $w_2, \dots, w_n, \dots$ lie in $w^{\perp}$, i.e., one can find a pair $w_1 < w_k$. Analogously, not all the words $w_{k+1}, \dots w_n, \dots$ lie in $w_k^{\perp}$, i.e., $w_k < w_m$ at a certain $m > k$, and so forth. This is the proof of the lemma..

6.2.19. The **proof** of lemma **6.2.17**. Let

$$I_1 < I_2 < \ldots I_m < \ldots$$

be an infinite strictly ascending chain of S-ideals in $F(V)$.Let us fix a basis $X = \{x_1, \ldots, x_n\}$ of a space V. From the elements of the set $I_m \setminus I_{m-1}$ let us choose a homogenous element z_m with a minimal senior word w_m in the lexicographical ordering of the words. By the Higman lemma, one can find numbers $i < j$, such that

$$w_j = y_1 \ldots y_m \qquad (y_k \in X, 1 \leq k \leq m)$$
$$w_i = y_{r_1} y_{r_2} \ldots y_{r_s} \qquad (1 \leq r_1 < r_2 < \ldots < r_s \leq m)^{\cdot}$$

Let us denote through h the word obtained from w_j by crossing out the letters standing in the positions $r_1, \ldots, r_s$. Let $w_i h = y_{\pi(1)} \ldots y_{\pi(m)}$. Then π is a permutation, in which case $w_j = (w_i h)^{\sigma}$, where $\sigma = \pi^{-1}$.

Let us write an element z_i as $w_i + \sum_w w \cdot \alpha_w$, where summation is carried out over monomials w which are less than w_i and α_w are from the basic field. (With no generality violated, one can assume that the senior coefficient is equal to the unit). We have

$$(z_i h)^{\sigma} = (w_i h)^{\sigma} + \sum (wh)^{\sigma} \alpha_w .$$

It should be remarked that the word $(w_i h)^{\sigma}$ is the senior word in the presentation of $(z_i h)^{\sigma}$. If $w < w_i$, then, by crossing out the letters occurring in the positions $r_1, \ldots, r_s$ in the monomials $(w_i h)^{\sigma}$ and $(wh)^{\sigma}$, we get the same word, h. Therefore, the first from the left difference between the words $(w_i h)^{\sigma}$ and $(wh)^{\sigma}$ is attained at one of the positions $r_1, \ldots, r_s$, i.e., where the words w_i and w are located, respectively.

Since $(w_i h)^{\sigma} = w_j$, then the senior word of the element $z_j - (z_i h)^{\sigma}$ is less than w_j. However, this element belongs to $I_j \setminus I_{j-1}$, which contradicts the choice of z_j. The theorem is proved.

6.2.20. Let h be an arbitrary element of $F\langle V\rangle$, G be a reductive group of linear transformations of a space V. Let M_h be a space generated in $F\langle V\rangle$ by the orbit of the element h, while N_h be a subspace generated by all the differences $h^g - h$, where $g \in G$. Since automorphisms from G do not change the degrees of elements, then the space M_h is finitely-dimensional. Its invariant subspace N_h has a co-dimensionality not greater than one, in which case the group G acts identically on the factor-space M_h / N_h. Moreover, the presentation $G \to GL(M_h)$ will be rational if in M_h we choose a basis consisting of the elements of the orbit $\bar{h}$.

Therefore, if if $M_h \neq N_h$, then, due to reductivity, one can find a fixed element $h^* \in M_h$ not lying in N_h. In any case we get a presentation

$$h = h^* + h', \tag{3}$$

where h^* is a fixed element from M_h, while h belongs to N_h.

6.2.21. Lemma. *Let A be a certain S-subalgebra in $F\langle V\rangle$; $f_1,\ldots, f_m$ be its homogenous elements. Then any element v from an S-ideal generated by the elements $f_1,\ldots, f_m$ has a presentation*

$$v = \sum_{i,j} (f_i h_{ij})^{\sigma_j}, \tag{4}$$

where h_{ij} are either homogenous elements from A, or elements of the field, while σ_j are suitable permutations.

The proof results immediately from the fact that any element of the type uf_iv, where u, v are words, is transferred through a permutation into f_iuv.

6.2.22. Lemma. *An intersection of an S-ideal generated in $F\langle V\rangle$ by homogenous invariants $f_1,\ldots, f_m$ of a reductive group G, with the algebra of invariants $\mathrm{Inv}G$ coinciding with an S-ideal generated by these elements in $\mathrm{Inv}G$.*

Proof. Let us denote by I an S-ideal generated by elements $f_1,\ldots, f_m$ in $F\langle V\rangle$, and by I_1 an S-ideal generated by these elements in $\mathrm{Inv}G$. The inclusion $I_1 \subseteq I \cap \mathrm{Inv}G$ is obvious. Let $v \in I \cap \mathrm{Inv}G$.

Then by lemma **6.2.21**, we get presentation (2). Let us consider a finite set $\{(i, j) \mid h_{ij} \neq 0\}$. Let us prove that $v \in I_1$ by induction over the number k of the elements of this set. At $k = 0$ we have nothing to prove.

In decomposition (4) let us single out one of the terms

$$v = (fh)^{\sigma} + y, \tag{5}$$

where y is a sum of all $k-1$ addends of the type $(fh)^{\sigma}$, i.e.,

$$y = \sum_B (f_i h_{ij})^{\sigma_j}, \tag{6}$$

in which case B contains $k-1$ elements. Let $g \in G$. As $[(fh)^{\sigma}]^g - (fh)^{\sigma} = (f(h^g - h))^{\sigma}$, we have

$$0 = v^g - v = (f(h^g - h))^{\sigma} + \sum_B (f_i(h_{ij}^g - h_{ij}))^{\sigma_j}. \tag{7}$$

For the element h let us consider a decomposition $h = h^* + h'$ by formula (3), where $h' = \sum_G (h^g - h)\alpha_g$ for some $\alpha_g \in F$. Multiplying equality (7) by α_g and summing them up over all g from G, we get

$$0 = (fh')^{\sigma} + \sum_B (f_i h'_{ij})^{\sigma_j}, \tag{8}$$

where h'_{ij} are certain homogenous elements from $F\langle V\rangle$.

According to (3) and (5), the invariant $v - (fh^*)^{\sigma}$ is equal to $(fh')^{\sigma} + y$. Writing y by formula (6), and substituting $(fh')^{\sigma}$ by formula (8), we get

$$v - (fh^*)^{\sigma} = \sum_B (f_i(h_{ij} - h'_{ij}))^{\sigma_j}.$$

Now we apply the inductive proposition to the right-hand part of the above equality, i.e., $v - (fh^*)^{\sigma} \in I_1$, but then $v \in I_1$, which is the required proof.

6.2.23. Lemma. *Let* A^+ *be an ideal of an* S*-subalgebra*

$A \subseteq F\langle V\rangle$ which consists of all the elements of A with a zero free term. Then, if A^+ is generated as an S-ideal in A by homogenous elements $f_1,\dots, f_m$, then A is generated as an S-subalgebra (with a unit) by the same elements.

The proof is carried out by induction over the degree of a homogenous element $v \in A^+$. If $v = \sum (f_i h_{ij})^{\sigma_j}$, where $h_{ij} \in A$, then the degree of h_{ij} is strictly less than that of v and the induction is obvious.

6.2.24. From the lemmas proved above one can easily deduce now theorem **6.2.16**. Let us consider an S-ideal J in $F\langle V\rangle$ generated by a set $(Inv\, G)^+$ of invariants with no free terms. By theorem **6.2.17**, one can find a finite set of homogenous invariants $f_1,\dots, f_m$ generating this S-ideal. By lemma **6.2.22**, the intersection $J \cap Inv\, G$ is generated by the same elements as the S-ideal in $Inv\, G$. Since this intersection is equal to $(Inv\, G)^+$, then, by lemma **6.2.23**, the algebra of invariants is generated by the set $f_1,\dots, f_m$ as an S-algebra, which is the required proof.

6.2.25. Remark. As was the case in the Nagata-Hilgebt theorem, reducibility of a group is necessary only in decomposition (3), the latter demanding a somewhat weaker condition, i.e., splitting of sequences of homomorphisms of finite-dimensional G-modules:

$$0 \longrightarrow N \longrightarrow M \longrightarrow M/N \longrightarrow 0,$$

where G acts rationally on M and trivially on M/N. This condition, for instance, works for some rational representations of groups $GL(h, F)$ and $SL(n, F)$ in $GL(V)$, if F is an algebraically closed field of a zero characteristic, as well as for rational representations of any semi-simple algebraic group $\Gamma \subseteq GL(n, k)$ over a field of a zero characteristic.

6.2.26. Hilbert series. The problem of describing groups $G \subseteq GL(V)$, the algebras of non-commutative invariants of which are finitely generated, was posed by S.Montgomery at the conference "Nöether Rings and Rings with Polynomial Identities" at Durham University in 1979, as well as by the authors of the "Dnestrovskaya Notebook" (problem N **2.88**). E.Formanek and W.Dicks have chosen a way to solve this problem which is different from the one presented here. Their approach is based on studying Hilbert series of the algebras of invariants.

It should be recalled that a *Hilbert series* of a graded F-algebra

$$R = F \oplus R_1 \oplus R_2 \oplus \dots$$

such that the dimensions of the components R_i are finite, is a series with integer coefficients

$$H(R) = 1 + \sum_{n\geq 1}(\dim R_n)t^n.$$

This series can be viewed as an element of a ring of the power series $Z[[t]]$, and as a function, provided t is considered to be running through a set of either real or complex numbers.

Let now $G \subseteq CL(V)$. Then, as we know, $R = InvG$ is a homogenous free subalgebra in $F\langle V\rangle$. In particular, one can find a homogenous subspace $U \subseteq InvG$, such that $InvG \cong F\langle U\rangle$. Let us determine an auxiliary series

$$P(U) = \sum_{n\geq 1}\dim(U \cap R_n)t^n.$$

From the obvious decomposition of graded spaces

$$F < U > \cong F \oplus (U \otimes F < U >)$$

we get $H(F(U)) = 1 + P(U)\, H(F\langle U\rangle)$. Hence, in the ring of series we have the relation

$$H(InvG)^{-1} = 1 - P(U).$$

In particular, this affords that $InvG$ is finitely generated iff its Hilbert series has the form $\frac{1}{1-f(t)}$, where f is a polynomial.

We know (theorem **6.2.4**) that, as a rule, a Hilbert series has no form of the kind. However, calculations of a Hilbert series in an explicit form gives an essential positive information on the algebra of invariants (unlike the very fact of being not finitely generated).

Hilbert series for the algebras of commutative invariants were calculated as far back as the last century. For instance, well known is the Molien's theorem published in 1897: if a field F has a zero characteristic and a group G is finite, then the Hilbert series of the algebra $F[V]^G$ has the form:

$$H(t) = \frac{1}{|G|}\sum_{g\in G}\frac{1}{\det(1-gt)}$$

An analogous result is also valid for the algebras of non-commutative invariants (see [2]).

6.2.27. Theorem. *Let F be a field of a zero characteristic, G be a finite subgroup in $GL(V)$. Then we have*

$$H(t) = \frac{1}{|G|} \sum \frac{1}{1 - tr(g)t},$$

where tr is a trace of a linear transformation.

In the works by G.Almkwuist, W.Dicks and E.Formaneck [3], this formula was transferred to the case of compact groups in $GL(V)$ for a field of complex numbers. The authors called their result

"The non-commutative Molien-Weyl theorem". Let G be a compact subgroup in $GL_n(C)$. In this case

$$H(InvG) = \int_G \frac{d\mu}{1 - t\ tr(g)},$$

where $d\mu$ is a normalized Haar measure on G and $|t| < \frac{1}{n}$.

REFERENCES

G.Almkvist [2];
G.Almkvist, W.Dicks, E.Formanek [3];
J.Dieudonne, J.Carrol [40];
Dnestrovskaya Notebook [41];
G.Higman [58];
V.K.Kharchenko [82,83];
A.T.Kolotov [87];
A.N.Koryukin [88];
T.Molien [109];
H.Weil [151];
M.C.Wolf [154];

An analogous range of problems for reduced-free algebras:
R.Guaralnick [54];
V.K.Kharchenko [82].
I.V.L'vov, V.K.Kharchenko [90];
S.Montgomery [112];
J.Fisher, S.Montgomery [122];

6.3 Relations of a Ring with Fixed Rings

The trend we are going to discuss in this quite a bulky section was initiated by the Bergman-Isaaks theorem (theorem **1.3.6**) and is associated with the fact that when studying finite groups of automorphisms of arbitrary rings, the most important notions of the classical Galois theory, such as finite dimension, separability, normality lose their primary sense and the problem of relations of a fixed ring R^G with the basic ring R gains a priority. The Bergman-Isaaks theorem shows that when in R there is no additive $|G|$-torsion, the nilpotency of R^G implies that of R, which fact opens perspectives for a positive solution of the basic problem of the trend arising here, i.e., the problem of transferring certain properties from a fixed ring to the basic one, and vice versa. At the same time, in the latest investigations the theorem itself has had a limited application since studies of M-groups have gained in importance, as, on the one hand, they embrace the case of an arbitrary finite group of automorphisms of a domain (or of a ring without nilpotent elements), while on the other hand, it embraces the case of infinite groups to the extent which is necessary for considering Galois closures of finite groups. The reader will see that the key moment in proving nearly all the results presented in this section is the use (in different variations) of the Shirshov local finiteness of a certain essential ideal over a fixed ring (see **5.10**).

When considering finite groups, the most natural restrictions (i.e., the absence of an additive $|G|$-torsion in the ring) can be somewhat attenuated (in literature one can find limitations of the kind: the existence of an element γ with a unit trace $t(\gamma)=1$; non-degenerality of the form of a trace; $Rt(R)\ni 1$, etc.) or even eliminated at all, as, for instance, when studying domains or rings without nilpotent elements.

We are going to consider theorems under most natural, in our opinion, limitations: when viewing the problems referring to subclasses of prime rings the limitation is the semiprimity of the algebra of a group; when considering the problems pertaining to all the rings, it is the absence of additive $|G|$-torsion or inversability of the order of a group. These limitations do not always correspond to maximum generality, and sometimes somewhat weaker limitations prove to be essentially more cumbersome, obscuring the essence. The reader interested in corresponding generalizations will easily find them in the references.

It is obvious that for every property of a ring one can pose a problem on transference of this property from a fixed ring onto a ring and vice versa. Therefore, perspective might prove a general approach to the problem: to describe in some terms the properties for which the problem posed has a positive solution.

A step in this direction is consideration of radical problems.

A. *Radicals of algebras.* It should be recalled that in the class of associative algebras over a field F, a *radical* is a mapping ρ setting every algebra R into correspondence to its ideal $\rho(R)$ and obeying the following properties:

(1) If $\varphi: A \to B$ is a homomorphism of the algebras, then $\varphi(\rho(A)) \subseteq \rho(\varphi(A))$.

(2) In any algebra R the ideal $\rho(R)$ is the biggest among all the ideals I, such that $\rho(I) = I$.

(3) For any algebra R the relation $\rho(R/\rho(R)) = 0$ is valid.

Any property of the algebras can be associated with a mapping correlating the biggest ideal (if any) to the algebra R, and obeying, as an algebra, this property. If such a mapping proves to be a radical, then the property is called *a radical property*.

Let us consider several well-known examples of radicals and radical properties.

(a) The Baer radical (see **1.1**).

(b) The Levitsky locally-nilpotent radical, which is determined by the property of local nilpotency.

(c) A locally-finite radical, which is determined by the property of local finiteness of an algebra.

(d) The Koethe or upper nil-radical, which is determined by the property of nilpotency of all the elements of an algebra.

(e) An algebraic kernel, which is determined by the property of algebraicity of all the elements of an algebra.

(f) The Jacobson radical, the most well-known, which is determined by the property of quasi-regularity of all the elements

$$\forall x \exists y \qquad xy + x + y = 0$$

All the radicals listed above (the Koethe radicals and algebraic kernel under the condition that F is an uncountable field) obey the additional condition, i.e., they are overnilpotent and strict. Let us recall the definitions.

6.3.1. Definition. A radical is called *overnilpotent* if every algebra with a zero multiplication is radical.

6.3.2. Definition. A radical is called *strict* if every one-sided radical ideal of the algebra is contained in the radical of this algebra.

Moreover, all the radicals listed, but the Jacobson radical, are *absolutely hereditary*, i.e., any subalgebra of a radical algebra is radical.

6.3.3. Theorem. *Let* ρ *be an overnilpotent strict absolutely* *hereditary radical in the class of algebras over a field* F. *Then, if a finite group* G *acts on an algebra* R, *and the order of this group is invertible in* F, *then we have*

$$\rho(R^G) = \rho(R) \cap R^G = \rho(R)^G,$$

and, besides, if R^G *is radical, then* R *is also radical.*

Let us fix a radical ρ, obeying the conditions of the theorem and prove some preliminary lemmas.

6.3.4. Lemma. *If an algebra* R *is radical, then the algebra of matrices* R_n *is also radical.*

Proof. Let $1 \le j \le n$. Let us denote by $R^{(j)}$ a set of all matrices with all their elements, but those in the j-th column, equal to zero. In this case $R^{(j)}$ is a left ideal in R_n. Besides, a set A of matrices from $R^{(j)}$ which have a zero everywhere but at the point (j, j), forms the ideal of $R^{(j)}$, in which case $A^2 = 0$. Since the radical ρ is overnilpotent, we have $\rho(R^{(j)}) \supseteq A$.

Then, the mapping which puts an element α_{jj} located at (j, j) into correspondence to the matrix α, sets an epimorphism $\varphi\colon R^{(j)} \to R$ with a kernel A, i.e., $R^{(j)} / A \cong R$. Therefore, $\rho(R^{(j)} / A) = R^{(j)} / A$, i.e., $\rho(R^{(j)}) = R^{(j)}$.

Since ρ is a strict radical, then $\rho(R_n)$ contains a sum of all left ideals $R^{(j)}$, $1 \le j \le n$, which is equal to R_n. The lemma is proved.

6.3.5. Lemma. *Let a left* R*-module* V *to be generated by* n *elements. In this case there is an epimorphism of a certain subalgebra* H *of the algebra of matrices* R_n *to the algebra of all endomorphisms of the module* V, *the images of which are contained in* RV, *i.e.,* $R_n \supset H \xrightarrow{\lambda} Hom(V, RV) \to 0$.

Proof. Let $R^{\#}$ be an algebra obtained from R by an external addition of a unit, it it is absent, and $R^{\#} = R$, if it exists. In this case the algebra of matrices R_n is naturally identified with the algebra

$$Hom_{R^{\#}}\Big(\sum_{\alpha=1}^{n} \oplus R^{\#}_{\alpha},\ \sum_{\alpha=1}^{n} \oplus R_{\alpha}\Big) \text{ where } R^{\#}_{\alpha} \cong R^{\#},\ R_{\alpha} \cong R.$$

Let now p be a homomorphism of a free $R^{\#}$-module $\sum_{\alpha=1}^{n} \oplus R^{\#}_{\alpha}$ onto V, which maps the element $\sum r_{\alpha}$ onto $\sum r_{\alpha} v_{\alpha}$, where $V = \sum R^{\#} v_{\alpha}$.

Let us consider the following diagram:

$$\begin{array}{ccc} Hom(\sum\oplus R^{\#}_{\alpha}, \sum\oplus R_{\alpha}) & \xrightarrow{\nu} & (\oplus R^{\#}_{\alpha}, RV) \longrightarrow 0 \\ \downarrow \lambda & & \uparrow \mu \\ Hom(V, RV) & ===== & Hom(V, RV) \\ & & \uparrow \\ & & 0 \end{array}$$

In this diagram the mapping ν transfers the homomorphism ξ into a superposition $\xi \bullet p$, while the mapping μ transforms the homomorphism β into a superposition $p \bullet \beta$. The line of the diagram is exact, because $\sum\oplus R^{\#}_{\alpha}$ is a free $R^{\#}$-module. The exactness of the column results from the fact that p is an epimorphism of modules.

Let $H = \nu^{-1}(\mu Hom(V, RV))$. Then there is a linear mapping $\lambda\colon H \to Hom(V, VR)$, such that $\lambda\mu = \nu$ provided it is determined by the formula $h\lambda = (h\nu)\mu^{-1}$. Now we have to remark that λ is a homomorphism of algebras. If $\xi\lambda = \alpha$ and $\eta\lambda = \beta$, then $\xi \bullet p = p \bullet \alpha$ and $\eta \bullet p = p \bullet \beta$. Therefore, $\xi \bullet \eta \bullet p = \xi \bullet p \bullet \beta = p \bullet \alpha \bullet \beta$, i.e., $(\xi \bullet \eta)\lambda = \alpha \bullet \beta$. The lemma is proved.

Let B be a certain subalgebra of the algebra R. Let us call, for simplicity, an element $a \in R$ *finite* provided $R^{\#} a \subseteq \sum_{i=1}^{n} Ba_i$ for certain $a_1, \ldots a_n \in R$, where $R^{\#}$, as above, is an algebra obtained from R by adding a unit. It is evident that a set $W_R(B)$ of all finite over B elements forms a two-sided ideal of the algebra B.

6.3.6. Lemma. *Let L be a finitely generated left ideal of a ring R contained in $W_R(B)$.. In this case there is an epimorphism of algebras $\pi\colon S \to L$ which is determined on a certain subalgebra S of the algebra of matrices B_n.*

Proof. By the condition, L is contained in a certain finitely generated left B submodule $\sum_{i=1}^{n} Bv_i$ of the module R. Let us consider a left B-submodule V of $R^{\#}$, which is generated by elements $1, v_1, \ldots, v_n$. It should be remarked that $VL = L + \sum B^{\#} v_i L \subseteq L \subseteq BV$, which means that the operator of the right multiplication $r(l)$ by any element $l \in L$ belongs to $Hom_B(V, BV)$ and we have a homomorphism of the algebras $r\colon L \to$

$\to Hom_B(V, BV)$. The kernel of this homomorphism is zero, since V contains the unit of the ring $R^{\#}$. If we now make use of lemma **6.3.5**, then the lemma is proved.

Proof of theorem 6.3.3. Let us assume that the algebra of invariants R^G is radical, but $\rho(R) \neq R$. According to condition (1) from the definition of a radical, $\rho(R)$ is an invariant ideal and the action of the group G is induced on the factor-algebra $R / \rho(R)$. In this case $(R / \rho(R))^G = R^G / \rho(R) \cap R^G$, since the invariance of the element a is determined by the condition $a = \frac{1}{|G|} \sum a^g$, which is preserved under the homomorphisms which act in agreement with the group G.

Therefore, we can assume $\rho(R) = 0$. Since ρ is an overnilpotent radical, then R is a semiprime ring and, by theorem **5.14.15**, the group G is a Maschke group.

Let us apply theorem **5.10.6**, on local finiteness in the Shirshov sense to the ring R. In this case the ideal $W_R(R^G)$ is other than zero. By lemma **6.3.6** and **6.3.4**, each finitely generated left ideal lying in $W_R(R^G)$ is radical. As ρ is a strict ideal, then $\rho(R)$ contains the whole ideal $W_R(R^G)$, which is a contradiction.

Therefore, since R^G is radical, R is also radical. The inclusion $\rho(R) \cap R^G \subseteq \rho(R^G)$ results from the absolute heredity of the radical ρ. Let us prove the inverse inclusion.

Let us consider a right invariant ideal $L = \rho(R^G) R^{\#}$ as an algebra on which the group G acts. It should be remarked that $(\rho(R^G)R)^G = \rho(R^G) \cdot R^G$: if $a = \sum_i \rho_i r_i$ and $a = \frac{1}{n} \sum a^g$, then $a = \sum_i \rho_i (\frac{1}{n} \sum r_i^g) \in \rho(R^G) R^G$. This implies that the fixed ring of the algebra L is radical. According to the above proved, the algebra L is also radical. Since ρ is a strict radical, then $L \subseteq \rho(R)$ and, hence, $\rho(R^G) \subseteq \rho(R)$, which is the required proof.

An analogous result is also valid for the Jacobson radical.

6.3.7. Theorem. *If a finite group the order of which is invertible in F acts on the algebra R, then we have*

$$J(R^G) = J(R) \cap R^G = J(R^G),$$

where J is a Jacobson radical. If R^G is quasi-regular, then the algebra R is also quasi-regular.

Proof. Let us first assume that $J(R^G) = R^G$ and make use of the characteristic of a Jacobson radical as an intersection of the kernels of all irreducible right R-modules. Going over to the algebra $R/J(R)$, we can assume that $J(R) = 0$ and, in particular, R is a semiprime algebra.

By lemma **1.3.4** (or by lemma **5.6.6**), the left annihilator R^G in R is zero. By the symmetry, the right annihilator of the left ideal RR^G in a ring $Q(R)$ is also zero and, by theorem **5.10.2** (or by corollary **5.10.5**), the set RR^G contains an essential ideal I of the ring R.

By theorem **5.10.1** we can find an essential ideal W which is locally-finite over R^G in the Shirshov sense. Let V be an irreducible right R-module the kernel of which contains no intersection $W \cap I$. In this case $VW \neq 0$ and, if $v \in V$, $a \in W$ are elements, such that $va \neq 0$, then, because of the irreducibility, $vaR = V$. However, $aR \subseteq a_1R^G + ... + a_nR^G$ for appropriate $a_1,..., a_n \in R$. Therefore, the right R^G-module V is generated by a finite number of elements $va_1,..., va_n$.

As any finite-generated module, V contains a maximal R^G-submodule $W \neq V$. The factor-module V/W is irreducible and, hence, its kernel contains the radical $J(R^G) = R^G$.

We now come to a contradiction with the fact that $VI = V$:

$$V = VI \subseteq VRR^G \subseteq VR^G \subseteq W.$$

Therefore, the quasi-regularity of R^G implies that of R.

The inclusion $\rho(R^G) \supseteq R^G \cap \rho(R)$ results from the fact that the latter intersection is a quasi-regular ideal of R^G: if $a \in R^G$ and $x \in R$, $ax + a + x = 0$, then, applying the operator $\frac{1}{n}t$, we get $a\frac{1}{n}t(x) + a + \frac{1}{n}t(x) = 0$, i.e., the element a is quasi-regular in R^G as well.

Since J is a strict radical, then, repeating the last paragraph of the proof of theorem **6.3.3** word per word, we get $J(R^G) \subseteq J(R)$. The theorem is proved.

By way of conclusion it should be remarked that theorem **6.3.7** is not valid for Maschke groups. This has been proved by Martindale [105] when he constructed an example, where $J(R^G) \neq J(R)^G$ for a finite group of automorphisms G of a domain R with no additive torsion. Nevertheless, the semisimplicity of R is equivalent to that of R in the case of Maschke groups also, we shall see it in theorem **6.3.18**.

REFERENCES
V.A.Andrunakievich, Yu.M.Ryabukhin [8];
K.I.Beidar [13];
K.I.Beidar, Ten [17];
V.K.Kharchenko [72];
W.Martindale [104];
S.Montgomery [115];
J.-L.Pascaud [133];
D.S.Passman [135].

Jacobson and Baer radicals for (Jordan) fixed rings of Jordan isomorphisms were considered by W.Martindale and S.Montgomery [106].

B. Units, semisimple artinian rings and essential one-sided ideals

When considering various properties associated with modules it would be convenient to assume that the initial ring contains a unit. If this is not the case, a unit can be added to from outside which could not result in complications if the ring of invariants had its own unit. Therefore, it would be useful to take into account the following result.

6.3.8. Lemma. *Let G be a Maschke group of automorphisms of a semiprime ring R. Then the unit of the fixed ring will be that of R.*

Proof. If e is the unit of R^G, then $(x - xe)R^G = 0$ at any $x \in R$. Lemma **5.6.6** gives the identity $x = xe$. Analogously, $x = ex$. The lemma is proved.

6.3.9. Theorem. *Let us consider an M-group G of automorphisms of a semiprime ring R. The fixed ring R^G is a semisimple artinian ring iff the ring R is also a semisimple artinian ring.*

Let us recall one of many characterizations of semisimple artinian rings, i.e., that these are rings with a unit having no proper essential left ideals. It should be also recalled that a left ideal is called essential if it has a nonzero intersection with any nonzero left ideal.

6.3.10. Lemma. *Let G be a reduced-finite group of automorphisms of a semiprime ring R.*

(a) *If A is an essential left ideal of the ring R, then for any trace form τ there can be found an essential left ideal $A' \subseteq A$, such that $\tau(A') \subseteq A$.*

(b) *If A is an invariant left ideal, then for any trace form τ there* *exists an essential ideal* I, *such that* $\tau(IA) \subseteq A$.

Proof. The form τ can be easily presented as $\sum c_i \psi_i x^{g_i} \varphi_i$, where $g_i \in G$, $c_i \in C$, φ_i belong to some components Φ_{h_i}, $\psi_i \in B(G)$. Or, in another expression,

$$\tau = \sum c_i \psi_i \varphi_i x^{g_i h_i}.$$

If the ideal A is essential, but not invariant, then we set $A_1 = \cap A^{h_i^{-1} g_i^{-1}}$, while if A is an invariant left ideal, then we set $A_1 = A$. Let I_1 is an ideal of F, such that $c_i \psi_i \varphi_i I_1 \subseteq R$. We shall find an ideal $I \in F$, such that $I^{h_i g_i} \subseteq I_1$. Then for any $r \in IA_1 \overset{df}{=} A'$ we have

$$\tau(r) = \sum c_i \psi_i \varphi_i r^{g_i h_i} \in \sum c_i \psi_i \varphi_i (IA_1)^{g_i h_i} \subseteq A.$$

Let us show that A is an essential left ideal (if A is essential). To this end it is sufficient to show that an intersection $A \cap A^g$ is essential for any $g \in G$. Let $v \in R$. Let us find an ideal $I \in F$, such that $Iv^{g^{-1}} \subseteq R$ and let $J^g \subseteq I$, $J \in F$. Then $IJv^{g^{-1}}$ is a nonzero ideal in R. Therefore, $0 \neq IJv^{g^{-1}} \cap A$, which affords $0 \neq (IJ)^g v \cap A^g \subseteq Iv \cap A^g \subseteq R$ and, hence, $((IJ)^g v \cap A^g) \cap A \neq 0$. The latter intersection is contained in $Rv \cap (A^g \cap A)$ and, hence, $A^g \cap A$, which affords that A_1 are also essential.

Finally, $0 \neq I(L \cap A_1) \subseteq L \cap IA_1$ for any nonzero left ideal L and $I \in F$. The lemma is proved.

6.3.11. It is now evident that the same statements are true for right ideals as well.

6.3.12. Lemma. *Let* G *be an* M*-group of automorphisms of a semiprime ring* R, A *be an essential one-sided ideal of* R. *Then* $A \cap R^G$ *is an essential one-sided ideal of the ring* R^G.

Proof. Let W be a left nonzero ideal of R^G. Let us choose an arbitrary principal invariant form $\bar{\tau}_e = \tau$ and by lemma **6.3.10** find an essential left ideal $A' \subseteq A$, such that $\tau(A') \subseteq A$. If $I \in F$ and $\tau(I) \subseteq R$, $Ie \subseteq R$, then $\tau(IW) = \tau(I)W \subseteq W$ and, hence, $\tau(A' \cap IW) \subseteq A \cap W$. If the latter intersection is zero, then, by lemma **5.6.8**, we get $0 = e \cdot (A' \cap IW) = A' \cap (eI)W$, i.e., $eIW = 0$ and $eW=0$. Since e is an arbitrary homogenous idempotent, then $W = 0$. The lemma is proved.

6.3.13. Proof of lemma 6.3.9. By lemma **6.3.8**, the unit of a ring R^G will be that of R. If A is an essential left ideal of the ring R, then, by lemma **6.3.12**, the intersection $A \cap R^G$ is an essential left ideal, i.e., by the condition it contains the unit and $A = R$, which is the required proof.

The inverse statement can be proved in many ways in view of plentiful characterizations of semiprime artinian rings. We shall make use of the following characterization: *if $S^2 = S$ and the semiprime ring S obeys the descending chains condition for left ideals L, such that $SL = L$, then S is a semisimple artinian ring.*

So, let R be a semisimple artinian ring. By theorem **5.5.8**, it suffices to consider the case when R is a simple artinian ring.

Let τ be a principal invariant form. Then $\tau(R)$ is an essential two-sided ideal of a semiprime ring R^G (see lemma **3.8.1** and theorem **3.6.1**). Using the above characterization, let us show that $S = \tau(R)$ is a semisimple artinian ring.

If $I_1 \supset I_2 \supset ... \supset I_n \supset ...$ is a strictly decreasing chain of left ideals in S, such that $SI_n = I_n$, then there is a number k, such that $RI_k = RI_{k+1}$. Applying to this equality the operator τ we get $\tau(R) \cdot I_k = \tau(R) \cdot I_{k+1}$, which is a contradiction. Then, by lemma **3.8.3**, $R\tau(R) = R$, and, applying τ to both parts, we get $S^2 = S$.

Thus, S is a semisimple artinian ring. Hence, S contains a unit, and this unit will be that in R^G as well, since S is an essential ideal. Therefore, $S = R^G$. The theorem is proved.

REFERENCES

M.Cohen [28];
M.Cohen, S.Montgomery [32];
V.K.Kharchenko [80];
D.S.Passman [135].

C. *Primitive Rings.* It should be recalled that a *left primitive ring* is a ring with an exact irreducible left module. A primitive ring is certain to be prime if V is an irreducible exact module, I, J are nonzero ideals, then $IJV = I(JV) = IV = V$ and, hence, $IJ \neq 0$.

The task is to elucidate a possibility of transferring the property of primitivity from the ring R to R^G and vice versa. As the prime dimension of a fixed ring is equal to the prime invariant dimension of the algebra of a group G (theorem 3.6.7), then it would be natural to consider the groups G with G-simple algebras $\mathbb{B}(G)$.

6.3.14. Theorem. *Let R be a left primitive ring, G be a reduced-finite group, with its algebra, $\mathbb{B}(G)$, being G-simple. Then the fixed ring R^G is left primitive.*

Proof. Let V be an exact irreducible R-module. Let us show that it is finitely-generated as a left R^G-module.

By theorem **3.5.6** we can find a nonzero element $a \in R$, such that $Ra \subseteq \sum_{i=1}^{n} R^G a_i$ for suitable $a_i \in R$. If v is an element from V, such that $av \neq 0$, then $V = Rav \subseteq \sum_{i=1}^{n} R^G(a_i v) \subseteq V$, and, therefore, V is finitely-generated.

Now we can find a maximal left R^G-module $W \neq V$. Let $S = \{s \in R^G: sV \subseteq W\} \neq 0$. Then SR is a nonzero $(R^G - R)$-bimodule. In this case its left annihilator in $\mathbb{B}(G)$ is an invariant ideal (as $S \subseteq R^G$) and, since $\mathbb{B}(G)$ is G-is simple, it equals zero. By corollary **3.7.4**, we find a two-sided nonzero ideal $I \subseteq SR$ and get $V = IV \subseteq SRV \subseteq SV \subseteq W$, which is a contradiction. Thus, $S = 0$ and, hence, V / W is an exact invertible R^G-module. The theorem is proved.

In the case of an arbitrary M-group an analogous result is valid. It should be recalled that the ring R is called *semiprimitive*, if it has an exact completely reducible left module of a finite length.

Let us begin with characterization of semiprimitive rings.

6.3.15. Proposition. The following conditions on a ring are equivalent:

(1) *R is a semiprimitive ring;*

(2) *R is a subdirect product of a finite number of primitive rings;*

(3) *R has an essential two-sided ideal, decomposing into a direct sum of a finite number of primitive rings.*

Proof. (1)$\rightarrow$(2). Let $V_1 \overset{\circ}{+} \dots \overset{\circ}{+} V_n$ be an exact left R-module, V_k, $1 \leq k \leq n$ be its irreducible submodules. Let us denote by N_k a kernel of the module V_k, i.e., $N_k = \{r \in R | rV_k = 0\}$. Then V_k is an exact R / N_k-module, in which case $\bigcap_{k=1}^{n} N_k = 0$. It means that R is a subdirect product of primitive rings R / N_k, $k = 1, 2, \dots, n$.

(2)→(3). Let R be a subdirect product of primitive rings $R_1, ..., R_n$ and $\pi_k: R \to R_k$, $1 \le k \le n$ be approximating epimorphisms. Let us assume that R is not approximated by a less number of primitive rings. Then $I_k = \bigcap_{i \neq k} N_i \neq 0$, where $N_i = \ker \pi_i$. Since

$$I_k \cap \sum_{j \neq k} I_j \subseteq (\bigcap_{i \neq k} N_i) \cap N_k = 0,$$

then the sum $I_1 + ... + I_n = I$ is direct.

Then, $\ker\pi_k \cap I_k = 0$ and, hence, $I_k \cong \pi_k(I_k)$ is an ideal of R_k, which affords that I_k are primitive rings. If now $vI = 0$, then, applying a projection π_k, we get $\pi_k(v) \cdot \pi_k(I_k) = 0$, i.e., $\pi_k(v) = 0$ for all k and $v = 0$. Thus, I is the sought essential ideal.

(3)→(1). Let $I = I_1 \oplus ... \oplus I_n$ be a decomposition of an essential ideal into a direct sum of primitive rings, and V_k be an exact irreducible module for the ring I_k, $1 \le k \le n$. Then $I_k V_k$ is, on the one hand, a left R-module, and, on the other hand, $I_k V_k = V_k$, i.e., V_k are invertible left R-modules. Now we have to remark that the sum $V_1 \dot{+} ... \dot{+} V_k$ is an exact module. If $rI_k V_k = 0$ for all k, then , as V_k is exact, we have $rI_k = 0$ and $rI = 0$, which affords $r = 0$. The proposition is proved.

Let us recall an interaction between iirreducible modules and left ideals of a ring. If V is an invertible module and $0 \neq v \in V$, then $Rv = V$ and, hence the module epimorphism $\varphi: R \to V$, which acts by the formula $r \to rv$, is determined. Therefore, $V \cong R/\lambda$, where $\lambda = ann\, v = \{ r \in R | \ rv = 0 \}$. It is evident that λ is a maximal left ideal of the ring R. Moreover, R has a *left unit over* λ, i.e., an element i, such that $xi - x \in \lambda$ for all $x \in R$: if $v = iv$, then $(xi - x)v = 0$ and, hence, $xi - x \in \lambda$. The left ideal λ is called *modular* if the ring R has a left unit over λ.

Inversely, if λ is a maximum modular left ideal, then $R/\lambda = V$ is an irreducible R-module. The fact that R/λ has no proper submodules results from maximality of λ, and the inequality $RV \neq 0$ results from modularity: if $RV = 0$, then $R^2 \subseteq \lambda$ and, hence, $x = (x - xi) + xi \in \lambda$ for all $x \in R$, which is impossible.

6.3.16. Theorem. *A fixed ring for an M-group of automorphisms of a semiprime ring is semiprimitive iff the initial ring is semiprimitive.*

Proof. Let R be semiprimitive. By point (3) of the preceding proposition, R has a finite prime dimension. Using the same point, we can use theorem **5.5.8** and consider R to be a primitive ring.

Let V be an exact left invertible R-module. As in the beginning of the proof of theorem **6.3.14**, we see that the module V is finitely-generated over R^G.

Let us decompose the algebra of the group G into a direct sum of minimal invariant ideals

$$\mathcal{B}(G) = e_1\mathcal{B}(G) \dot{+} \ldots \dot{+} e_n\mathcal{B}(G).$$

Let I be a nonzero ideal of the ring R, such that $e_iI, Ie_i \subseteq R$. Let us consider left R^G-submodules $V_i = (e_iI)V$ in V. Since $IV = V$, we get $V = V_1 + .. + V_n$. Let us prove that this sum is direct. We have

$$(Ie_1)(\sum_{i>1} V_i) \subseteq \sum_{i>1} [Ie_1e_iI]V = 0,$$

while if $0 \neq v_1 \in V_1$, then $ae_1v_1 = av_1$ for all $a \in I$ and, hence, $Ie_1v_1 \neq 0$. Therefore, $V_1 \cap \sum_{i>1} V_i = 0$ and, by symmetry, we get a direct decomposition $V = V_1 \dot{+} \ldots \dot{+} V_n$. This, in particular, implies that all V_i are finitely-generated over R^G. In every V_i let us choose a maximum submodule W_i. Let $S = \{s \in R^G : sV_i \subseteq W_i$ for all $i\}$. Let us, for instance, suggest that $e_1S \neq 0$. In $\mathcal{B}(G)$ the annihilator of e_1S contains $(1 - e_1)\mathcal{B}(G)$ and is an invariant proper ideal, i.e., this annihilator is exactly equal to $(1 - e_1)\mathcal{B}(G)$. By theorem **3.7.3** we have $e_1J \subseteq e_1SI$ for a certain $J \in \mathcal{F}$. Therefore, we have

$$V_1 = e_1IV \supseteq e_1IJV = e_1I(JV) = e_1IV = V_1.$$

Hence,

$$V_1 = e_1IJV \subseteq (e_1SI)V = Se_1IV = SV_1 \subseteq W_1,$$

which is a contradiction that shows that $e_iS = 0$ for all i and, hence, $S = 0$.

Thus, the direct sum

$$V_1/W_1 \dot{+} \ldots \dot{+} V_n/W_n$$

is the sought completely reducible exact module.

Inversely, by proposition **6.3.15**, a fixed ring has a finite prime dimension. By theorem **5.6.7**, the ring R also has a finite prime dimension. Now theorem **5.5.10** and proposition **6.3.15** show that it would be enough to consider the case when R is prime. Let us prove a somewhat stronger statement which shows that for the ring R to be primitive, it is sufficient that one of the prime addends of the essential ideal of a fixed ring, the existence of which is stated by lemma **3.6.6.**, point (2), be primitive.

6.3.17. Lemma. *Let G be an M-group of automorphisms of a prime ring R. Then if R^G has a nonzero ideal which is a primitive ring, then R is primitive.*

Proof. Let $0 \neq T$ be an ideal of R^G and V be an irreducible exact T-module. Then $V = TV$ is an irreducible R^G-module. Let λ be a maximal modular left ideal of the ring R^G corresponding to it.. It should be remarked that λ contains no essential ideal of the ring R^G: if $A \subseteq \lambda$, then $ATV \subseteq ATv \subseteq \lambda v = 0$, where $\lambda = ann^{(1)}v$ and, hence, $AT = 0$.

Then, by lemma **3.8.3**, the left ideal RR^G contains a nonzero ideal of the ring R, which will be denoted by I.

Let $L_0 = R\lambda + \lambda$. Let us show that $L_0 \cap R^G = \lambda$.

If this is not the case, then, since λ is maximal, we get $L_0 \supseteq R^G$, which affords $L_0 \supseteq RR^G \supseteq I$. Let τ be a principal trace form, J be a nonzero ideal, such that $\tau(J) \subseteq R$. Then

$$\tau(IJ) \subseteq \tau(JL_0) \subseteq \tau(J\lambda) = \tau(J)\lambda \subseteq \lambda,$$

which, however, contradicts the choice of λ, since $\tau(JI)$ is an essential ideal of the ring R^G (see lemma **3.8.1**).

We now see that a set $\mathcal{M}$ of all the left ideals L of the ring R which are contained in $R' = RR^G + R^G$ and such that $L \cap R^G = \lambda$, is non-empty. It is evident that $\mathcal{M}$ is inductive by inclusion, i.e., by the Zorn lemma, there is a maximal element $L \in \mathcal{M}$. Let us prove that $V = R' / L$ is an exact irreducible R-module.

If $R' \supseteq L_1 \supset L$, then $L_1 \cap R^G \supset \lambda$ and, hence, $L_1 \supseteq R^G$, which affords $L_1 \supseteq RR^G + R^G = R'$. Consequently, V contains no proper submodules. Then, the kernel K of the module V is a two-sided ideal contained in L, in which case $\lambda = L \cap R^G \supseteq K^G$ and, if $K \neq 0$, then K^G is an essential ideal of the fixed ring, which contradicts the choice of λ. Analogously, $L \not\supseteq R^2$, i.e., $RV \neq 0$. The theorem is proved.

6.3.18. Theorem. *A fixed ring of an M-group of automorphisms of a semiprime ring is semisimple (in the sense of the Jacobson radical) iff the initial ring is semisimple.*

Proof. Let R be semisimple. Since a set W of elements w of the ring R, such that $Rw \subseteq \sum_{i=1}^{n(a)} R^G a_i$ forms an essential two-sided ideal of the ring R, then $W \cap R^G$ is an essential ideal of the fixed ring (lemma 5.6.6). In particular, if R^G is not semisimple, then we can find a fixed element $w \in W$ annihilating all the irreducible left modules of the ring R^G.

Let us consider a bimodule $R^G wR$, and by **5.10.3** find an idempotent $e \in B(G)$, such that $R^G wR \supseteq eI$, $ew = w$, $I \in F$. Since $Iw \neq 0$, then we can find an irreducible left R-module V, such that $IwV \neq 0$, and, hence, $IV = V$, $wV \neq 0$ and $V = IwV \subseteq \sum R^G a_i V \subseteq V$ is a finitely-generated left R^G-module. We have a decomposition $V = eIV \oplus (1-e)IV$ and, hence, eIV is a nonzero (if $eIV = 0$, then $wV = wIV = ewIV = weIV = 0$) finitely-generated module over R^G. If W is its maximal submodule, then eIV/W is irreducible, i.e., $W \supseteq weIV = wIV = wV$, which affords $R^G wRV \subseteq W$ or $eIV \subseteq W$, which is a contradiction.

Inversely, let $J(R^G) = 0$. It should be remarked that $J(R) \cap R^G \subseteq J(R^G)$. Indeed, if a is a quasi-regular fixed element in R, then it will also be quasi-regular in R^G: the equalities $a \bullet x \overset{df}{=} ax + a + x = x \bullet a = 0$ afford $a \bullet x^g = 0$ and, hence, $x = x \bullet (a \bullet x^g) = (x \bullet a) \bullet x^g = x^g$.

Therefore, $J(R) \cap R^G = 0$. Let τ be an arbitrary principal trace form. As was the case in lemma **6.3.10**, let us present it as $\tau(x) = \sum (b_i x)^{g_i}$. Let T, I be essential ideals of the ring R, such that $T \subseteq I^{g_i} \subseteq R$ for all i. Then $(IJ(R))^{g_i}$ is isomorphic , as a ring, to a quasi-regular ideal $IJ(R)$, in which case $T(IJ(R))^{g_i} < (IJ(R))^{g_i}$ since $T^{g_i^{-1}} \subseteq I$. It means that $T(IJ(R))^{g_i}$ is a quasi-regular left ideal of the ring R, i.e., $T(IJ(R))^{g_i} \subseteq J(R)$.

Let now I_1 be an ideal from F, such that $b_i I_1 \subseteq I$ for all i. Then $T\tau(I_1 J(R)) \subseteq J(R)$ and, hence, $T^G \tau(I_1 J(R)) \subseteq J(R) \cap R^G = 0$. Now , by lemma **5.6.6**, we have $\tau(I_1 J(R)) = 0$. Since τ is arbitrary, proposition **5.6.8** affords $I_1 J(R) = 0$ and, hence, R is semisimple. The theorem is proved.

REFERENCES

K.I.Beidar [13];
S.Montgomery, D.Passman [118];
J.-L.Pascaud [133].

D.*Quite primitive rings.* These rings have already been considered in relation with the Martindale theorem (**1.14, 1.15**). Since while transferring to a fixed ring the prime dimension can increase, then, as above, we should make this notion broader. A ring R will be called *quite semiprimitive* if it is a subdirect product of a finite number of quite primitive rings. Let us start with characterizing this class of rings.

6.3.19. Proposition. The following conditions on the ring R are equivalent:

(1) *R is a quite semiprimitive ring.*

(2) *R has an essential two-sided ideal decomposing into a direct sum of a finite number of quite primitive ring.*

(3) *R has the least essential ideal which decomposes into a direct sum of a finite number of simple rings coinciding with their socles.*

(4) *R has an exact completely reductive module of a finite length which is isomorphic to a left ideal of the ring R.*

Proof. (1)$\rightarrow$(2). One should repeat the proof of implication (2)$\rightarrow$(3) of proposition **6.3.16** substituting the word "primitive" with "quite primitive".

(2)$\rightarrow$(3). Let $I = I_1 \oplus ... \oplus I_n$ be a decomposition of an essential ideal into a direct sum of quite primitive rings and H_k be a socle of I_k, $1 \le k \le n$. Then $H = H_1 \oplus ... \oplus H_n$ is the sought ideal.

(3)$\rightarrow$(4). Let $H = H_1 \oplus ... \oplus H_n$. In every ring H_k let us choose a primitive idempotent e_k. Then $V_k = H_k e_k$ is an irreducible exact H_k-module (lemma **1.14.2**). The sum $V = V_1 + ... + V_n$ will be a left ideal of R, as $HV_k = V_k$. Besides,as left R-modules, the left ideals V_k are irreducible. Therefore, V is a completely reducible R-module. If, now, $rV = 0$, then $(rH)V = 0$ and, hence, $(rH_k)V_k = 0$, then $rH_k = 0$ at all k. Thus, $rH = 0$ and $r = 0$.

(4)$\rightarrow$(1). Let $V_1,... V_n$ be a set of minimal left ideals, the intersection of whose kernels (i.e., left annihilators in R) is zero. Let $N_1,..., N_n$ be corresponding kernels. Then V_k is an exact irreducible R/N_k-module. In particular, R/N_k is a prime ring, while R is a semiprime ring as a subdirect product of prime ones: $R = \overset{n}{\underset{k=1}{S}} R/N_k$.

Well-known is the fact that every non-nilpotent minimal one-sided ideal is generated by a primitive idempotent: if $V_k v_k \neq 0$, $v_k \in V_k$, then $V_k v_k = V_k$ and $e_k v_k = v_k$ for a certain $e_k \in V_k$, in which case $(e_k^2 - e_k)v_k = 0$, but an intersection of the left annihilator of v_k with V_k is a left ideal and, hence, it equals zero; if now I is a nonzero left ideal in $e_k Re_k$, then $V_k = RI$ and, hence, $e_k Re_k = e_k V_k = e_k Re_k I \subseteq I$. Therefore, $e_k Re_k$ is a sfield.

Since $V_k^2 \neq 0$, then $V_k \cap N_k = 0$ and, hence, $e_k + N_k$ is the sought primitive idempotent in the factor-ring R/N_k. The proposition is proved.

It should be remarked that the least essential ideal, arising in (3) is generated by all primitive idempotents of the ring R and, hence, it would be natural to call it *a socle* . When the ring R is arbitrary, a *left socle* is a sum of all minimal left ideals. If the ring is not semiprime, then the left socle is possible not to be generated by primitive idempotents (since minimal left ideal with zero multiplication can arise) and not to coincide with the right sockle. In the case of a semiprime ring, the left socle coincides with the right one and is generated by primitive idempotents.

6.3.20. Theorem. *A fixed ring of an M-group of automorphisms of a semiprime ring is quite semiprimitive iff the initial ring is quite semiprimitive.*

Proof. Let R be quite semiprimitive. According to (3) of the preceding proposition, R has a finite prime dimension. Therefore, we can make use of theorem **5.5.10** and assume R to be quite primitive.

Let H be the sockle of the ring R, τ be the principal trace form. As H is the least ideal, then $\tau(H) \subseteq H$, in which case $\tau(H)$ is the least essential ideal of R^G (lemma **3.8.1**). If

$$\mathcal{B} = e_1 \mathcal{B} + \ldots e_n \mathcal{B}.$$

is a decomposition of the algebra of the group into a direct sum of G-simple components, then $\tau(H)e_i < R^G$ and

$$\tau(H) = \tau(H)e_1 + \ldots + \tau(H)e_n$$

is a decomposition of $\tau(H)$ into a direct sum of nonzero prime (theorem **3.6.7**) rings. The task now is to show that each of the ideals $\tau(H)e_k$ has a minimal left ideal.

Let us assume that in $\tau(H)e_1$ such an ideal cannot be chosen. Then we can find a sequence of elements $s_1, s_2, \ldots s_m, \ldots$ from $\tau(H)e_1$, such that

$$\tau(H)s_1 \underset{\neq}{\supset} \tau(H)s_2 \underset{\neq}{\supset} \ldots \tau(H)s_m \underset{\neq}{\supset} \ldots$$

It should be remarked that $H\tau(H) = H$ (lemma **3.8.3**) and, hence, multiplying all the terms of the chain by H from the left, we get

$$H s_1 \supseteq H s_2 \supseteq \ldots \supseteq H s_m \supseteq \ldots$$

All the inclusions in this chain cannot be strict, since $s_1 \in H$ and $H s_1$ is a completely reducible module of a finite length. If $H s_k = H s_{k+1}$, then, applying τ, we get $\tau(H)s_k = \tau(H)s_{k+1}$, which is a contradiction.

Inversely, by theorem **5.6.7**, the ring R has a finite prime dimension. Theorem **5.5.10** and proposition **6.3.18** show that it would be sufficient to consider the case when R is a prime ring.

Let e be a primitive idempotent from R^G. Then eR^Ge is a sfield, in which case the group G is induced on the ring eRe and we have $(eRe)^G \subseteq eR^Ge \subseteq (eRe)^G$, i.e., $(eRe)^G$ is a sfield. If we now make use of theorem **6.3.9**, we see that eRe is a semisimple artinian ring. In particular, a primitive idempotent f can be found in it, in which case $fRf = feRef$ is a sfield and f is a primitive idempotent of R. The fact that theorem **6.3.9** can be used here, results from the following lemma.

6.3.21. Lemma. *Let G be an M-group of automorphisms of a prime ring R and e be a fixed idempotent. Then a group $\bar{G}$ induced by the group G on the ring eRe will be an M-group, the algebra of which is isomorphic to $e\mathcal{B}(G)$.*

Proof. First, we should remark that eRe is a prime ring: if $J, I \lhd eRe$, $IJ = 0$, then $IRJ = IeReJ = 0$. Second, it should be also remarked that $\bar{G} \subseteq A(eRe)$. If $J \subseteq I^g \subseteq R$, then $eJe \subseteq (eIe)^g \subseteq eRe$, in which case eJe, eIe are nonzero ideals in eRe. Third, $e\mathcal{B}(G)$ (and even eQe) is naturally embedded into $Q(eRe)$: if $Ieqe \subseteq R$, then $eIe \cdot eqe \subseteq eRe$.

Then, if b is an invertible element from $\mathcal{B}(G)$ and $\tilde{b}$ is an automorphism from G corresponding to it, then eb is invertible in $e\mathcal{B}$ and $e\tilde{b}$ is induced by $\tilde{b}$, i.e., $\mathcal{B}(\bar{G}) \supseteq e\mathcal{B}(G)$.

Finally, if $\xi \in \Phi_{\bar{g}}$, $\bar{g} \in \bar{G}$, then $exe\xi = \xi ex^g e$ for all $x \in R$. In

eRe let us find a nonzero ideal A, such that $\xi A, A\xi \subseteq eRe$. Let $a \in A$. Then $ax(\xi a) = (a\xi) x^g a$ is an identity on R. By theorem **2.2.2**, we see that $g = \tilde{b}$ is an inner automorphism and $a \otimes \xi a = a\xi b^{-1} \otimes ba$, i.e., in the generalized centroid C one can find an element λ, such that $\lambda a = (a\xi)b^{-1}$. From here we get a relation in the ring $Q(eRe)$ of the type $a \cdot \lambda e \cdot (b^{-1}e) = a\xi$ or $A(\xi - (\lambda e)(b^{-1}e)) = 0$ and, hence, $\xi = \lambda b^{-1} \cdot e \in B(G)e$, which is the required proof.

REFERENCES

M.Hacque [55];
N.Jaconson [65].

E. Goldie Rings. The conditions which determine this class of rings emerged in the famous works by A.Goldie on the orders of simple and semisimple artinian rings. As we have no possibility of presenting here the basic notions of this theory, we shall limit ourselves with getting acquainted with the main definitions and facts.

A *left Goldie dimension* ($Gol-\dim V$) of the left module V is the biggest number n, such that V contains a direct sum of n non-zero submodules. This dimension is also often called *uniform*.

One can prove (it is not obvious) that if a module does not contain a direct sum of an infinite number of nonzero submodules, then it has a finite Goldie dimension (see **6.3.55** below).

A ring R is called a *left Goldie ring* if in R the condition of maximality is fulfilled for the left annihilator ideals, and R has no infinite direct sums of nonzero left ideals (i.e., as has been noted above, $Gol-\dim R < \infty$).

The following statement presents the most important property of semiprime Goldie rings.

6.3.22. Proposition. *Any left essential ideal of a semiprime left Goldie ring has a regular element.*

It should be recalled that an element $r \in R$ is called a regular element if $sr \neq 0$, $rs \neq 0$ for any $s \in R$, $s \neq 0$.

6.3.23. Definitions. Let R be a subring of a ring S. The ring S is called a *classical left ring of quotients of the ring* R, and the ring R is called a *left order* in S iff the following conditions are met:

(1) all regular element of the ring R are invertible in the ring S;

(2) all elements of the ring S have the form $a^{-1}b$, where

$a, b \in R$ and a is a regular element of the ring R.

It is not any ring that has a classical left ring of quotients. Indeed, if a, b are elements of R and b is a regular element, then $ab^{-1} = c^{-1}d$, where $c, d \in R$ and c is a regular element. Hence, $ca = db$ and we come to the necessity of the following condition.

The left Ore condition. For any elements $a, b \in R$, where b is a regular element, one can find elements c, d, where c is a regular element, such that $ca = db$.

This condition is known to be sufficient for the existence of a left classical ring of quotients (the Ore theorem). Moreover, the Ore condition guarantees the uniqueness of the left classical ring of quotients, this ring denoted by $Q_{cl}(R)$.

6.3.24. Goldie theorem. *A ring R is a left order in a semisimple artinian ring iff R is a semiprime left Goldie ring.*

Semiprime left Goldie rings have some interesting characterizations. Let us recall one of them which is required here. A ring R is called *left nonsingular*, if every essential left ideal has a zero right annihilator.

6.3.25. Johnson theorem. *A semiprime ring R is a left Goldie ring iff it is nonsingular and contains no infinitive direct sums of nonzero left ideals.*

Now we go over to the basic topic.

6.3.26. Theorem. *A ring of invariants of an M-group G of automorphisms of a semiprime ring R is a left Goldie ring iff R is a left Goldie ring. In this case automorphisms of the ring R are extended onto $Q_{cl}(R)$ and $Q_{cl}(R)^G = Q_{cl}(R^G)$.*

Proof. A left Goldie ring is evident to have a finite prime dimension. Thus, theorem **5.5.10** and proposition **6.3.15** make it possible to limit ourselves with considering only a prime ring R (here one should pay attention to the fact that the Goldie conditions can be evidently transferred onto essential two-sided ideals and vice versa).

Let R be a prime left Goldie ring. The condition of maximality for annihilators is preserved when going over to subrings and, hence, it is fulfilled in R^G as well.

If A, B are left ideals in R^G and $A \cap B = 0$, then for a principal invariant form τ we have

$$\tau(IA \cap IB) \subseteq \tau(I)A \cap \tau(I)B = 0 ,$$

where the nonzero ideal I of the ring R is chosen in such a way that $\tau(I) \subseteq R$. By lemma **3.8.1** we get $IA \cap IB = 0$. This remark shows that if

$$A_1 \dot{+} A_2 \dot{+} \dots A_k \dot{+} \dots$$

is a direct sum of left ideals R^G, then

$$IA_1 \dot{+} IA_2 \dot{+} \dots \dot{+} IA_k \dot{+} \dots$$

is a direct sum of left ideals of R and, hence, $Gol-\dim R^G < \infty$, which is the required proof.

The inverse statement is based on the following lemma.

6.3.27. Lemma. *Let G be an M-group of automorphisms of a semiprime ring R. Then, if R^G is a left Goldie ring, then the left R^G-module R is embeddible into a finite direct sum of copies of the regular module R^G.*

Proof. Let us make use of theorem **5.10.4**. By this theorem, a set W of elements $a \in R$, such that Ra is embeddible into a finite direct sum of copies of the module R^G, contains an essential ideal. By lemma **6.3.12**, an intersection $W \cap R^G$ has a regular element a of the ring R^G (see **6.3.22**). The element a will be a regular element in R as well: if $va = 0$, then for any principal invariant form $\bar{\tau}_e$ and an ideal $I \in F$, such that $\bar{\tau}_e(I) \subseteq R$, we get $\bar{\tau}_e(Iv)a = \bar{\tau}_e(Iva) = 0$, i.e., by lemma **5.6.8** we have $Iev = 0$ or $ev = 0$ and $v = 0$.

Now we have to remark that the mapping $r \to ra$ implements the isomorphism of the left R^G-modules $R \cong Ra$. The lemma is proved.

Let us go on proving the theorem. It should be remarked that the modules A, B have no infinite direct sums of nonzero submodules, then their direct sum $A \dot{+} B$ also obeys this property. If $V = V_1 \dot{+} \dots \dot{+} V_n \dot{+} \dots$ is an infinite direct sum in $A \dot{+} B$, then we can group it into an infinite direct sum of infinite direct sums:

$$V = W_1 \dot{+} \dots \dot{+} \dots ; \qquad W_m = V_{m1} \dot{+} \dots \dot{+} V_{mn} \dot{+} \dots \quad .$$

Only a finite number of modules $W_1, \dots, W_m$ have nonzero intersections with A. Let, for instance, $W_1 \cap A = 0$. Then the module W_1 is embeddible into $(A \dot{+} B)/A \cong B$, which is impossible.

Now lemma **6.3.27** shows that R has no infinite direct sums of nonzero left R^G-submodules and, moreover, nonzero left ideals.

Let us check if R is nonsingular. If $La = 0$, where L is an essential left ideal of the ring R, then, by lemma **6.3.12**, an intersection

$L \cap R^G$ is an essential left ideal of R^G. Proposition **6.3.22** results in a regular element $t \in L^G$. And now property **6.3.25** shows that R is a Goldie ring.

Let us prove, finally, that $Q_{cl}(R)^G = Q_{cl}(R^G)$. For this purpose let us consider a set T of regular elements in R^G. We have already paid attention to the fact that nonzero divisors of the ring R^G will be those in R as well. Therefore, T consists of invertible elements of the ring $Q_{cl}(R)$. Let us show that $T^{-1}R = Q_{cl}(R)$, i.e., any element from $Q_{cl}(R)$ has the form $t^{-1}r$, where $t \in T$, $r \in R$.

Let a be a regular element from R. In this case the left ideal Ra will be essential: if $V \cap Ra = 0$, then the sum $V + Va + Va^2 + \dots$ is infinite and direct. By lemma **6.3.12**, an intersection $Ra \cap R^G$ will be an essential ideal of the fixed ring, i.e., $Ra \cap T \neq \emptyset$. Let $t = ra$. Then $a^{-1} = t^{-1}r \in T^{-1}R$. If now $a^{-1}b$ is an arbitrary element from $Q_{cl}(R)$, then $a^{-1}b = t^{-1}rb = T^{-1}R$.

Now we have got an evident extension of the group G: $(t^{-1}r)^g = t^{-1}r^g$. Strictly speaking, the right-hand part of this equality is determined (by r) only on a certain essential ideal I of the ring R, but I contains a regular element t_0 from T and, hence, $T^{-1}I \supseteq T^{-1}t_0R = T^{-1}R$. Besides, it is necessary to check the correctness of the extension and the fact that it results in automorphisms.

If $t_1^{-1}r_1 = t_2^{-1}r_2$, then, by the Ore condition, we can find elements $s_1 \in R^G$, $s_2 \in T$, such that $s_1t_1 = s_2t_2 \overset{df}{=} t$. We have $t_1^{-1}r_1^g - t_2^{-1}r_2^g = = t^{-1}s_1r_1^g - t^{-1}s_2r_2^g = t^{-1}(s_1r_1 - s_2r_2)^g = 0$, since $s_1r_1 - s_2r_2 = t\cdot(t_1^{-1}r_1 - -t_2^{-1}r_2) = 0$, which implies correct-ness. Now, if $t_1^{-1}r_1$ is not obligatory equal to $t_2^{-1}r_2$, then , preserving the notations, we get $(t_1^{-1}r_1 + t_2^{-1}r_2)^g = {}^{-1}(s_1r_1 + s_2r_2)^g = t_1^{-1}r_1^g + t_2^{-1}r_2^g$. Besides, if $r_1t_2^{-1} = t_3^{-1}s_3$, then $t_3r_1 = s_3t_2$ and $t_3r_1^g = s_3^gt_2$, which affords $r_1^gt_2^{-1} = t_3^{-1}s_3^g$ and, hence,

$$(t_1^{-1}r_1t_2^{-1}r_2)^g = (t_1^{-1}t_3^{-1}s_3r_2)^g = t_1^{-1}t_3^{-1}s_3^g r_2^g = t_1^{-1}r_1^g t_2^{-1}r_2^g,$$ i.e., g is an automorphism.

Finally, an element $t^{-1}r$ will be fixed iff r is fixed, i.e., $Q_{cl}(R)^G = T^{-1}R^G$. The theorem is completely proved.

6.3.28. In the course of the proof we have established an interesting fact that when calculating a ring of quotients $Q_{cl}(R)$ it is sufficient to make invertible only fixed elements $Q_{cl}(R) = T^{-1}R$.

REFERENCES

M.Cohen [28];
K.Faith [43];
J.Fisher, J.Ostenburg [48];
V.K.Kharchenko [69,80,81];
M.Lorentz, S.Montgomery, L.Small [99];
S.Montgomery [111];
A. Page [130];
D.Passman [135];
H.Tominaga [149].

The Goldie theory is presented in the following monographs:
I.Herstein [57];
N.Jacobson [65];
I.Lambeck [91].

F. Noetherian Rings. It should be recalled that a module is Noetherian if there are no infinite ascending chains of submodule is generated by a finite number of elements. It is evident that a submodule of a Noetherian module will be Noetherian. The ring R is called *left Noetherian* if such is the left regular module $_R R$. Well known is the fact that a finitely-generated module over a Noetherian ring is also Noetherian.

6.3.29. Theorem. *If a fixed ring of an M-group of automorphisms of a semiprime ring is left Noetherian, then such is the initial ring.*

Proof. Let us denote by W a set of all elements $a \in R$, such that $Ra \subseteq \sum_{i=1}^{n} R^G a_i$ for some $a_1, \ldots a_n$. Theorem **5.10.1**, applied to a ring antiisomorphic to R, shows W to be an essential ideal. A Noetherian ring is certain to be a Goldie ring. By theorem **6.3.26** and proposition **6.3.22** we can find in W a regular element a. Then $R \cong Ra \subseteq \sum_{i=1}^{n} R^G a_i$ are Noetherian R^G-modules. The theorem is proved.

6.3.30. Theorem. *Let* $G \subseteq Aut(R)$ *be a finite group of automorphisms of a left Noetherian ring* R, *which has an invertible in* R *order* $|G|^{-1} \in R$. *In this case the fixed ring* R^G *is left Noetherian.*

Proof. Let

$$I_1 \subset I_2 \subset \ldots \subset I_n \subset \ldots$$

be a strictly ascending chain of left ideals in R^G. Then, since R is a Noetherian ring, there can be found a number k, such that $RI_k = RI_{k+1}$.

Let us apply the operator $t = |G|^{-1} \sum_{g \in G} g$ to both parts of this equality. As $t(s) = s$ for any fixed element s, we get a contradiction $I_k = I_{k+1}$. The theorem is proved.

The latter theorem stops being valid for M-groups. Moreover, the condition $|G|^{-1} \in R$ cannot be changed with that of the absence of additive torsion. Let us give here an example by Chang and Lie which is based on a Nagarajan example.

6.3.31. Example. Let $A = \mathbf{Z}[a_1, b_1, a_2, b_2, \ldots]$ be a ring of polynomials with integer coefficients. Let us denote by K the localization of A relative $2A$, i.e., $K = \{\frac{g}{f} | g \in A, f \in 2A\}$ and let us consider a ring $R = K[[x, y]]$ of the power series on variables x, y. Since K is a principal ideal ring, then R is Noetherian. Let us determine an automorphism g on R by the formulas:

$$x^g = -x, \; y^g = y, \; a_i^g = -a_i + (a_{i+1}x + b_{i+1}y)y,$$

$$b_i^g = b_i + (a_{i+1}x + b_{i+1}y)x.$$

In this case the group G generated by g has the order of 2, and the ring R has no additive torsion. One can prove that R^G is not a Noetherian ring. The proof is based on the result by Nagarajan, who claims that in the ring $(R/2R)^G$ there is a strictly ascending chain of ideals

$$(P_1) \subset (P_1, P_2) \subset \ldots \subset (P_1, \ldots, P_n) \subset \ldots$$

where $P_i = a_{i+1}x + b_{i+1}y + 2R$.

REFERENCES
K.L.Chung, P.G.Lee [26];
M.Cohen, S.Montgomery [33];
D.Farkash, L.Snaider [44];
J.Fisher. J.Ostenburg [48];
S.Montgomery, L.Small [91];
S.Montgomery [111,113];
K.Nagarajan [123];
J.-L.Pascaud [133];

G.Prime and subdirectly undecomposible rings. The problem of invariant simple rings has been considered in **3.6**, having showed that a ring of invariants of a finite group G is a finite direct sum of simple rings, provided there is no additive $|G|$-torsion in the initial ring. This result stops being valid when transferring to M-groups. Let us describe an example by Osterburg, which is based on an example by Zalesskii and Neroslavskii of a simple ring with a unit of a characteristic 2, having an outer automorphism of the order 2 with a non-simple ring of invariants.

6.3.32. Example. Let k be a field of characteristic 2. Let us consider an algebra $R_1 = k(y)[x, x^{-1}]$ over a field of rational functions of a variable y. Let g be an automorphism of this algebra which transfers x into xy and let G be a finite cyclic group generated by the automorphism g. Let us assume that $R_2 = R_1 * G$ is a skew group ring and let us consider an automorphism h of the algebra R_2, determined by the formulas $h(x) = x^{-1}$ and $h(g) = g^{-1}$. One can show that R_2 is a simple ring, h is an outer automorphism, and the ideal $t(R_2) = \{x + x^h | x \in R_2\}$ of the fixed ring contains no unit, i.e., $R_2^{<h>}$ is not simple.

Therefore, for characterizing the invariants of M-groups of simple algebras, we should somewhat extend the class of the rings under consideration.

It should be remarked that if a simple ring R contains a unit, then $R = R_F$. If R has no unit, then $R \neq Q(R)$, but, however, R will be an ideal in $Q(R)$, in which case this ideal will be contained in any other nonzero ideal of the ring $Q(R)$, i.e., R is a heart of $Q(R)$.

It should be recalled that the rings with hearts are called *subdirectly undecomposable*. It is known that any ring can be presented as a subdirect product of subdirectly undecomposable rings. In this case the hearts of cofactors, as the least ideals, either have zero multiplication or prove to be simple rings. It is evident that the heart of a subdirectly undecomposable ring will be simple iff the ring is prime.

It would be now natural to pay our attention to simple rings as hearts of subdirectly undecomposable prime rings. As when transferring to a ring of invariants the prime dimension can increase, we come to a natural generalization.

6.3.33. Lemma-definition. A ring R will be called *almost simple* if it obeys the following equivalent conditions.

(1) R has the least essential ideal decomposing into a direct sum of a finite number of simple rings, which will be called a *semi-heart.*

(2) the ring R is a subdirect product of a finite number of subdirectly undecomposable prime rings.

Proof. $(1)\to(2)$. Let $\sigma = \sigma_1 \oplus ... \sigma_n$ be the semi-heart, and $T_k = ann_R \sigma_k$. Then $\sigma_k \cong \sigma / T_k$ will be the heart of the ring R / T_k and, hence, R / T_k is a subdirectly undecomposable prime (since the heart is prime) ring . It is evident that $\cap T_k = 0$ and $R = \underset{k=1}{\overset{n}{S}} R / T_k$.

$(2)\to(1)$. Let R be a subdirect product of rings $R_1,... R_n$ with simple hearts $\sigma_1,... \sigma_n$ and π_k: $R \to R_k$, $1 \le k \le n$ be approximating projections. Let us assume that n are the least possible numbers. Then $I_k \overset{df}{=} \bigcap_{i\neq k} \ker \pi_i \neq 0$. The sum of ideals $I_1 + ... + I_n$ will be direct, as

$$I_k \cap \sum_{i\neq k} I_i \subseteq \bigcap_{i\neq k} \ker \pi_i \cap \ker \pi_k = 0$$

Then, $\ker \pi_k \cap I_k = 0$ and, hence, $I_k \cong \pi_k(I_k)$ is an ideal of R_k, and, in particular, I_k contains an ideal $\sigma_k = \pi_k^{-1}(\sigma_k) \cong \sigma_k$. Let us show that $\sigma_1' + ... + \sigma_n'$ is the semiheart of R.

If $x\sigma = 0$, then $0 = \pi_k(x\sigma) = \pi_k(x)\sigma_k$, i.e., $\pi_k(x) = 0$ for all k $x = 0$ is a nonzero ideal of σ_k' and, hence, $I\sigma_k' = \sigma_k'$. It affords $I \supseteq \sum I\sigma_k' = \sigma$. The lemma is proved..

6.3.34. Theorem. *A ring of invariants of an* M*-group of automorphisms of a semiprime ring is almost simple iff such is initial ring. In this case the semiheart of the ring* R *is generated as a one-sided ideal by a semiheart of* R^G.

Proof. Let R be an almost simple ring. Then it has a finite prime dimension and, by theorem **5.5.10**, it is sufficient to consider the case when

R is a prime ring with a heart σ.

Let τ be a principal trace form. Then $\tau(\sigma) \subseteq R$. Let us show that $\tau(\sigma)$ is a semiheart of the fixed ring.

If $a \in R^G$ and $a\tau(\sigma) = 0$, then $\tau(a\sigma) = 0$ and, hence, $a\sigma = 0$, i.e., $\tau(\sigma)$ is an essential ideal in R^G.

If I is an arbitrary essential ideal of the fixed ring, then $I\sigma$ will be an $(R^G - R)$-bimodule with a zero left annihilator, i.e., this bimodule contains a nonzero ideal of the ring R (see **3.8.3**) and, moreover, $I\sigma \supseteq \sigma$. Therefore,

$I \supseteq I\tau(\sigma) = \tau(I\sigma) \supseteq \tau(\sigma)$.

As R^G has a finite prime dimension, then a certain essential ideal I of R^G is decomposed into a direct sum of prime rings $I = I_1 \oplus .. \oplus I_n$. We have $\tau(\sigma) = I_1\tau(\sigma) \oplus ... \oplus I_n\tau(\sigma)$. We now have to remark that $I_k\tau(\sigma)$ are simple rings. If, for instance, $0 \neq J$ is an ideal of $I_1\tau(\sigma)$, then $J \oplus I_2\tau(\sigma) \oplus ... \oplus I_n\tau(\sigma)$ is an essential ideal in R^G and, by the above proved, it contains $\tau(\sigma)$. Hence, $J = I_1\tau(\sigma)$.

Inversely, let σ be a semiheart of the ring R^G. By theorem **5.6.7**, the ring R has a finite prime dimension and, by theorem **5.5.10**, we can limit ourselves with considering only the prime ring R.

By lemma **3.8.3**, the left ideal $R\sigma$ contains a nonzero ideal I of the ring R. If J is an arbitrary nonzero ideal of the ring R, then $J \cap R^G$ is an essential ideal in R^G (lemma **3.8.4**) and, hence, $J \supseteq \sigma$. This affords $J \supseteq R\sigma \supseteq I$, which means that $I = R\sigma$ is a heart of the ring R. The theorem is proved.

For simple rings with a unit we can now give a certain necessary and sufficient condition of decomposition of a fixed ring into a direct sum of simple rings.

6.3.35. Corollary. *Let R be a simple ring with a unit, G be an M-group of its automorphisms. Then R^G will be a direct sum of simple rings iff for a certain invariant form τ there is an element $\gamma \in R$, such that $\tau(\gamma) = 1$.*

Proof. If $\tau(R) \ni 1$ and σ is a semiheart of the ring R^G, then $R\sigma = R$ and $R^G = \tau(R) = \tau(R)\sigma \subseteq \sigma$. Inversely, if τ is a principal form, then $\tau(R)$ is a semiheart of R^G, i.e., $\tau(R) = R^G \ni 1$, which is the required proof.

Returning to the case when G is a finite group and R has no $|G|$-torsion, it should be remarked that theorem **3.6.9** is also valid for finite direct sums of simple rings (by theorem **5.5.10**). As to the inverse statement, one can easily give examples proving that it is not valid. At the same time, in this case the Bergman-Isaaks theorem makes it possible to calculate a semiheart more exactly.

6.3.36. Corollary. *Let a fixed ring of a finite group of automorphisms of a semiprime ring R be a direct sum of a finite number of simple rings. Now, if there is no additive $|G|$-torsion, then a certain power of the ring is decomposed into a direct sum of simple rings.*

Proof. By theorem **6.3.4**, the ring R is almost simple. Let σ be its semiheart. Then σ is an invariant relative $AutR$ ideal, since it is equal to the intersection of all essential ideals. Moreover, in the factor-ring R/σ there is no additive $|G|$-torsion : $|G|\sigma$ is an essential ideal in R and, hence, $|G|\sigma=\sigma$; if $|G|r\in\sigma$, then $|G|r=|G|s$, where $s\in\sigma$, i.e., $r=s\in\sigma$. Now the group G is induced onto R/σ.

Since the intersection $\sigma\cap R^G$ is an essential ideal of the ring of invariants, then $\sigma\supseteq R^G$, i.e., there are no nonzero fixed elements in the factor-ring R/σ. By the Bergman-Isaaks theorem we get $R^m=\sigma$, which is the required proof.

REFERENCES

V.K.Kharchenko [127];
J.Osterburg [128,129];
T.Sundström [147];
A.I.Zalessky. O.M.Neroslavsky [157].

H.Prime Ideals, Montgomery Equivalence. Here we shall consider relations between prime ideals of a given ring R and a ring of invariants R^G for the case when G is a finite group of an invertible in R order. Since in this case fixed elements are determined by the equality $x=t(x)$, where $t=\frac{1}{|G|}\sum_G g$, then we can go over to considering factor-rings with respect to invariant ideals. Here we have $(R/I)^G=R^G/I^G$, where I is an invariant ideal. In particular, if P is a prime ideal of the ring R, then $\bigcap_{g\in G} P^G$ is an invariant ideal and we can go over to considering a factor-ring $R/\cap P^g$, which is a subdirect product of a finite number of prime rings. Let us start with presenting the well-known properties of such rings and general properties of prime ideals.

6.3.37. Lemma. *A semiprime ring is presented as a subdirect product of* n *prime rings iff its prime dimension* $\le n$ (see **3.6.6**).

Proof. Let $R = \overset{n}{\underset{i=1}{S}} R_i$ and $\pi_i: R \to R_i$ be approximating epimorphisms. Let $I_1 \oplus ... \oplus I_{n+1}$ be a direct sum of ideals in R. Then $\pi_k(I_1) \cdot ... \cdot \pi_k(I_{n+1}) = 0$ and, hence, the kernel of π_k contains all the ideals $I_1,... I_{n+1}$ with, possibly, only one excluded. It means that the intersection of all kernels of π_i contains at least one of the ideals $I_1,..., I_{n+1}$, i.e., the prime dimension of R is not greater than n.

Inversely, by lemma **3.6.6**, let us find in R an essential ideal I, which decomposes into a direct sum of prime rings $I = I_1 \oplus ... \oplus I_m$, $m \le n$. Let P_k be an annihilator of the ideal I_k in R. Then $\bigcap_k P_k = 0$. Let us check if P_k is a prime ideal. If $aRb \subseteq P_k$, then $I_k aRb = 0$ and, moreover, $I_k a \cdot I_k \cdot I_k b = 0$. However, I_k is a prime ring and, therefore, we have either $I_k a = 0$ or $I_k b = 0$, which is the required proof.

6.3.38. Lemma. *Any prime ideal* P *contains a minimal prime ideal* $P' \subseteq P$.

Proof. It should be remarked that an intersection of any chain of prime ideals is a prime ideal: if $AB \subseteq \cap P_\alpha$, $A \not\subseteq \cap P_\alpha$, then for a certain β we have $A \not\subseteq P_\beta$ and, hence, for all $\gamma > \beta$ we get $A \not\subseteq P_\gamma \subset P_\beta$, i.e., $B \subseteq P_\gamma$ and, hence, $B \subseteq \cap P_\gamma = \cap P_\alpha$. If we now use the Zorn lemma, we get the required proof.

6.3.39. Proposition. *Let the prime dimension of a semiprime ring* R *be equal to* n. *Then the following statements are valid.*

(a) Any irreducible presentation of a ring R *as a subdirect product of prime rings contains exactly* n *cofactors.*

(b) A presentation of R *as a subdirect product of* n *prime cofactors is unique, i.e., if* $R = \overset{n}{\underset{i=1}{S}} R_i = \overset{n}{\underset{i=1}{S}} R'_i$, *then there is a permutation* σ, *such that* $\ker \pi'_i = \ker \pi_{\sigma(i)}$, *where* π_i, π'_i *are the corresponding approximating projections.*

(c) The ring R *has exactly* n *minimal prime ideals and their intersection equals zero.*

Proof. It should be recalled that a subdirect product is called *irreducible* if we cannot reduce a single cofactor without the rest projections to remain an approximating family. In terms of the kernels of projections it means that $\bigcap_{i \ne k} \ker \pi_i \ne 0$ for all k. Certainly, not any prime ring has an invertible presentation, but if its prime dimension is finite, then the existence of such a presentation becomes obvious.

So, let $R = \underset{\alpha \in A}{S} R_\alpha$ be an irreducible product, $\pi_\alpha : R \to R_\alpha$ be approximating projections. Let $I_\alpha = \bigcap_{\beta \neq \alpha} \ker \pi_\beta$. Then

$$I_\alpha \cap \sum_{\beta \neq \alpha} I_\beta \subseteq \bigcap_{\beta \neq \alpha} \ker \pi_\beta \cap \ker \pi_\alpha = 0 .$$

Therefore, $\sum I_\alpha$ is a direct sum of nonzero ideals and, hence, the power of A does not exceed the prime dimension n. By lemma **6.3.37**, n, on the contrary, does not exceed the power of A.

If $R = \overset{n}{\underset{i=1}{S}} R_i = \overset{n}{\underset{i=1}{S}} R'_i$ and π_i, π'_i are the corresponding projections, then for any i we have

$$\ker \pi'_1 \cdot \ker \pi'_2 \cdot \ldots \cdot \ker \pi'_n \subseteq \bigcap_{j=1}^{n} \ker \pi'_j = 0 \subseteq \ker \pi_i .$$

Since R_i is prime, it means that there is a number j, such that $\ker \pi'_j \subseteq \ker \pi_i$. Analogously, by the number j one can find a number k, such that $\ker \pi_k \subseteq \ker \pi'_j$. If $k \neq i$, then $R = \underset{\alpha \neq i}{S} R_\alpha$, which contradicts lemma **6.3.37**. Therefore, $\ker \pi_i = \ker \pi'_j$ and the mapping $\sigma: i \to j$ implements the required permutation.

Finally, by lemma **6.3.38.** we can find n minimal prime ideals $P_1, \ldots, P_n$ the intersection of which is zero. If P is any other prime ideal, then $P \supseteq P_k$ for a certain k, i.e., P cannot be minimal. The proposition is proved.

6.3.40. Lemma. *If R is semiprime and has a finite prime dimension, then any not minimal prime ideal is essential.*

Proof. If Q is not minimal, and $Q \cdot U = 0$, then for any minimal prime ideal P we have $Q \cdot U \subseteq P$ and, hence, $U \subseteq P$. However, the intersection of all minimal ideals is zero. The lemma is proved.

6.3.41. ***Spectrum.*** We have already encountered this notion when considering a commutative self-injective regular ring (see **1.9.3**). For an arbitrary case on a set $Spec\ R$ of all prime ideals of the ring R one can also set a topology using the same closure operation

$$\overline{\{P_\alpha\}} = \{P \in Spec\ R \mid P \supseteq \cap P_\alpha\} .$$

It is the obtained topological space that is called *spectrum* (or a *prime spectrum*) of the ring R. The above formulated lemmas are quite obvious from the viewpoint of this topological space. For instance, a decomposition of a semiprime ring into a subdirect product of prime rings means that a set of kernels of the corresponding projections is dense in $Spec\, R$, and the prime dimension is the least power of a dense subset, and so on.

If a certain group G acts on the ring R, the action of this group is induced on the space $Spec\, R$. In this case there arises a space of orbits $Spec\, R\,/\,G$ which is determined as a factor-space $Spec\, R$ relative the equivalence relation $P \approx Q \leftrightarrow \exists\, g \in G: P^g = Q$.

6.3.42. Theorem (Montgomery). *Let G be a finite group of automorphisms of the ring R, $G \subseteq Aut\, R$ with its order invertible in R. In this case the following statements are valid.*

(a) *If P is a prime ideal of the ring R, then $P \cap R^G = \rho_1 \cap .. \cap \rho_n$, where $n \leq |G|$, $\rho_1,.., \rho_n$ are prime ideals of the ring R^G uniquely determined by the condition that $\rho_i / P \cap R^G$ are minimal prime ideals in $R^G / P \cap R^G$.*

(b) *If P, Q are prime ideals of R and $P \cap R^G = Q \cap R^G$, then $Q = P^g$ for a certain $g \in G$.*

(c) *If ρ is a prime ideal of the ring R^G, then there is a prime ideal P of the ring R, such that ρ is minimal over $P \cap R^G$, i.e., $\rho / P \cap R^G$ is a minimal prime ideal of the ring $R^G / P \cap R^G$. In this case if Q is another prime ideal of the ring R with the same property, then $Q = P^g$ for a certain $g \in G$. If I is an ideal of the ring R, such that $I \cap R^G \subseteq \rho$, then $I \subseteq P^g$ for a certain $g \in G$.*

Proof.

(a) Let us consider an invariant ideal $J = \bigcap_{g \in G} P^g$. We have $P \cap R^G = J \cap R^G$. Going over to a factor-ring R / J and allowing for the fact that $(R / J)^G = R^G / J \cap R^G$, we can assume that $J = 0$.

By theorems **3.6.7** and **5.5.10**, the ring R^G has a finite prime dimension $\leq |G|$ and we are now to use point (c) of proposition **6.3.39.**

(b) If $P \cap R^G = Q \cap R^G$, then in the factor-ring $\bar{R} = R / \cap P^g$ we have $\bar{Q} \cap \bar{R}^G = \bar{0}$. Hence, the invariant ideal $\bigcap_G \bar{Q}^g$ of a semiprime ring $\bar{R}$ has no fixed elements. This does not contradict the Bergman-Isaaks theorem only in the case when $\cap \bar{Q}^g = \bar{0}$, i.e., $\cap Q^g \subseteq \cap P^g$. Analogously, $\cap P^g \subseteq \cap Q^g$. Now the factor-ring $\bar{R}$ has two presentations in the form of a subdirect product of prime rings. Cancelling spare cofactors from every presentation and making use of proposition **6.3.39.** (a,b), we get $P^g = Q^h$ for certain $h, g \in G$, i.e., $Q = P^{gh^{-1}}$, which is the required proof.

(c) Let us consider a set $\mathcal{M}$ of all ideals I of a ring R, such that $I^G \overset{df}{=} I \cap R^G \subseteq \rho$. This set is nonempty and inductive. By the Zorn lemma, there are maximal elements in $\mathcal{M}$. Let us show that all maximal elements of $\mathcal{M}$ are prime ideals and lie in one orbit of the spectrum.

If P is a maximal element, and $A \cdot B \subseteq P$, in which case $A \supset P$, $B \supset P$, then $A^G \not\subseteq \rho$ and $B^G \not\subseteq \rho$. Since the ideal ρ is prime, we get $A^G \cdot B^G \not\subseteq \rho$ and, moreover, $(AB)^G \not\subseteq \rho$. However, $(AB)^G \subseteq P^G \subseteq \rho$, which is a contradiction.

Let Q be another maximal element from $\mathcal{M}$. Let us consider invariant ideals $I = \cap P^g$, $J = \cap Q^g$. We have

$$(I + J)^G = t(I + J) = t(I) + t(J) \subseteq P^G + Q^G \subseteq \rho,$$

i.e., $I + J \in \mathcal{M}$ and, by the Zorn lemma, we can find a maximal element $S \in \mathcal{M}$ containing $I + J$. We have $\cap P^g \subseteq S$, i.e., since S is prime, we can find a $g \in G$, such that $P^g \subseteq S$ and $P \subseteq S^{g^{-1}}$, and, due to maximality, $P = S^{g^{-1}}$. Analogously, $Q = S^{h^{-1}}$. Now $Q = P^{gh^{-1}}$, which is the required proof.

Let now $I = \cap P^g$ be an intersection of all maximal elements of $\mathcal{M}$. Then $I^G = P \cap R^G$, and I is an invariant ideal. Going over to a factor ring R / I and allowing for the fact that $(R / I)^G = R^G / I^G$, we can assume that $I = 0$. We are now to show that in this case ρ is a minimal prime ideal.

If ρ is not minimal, then by lemma **6.3.9** it will be essential. By lemma **5.10.5**, the left ideal $R\rho$ contains a two-sided essential ideal W of the ring R. We have

$$W \cap R^G \subseteq t(W) \subseteq t(R\rho) \subseteq t(R)\rho \subseteq \rho.$$

This means that $W \in \mathcal{M}$. By the Zorn lemma, let us find a maximal element P^h of the set $\mathcal{M}$ containing W. Now we come to a contradiction with the fact that W is an essential ideal:

$$\cap W^g \subseteq \cap P^g = 0.$$

Let Q be another prime ideal, such that ρ is minimal over Q^G. Then $Q \in M$ and we can imbed Q into a maximal element $Q \subseteq P^h$. Let us consider a factor-ring $\bar{R} = R / \cap Q^g$. If $P^h \neq Q$, then $\bar{P}^h$ is not a minimal prime ideal of $\bar{R}$. By lemma **6.3.40**, it means that $\bar{P}^h$ is an essential ideal of $\bar{R}$. An intersection $\bar{P}^h \cap \bar{R}^G = \bar{\rho}$ will be also essential, but this is impossible since $\bar{\rho}$ is a minimal prime ideal in $\bar{R}^G = R^G / Q^G$. The theorem is proved.

This theorem proves that every prime ideal ρ of a ring of invariants uniquely determines a semiprime ideal $\underline{\rho}$, such that ρ is minimal over $\underline{\rho}$, and $\underline{\rho}$ has the form P^G for a certain prime ideal P of the ring R.

6.3.43. Definitions. Prime ideals ρ, μ of a fixed ring R^G are called *Montgomery equivalent*, $\rho \overset{M}{\sim} \mu$, if $\underline{\rho} = \underline{\mu}$.

Each class of equivalence contains not more than $n = |G|$ elements. Theorem **6.3.41**, shows the mapping f, which puts the class of equivalence of minimal over P^G prime ideals into correspondence to the orbit $\bar{P} \in Spec\, R / G$, to be one-to-one. Moreover, by point (b), the orbit $\bar{P}$ will be uniquely determined by both the invariant ideal $\bigcap_{g \in G} P^g$ and the intersection $P \cap R^G = P^G$. It enables one to determine on the orbit space and on the space $Spec\, R^G / \overset{M}{\sim}$ the order relation $\bar{P} \le \bar{Q} \overset{df}{\Leftrightarrow} \cap P^g \subseteq \cap Q^g$ and $\rho / \overset{M}{\sim} \le \mu / \overset{M}{\sim} \overset{df}{\Leftrightarrow} \rho \le \mu$.

6.3.44. Theorem. *The mapping f is a homeomorphism of the topological spaces preserving the order:*

$$f \colon Spec\ R / G \equiv Spec\ R^G / \overset{M}{\sim} .$$

Proof. We should check the fact that if the prime ideal Q contains an intersection $\bigcap_{\alpha} P_{\alpha}^G$, where P_{α} are prime ideals, then $Q \supseteq \bigcap_{\alpha, g} P_{\alpha}^g$. Let $A = \bigcap_{g} Q^g \not\supseteq \bigcap_{\alpha, g} P_{\alpha}^g = B$. Then in the factor-ring $\bar{R} = R / A$ we have $\bar{B} \neq 0$ and, by the Bergman-Isaaks theorem, we get $\bar{B}^G \neq \bar{0}$, i.e., $\bigcap_{\alpha} P_{\alpha}^G = B^G \not\subseteq A$, which is a contradiction. The theorem is proved.

In relation with the notion of Montgomery equivalence there arises a number of interesting problems which can be generally formulated in the following way: to study the properties of the points of the spectrum of the fixed ring, which are stable relative the Montgomery equivalence. We have already encountered one of these properties, i.e., primitivity (it should be recalled that an ideal P of the ring R is called *primitive* if the factor-ring R/P is primitive).

Indeed, if ρ is a primitive ideal of the fixed ring, and P is its corresponding prime ideal of the ring R, then, going over to a factor-ring $\bar{R} = R/\cap P^g$, we can assume that ρ is minimal and, hence, not essential. Its annihilator will be a nonzero ideal and a primitive ring. By lemma **6.3.17** and theorem **5.5.10**, the ring $\bar{R}$ is semiprimitive, and by theorem **6.3.16**, $\overline{R^G}$ is also semiprimitive. Since the minimal decomposition into a subdirect sum of prime rings is unique, we say that all the ideals which are Montgomery equivalent to the ideal ρ, will be primitive.

Another example of a Montgomery-stable characteristic is given by the monotonicity of the mapping f. It should be recalled that the *height* of the ideal ρ is the maximum length of a strictly descending chain of prime ideals $\rho \supset \rho_2 \supset \dots \supset \rho_n$. The Montgomery-equivalent points of the spectrum are evident to have the same height.

One of Montgomery-unstable (or, to be more exact, not always stable) characteristics is the *depth* of an ideal, i.e., the maximal length of a strictly ascending chain of prime ideals $\rho \subset \rho_2 \subset \dots \subset \rho_n$.

6.3.45. Example. Let R be a ring of all linear transformations of a countable-dimension space V over a field F with a zero characteristic. This ring is naturally identified with a ring of infinite finite-row matrices (**1.14.4**). Let us consider a conjugation g by a diagonal matrix $diag(-1, 1, 1, \dots)$. In this case the ring of invariants of a group $G = \{g, 1\}$ is isomorphic to $F \oplus R$, i.e., we have two Montgomery-equivalent ideals $P_1 = F \oplus 0$, $P_2 = 0 \oplus R$, one of them being maximal, another not maximal.

The same example shows the condition " R/P is a Goldie ring" to be Montgomery-unstable, in spite of the fact that it is transferred onto rings of invariants and vice versa.

6.3.46. Morita contexts. Montgomery-equivalence proves to be closely related with attenuated Morita-equivalence. Let us consider the properties of the points of a spectrum, which are determined by factor-rings. Setting such a property is equivalent to singling out a class $\mathbb{R}$ of prime rings. In this case Montgomery-equivalence is set by the relation between prime rings: $R \overset{M}{\sim} S$ iff there exists a prime ring A and a group G of an invertible order, such that $R \cong A^G/\rho$, $S \cong A^G/\rho_1$ for certain minimal prime ideals ρ, ρ_1 of a fixed ring A^G.

It should be recalled that a *Morita context* is a sequence (R, V, W, S), where R, S are rings; V is an (R, S)-bimodule; W is an

(S, R)-bimodule, and determined are the multiplications $V \otimes_S W \to R$, $W \otimes_R V \to S$, such that a set of all the matrices of the type $\begin{pmatrix} R & V \\ W & S \end{pmatrix}$ forms an associative ring relative the matrix operations of multiplication and addition. The rings R nd S with unity are called *Morita-equivalent* if there is a Morita context (R, V, W, S), where $VW = R$, $WV = S$. This notion has been studied in detail and its essence is that the categories of left modules over R and S are equivalent iff R and S re Morita-equivalent. The Morita context (R, V, W, S) is called *prime* if the corresponding ring $\begin{pmatrix} R & V \\ W & S \end{pmatrix}$ is prime. If in this case the rings R and S contain $\frac{1}{2}$, then, as in example 6.3.45, we can consider a conjugation g onto the element $\begin{pmatrix} 1 & 0 \\ 0 & -1 \end{pmatrix}$. For the group $G = \{1, g\}$ we have $M^G = R \oplus S$, i.e., the rings R and S are Montgomery-equivalent . The following theorem shows that the inverse statement is also , to a certain extent, valid.

Let us consider two conditions on the class of prime rings $\mathcal{R}$.

Inv. If R is a prime ring and $G \subseteq \operatorname{Aut} R$ is a group of its automorphisms of an invertible order, such that R^G is prime, then $R^G \in \mathcal{R}$ iff $R \in \mathcal{R}$.

Mor. If the nonzero prime rings R and S are related by a prime Morita context and $R \in \mathcal{R}$, then $S \in \mathcal{R}$.

6.3.47. Theorem. *If a class of prime rings $\mathcal{R}$ obeys conditions Inv and Mor, then it is stable under Montgomery equivalence.*

Proof. Let us first remark that condition *Mor* shows that the class $\mathcal{R}$ together with the ring R contains all its two-sided ideals. Indeed, if I is an ideal, then we have a prime Morita context (R, I, I, I).

Let R be a prime ring, G a group of its automorphisms of an invertible order. Let us decompose the algebra $\mathcal{B}(G)$ of the group G into a direct sum of G-simple components

$$\mathcal{B}(G) = e_1 \mathcal{B}(G) \oplus \ldots \oplus e_n \mathcal{B}(G).$$

We can find a nonzero ideal J of R, such that $e_i J$, $Je_i \subseteq R$. Let us set $I = J^2$. Then $e_i I e_j \subseteq R$ for any $i, j, 1 \le i, j \le n$. Since e_i, e_j are fixed idempotents, then the group G acts on all components $e_i I e_j$, in which case

$$I^G = (e_1 I e_1)^G \oplus \ldots \oplus (e_n I e_n)^G .$$

The next step will be a remark that $(e_i I e_i)^G$ is a prime ring, so that the algebra of an induced group is equal to $e_i \mathcal{B}(G)$ (see lemma **6.3.20**). Therefore, in the ring of invariants R^G we have an essential ideal I^G decomposing into a direct sum of prime nonzero rings $(e_i I e_i)^G = e_i I^G$.

Let ρ be a minimal prime ideal of a fixed ring. Then ρ is not an essential ideal and, hence, it has cannot have nonzero intersections with all the ideals $e_i I^G$. Let, for example, $\rho \cap e_1 I^G = 0$. Then $e_1 I^G \cong e_1 I^G + \rho / \rho$ is an ideal of R^G / ρ, i.e., $R^G / \rho \in \mathcal{R}$ iff $e_1 I^G \in \mathcal{R}$.

Therefore, we are to show that if $e_1 I^G \in \mathcal{R}$, then all the components $e_i I^G$ also lie in $\mathcal{R}$.

Since $(e_1 I e_1)^G = e_1 I^G$, then, by property Inv we get $e_1 I e_1 \in \mathcal{R}$. Then, we have a prime Morita context $(e_1 I e_1, e_1 I e_i, e_i I e_1, e_i I e_i)$ and, hence, $e_i I e_i \in \mathcal{R}$. By property Inv we get $e_i I^G = (e_i I e_i)^G \in \mathcal{R}$. The theorem is proved.

The results obtained in sections ***A* -*G*** and in the above theorem imply that Montgomery-stable are the classes of semisimple prime rings for radicals considered in section ***A***. the class of prime subdirectly nondecomposable rings, the class of primitive rings and that of quite primitive rings. Montgomery- unstable is the class of simple artinian rings, Goldie rings, Noetherian and simple rings.

Here one essential remark should be made. When considering Montgomery-equivalence not on the whole union of prime rings but on its certain parts, unstable properties can be come stable.

For instance, the Gel'fand-Kirillov dimension becomes a stable characteristic in the class of PI-rings (S.Montgomery, L.Small). The classical Crull dimension is stable in the class of affine PI-rings (J.Alev), as well as in the class of Noetherian PI-rings (S.Montgomery, L.Small).

REFERENCES

J.Alev [1];
M.Artin, W.Shelter [10];
K.I.Beidar [13];
M.Lorenz, S.Montgomery, L.Small [99];
S.Montgomery [110,114];
S.Montgomery, D.Passman [116];
S.Montgomery, L.Small [119,120,121].

I. Modular Lattices. Many properties studied in the theory of rings can be set in terms of the lattice of left (right) or two-sided ideals (for instance, Noetherity, artinity, Crull dimension, Goldie dimension, etc.). If a ring of invariants R^G obeys such a property, then this fact gives an information only on certain invariants (left or two-sided) ideals and there naturally arises a problem of an interrelation of the lattice of all the ideals with that of invariant ideals. As the lattice of ideals is modular (i.e., $I+(J\cap K)=(I+J)\cap K$ if $I\subseteq K$), then we come to the problem of studying fixed points of finite groups acting on modular lattices.

The following two theorems proved by J.Fisher [45] and P.Grzesc-zuk, E.P.Puchilovsky [33] give us a bulky information on relations between L and L^G.

It should be recalled that a *lattice* is an algebraic system L with two binary commutative associative operations $\wedge$ and $\vee$, interrelated by the equalities:

P 1 $(x\vee y)\wedge x=x$

P 2 $(x\wedge y)\vee x=x$

P 3 $x\wedge x=x$, $x\vee x=x$.

On the lattice one can introduce a relation of a partial order $a\leq b\Leftrightarrow a\wedge b=a$. From the point of view of this order the operation $\wedge$ calculates the exact lover boundary, while the operation $\vee$ calculates the exact upper boundary of the two elements. The mapping of the lattices $f\colon L\to L_1$ is called *strictly monotonous* if the inequality $a\underset{\neq}{<}b$ implies $f(a)\underset{\neq}{<}f(b)$. A strictly monotonous mapping can not always preserve the exact lower and upper boundaries of incomparable elements. The lattice L is called *modular* if for all $a\leq c$ from L the following equality holds:

M1. $a\vee(b\wedge c)=(a\vee b)\wedge c.$

Since $a\leq a\vee b$, $a\leq a\vee c$, then the above conditional identity yields the following identities:

M2. $a\vee(c\wedge(a\vee b))=(a\vee c)\wedge(a\vee b)=$
$a\vee(b\wedge(a\vee c))$.

These two formulas result in the *rule of substitution*:

$$\textit{if } a \wedge b = 0 \textit{ and } c \wedge (a \vee b) = 0, \textit{ then } b \wedge (a \vee c) = 0,$$

where 0 is the least element of the lattice.

Indeed, if $c \wedge (a \vee b) = 0$, then $a = a \vee (c \wedge (a \vee b)) = a \vee (b \wedge (a \vee c))$, i .e., $b \wedge (a \vee c) \leq a$ and, hence, $b \wedge (a \vee c) \leq b \wedge a = 0$.

In the theory of modular lattices the latter rule plays approximately the same role as the lemma on substitution in the theory of linear spaces which results in the theorem of the existence of a basis.

The following condition which readily follows from M1, plays the same role in the theory of modular lattices as the theorem of homomorphisms.

M3. If $a \leq x \leq a \vee b$, then $a \vee (x \wedge b) = x$.

This immediately affords that the mapping $f_b(x) = x \wedge b$ sets an isomorphism $[a, a \vee b] \cong [a \wedge b, b]$.

It should be recalled that for elements $x \leq y$ the interval $\{z \mid x \leq z \leq y\}$ is denoted by $[x, y]$. A direct product of the lattices is determined in a natural way:

$$(a_1, ..., a_n) \vee (b_1, ..., b_n) = (a_1 \vee b_1, ..., a_n \vee b_n);$$

$$(a_1, ..., a_n) \wedge (b_1, ..., b_n) = (a_1 \wedge b_1, ..., a_n \wedge b_n).$$

6.3.48. Theorem (J.Fisher). *Let G be a finite group of automorphisms of a modular lattice L. Then there is a strictly monotonous mapping $f: L \rightarrow L^G \times ... \times L^G$ of the lattice L into a direct product of a finite number of copies of a sublattice L^G of fixed elements.*

Proof. Let us denote through F a set of all unary terms in a signature $(G, \wedge, \vee)$. The elements of F will be called *G-lattice polynomials*. Examples of such polynomials can be the terms $\bigvee_{s \in S} (\bigwedge_{t \in T} x^t)^s$, where S, T are subsets of G. Each of such polynomials can be viewed as a mapping $f: M \rightarrow M$. By induction over the construction of the term f one can easily prove that the corresponding to it mapping will be monotonous (but, possibly, not strictly monotonous): $a \leq b \rightarrow f(a) \leq f(b)$.

A G-lattice polynomial f is called *invariant* if all its values lie in L^G (for instance, $\wedge_G x^g$, $\vee_G x^g$ are G-invariant). In order to prove the Fisher theorem, it is sufficient to construct a finite set of invariant G-lattice polynomials $f_1, ..., f_k$, such that for $a \leq b$ the equalities $f_i(a) = f_i(b)$, $1 \leq i \leq k$ imply $a = b$. Therefore, we shall say that f *trivializes the couple*

(a, b) if $a \le b$ and $f(a) = f(b)$.

6.3.49. Lemma. *Let* S *be a subset of* G, *and* $g \in G \setminus S$. *If the couple* (a, b) *is trivialized by each of the following polynomials*

$$\bigwedge_{t \in S \cup \{g\}} x^t, \tag{1}$$

$$\vee \{(\bigwedge_{s \in S} x^s)^{t^{-1}} \mid t \in S \cup \{g\}\}, \tag{2}$$

then the couple (a, b) is trivialized by the polynomial $\bigwedge_{s \in S} x^s$.

Proof. By the condition we have:

$$a^g \wedge \bigwedge_{s \in S} a^s = b^g \wedge \bigwedge_{s \in S} b^s, \tag{3}$$

$$\begin{aligned} &\bigvee_{t \in S} (\bigwedge_{s \in S} a^s)^{t^{-1}} \vee (\bigwedge_{s \in S} a^s)^{g^{-1}} = \\ &= [\bigvee_{t \in S} (\bigwedge_{s \in S} b^s)^{t^{-1}}] \vee (\bigwedge_{s \in S} b^s)^{g^{-1}}. \end{aligned} \tag{4}$$

It should be remarked that for every $t \in S$ we have $(\bigwedge_{s \in S} a^s)^{t^{-1}} \le a$ and, hence, $\bigvee_{t \in S} (\bigwedge_{s \in S} a^s)^{t^{-1}} \le a$, i.e., applying to equality (4) an automorphism g and neglecting all the terms in square brackets in the right-hand part, we get an inequality

$$a^g \vee (\bigwedge_{s \in S} a^s) \ge \bigwedge_{s \in S} b^s \tag{5}$$

and, moreover, $b^g \vee (\bigwedge_{s \in S} a^s) \ge \bigwedge_{s \in S} b^s$.

Let us now make use of the fact that $\bigwedge_{s \in S} a^s \le \bigwedge_{s \in S} b^s$ and that the lattice is modular, we get:

$$\bigwedge_{s \in S} b^s = \bigwedge_{s \in S} b^s \wedge (b^g \vee \bigwedge_{s \in S} a^s) = (\bigwedge_{s \in S} b^s \wedge b^g) \vee \bigwedge_{s \in S} a^s.$$

By equality (3) we find

$$\bigwedge_{s \in S} b^s = (\bigwedge_{s \in S} a^s \wedge a^g) \vee \bigwedge_{s \in S} a^s = \bigwedge_{s \in S} a^s.$$

The lemma is proved.

Let us arbitrarily order the elements of the group $1 = g_1, \ldots, g_n$. Let I be a set of all sequences $1 = i(1) < \ldots < i(m) \leq n$, where $m \geq 1$. By $i(m)$ we shall denote the last term of the sequence i; while i' will denote a sequence without this senior term, $i' = (i(1), \ldots, i(m-1))$, if $i \neq (1)$. By induction let us determine lattice G-polynomials $f(i; \wedge; x)$ and $f(i; \vee; x)$:

$$f((1); \wedge; x) = f((1); \vee; x) = x, \tag{6}$$

$$f(i; \wedge; x) = \bigwedge_{j=1}^{i(m)} f(i'; \vee; x)^{g_j}, \tag{7}$$

$$f(i; \vee; x) = \bigvee_{j=1}^{i(m)} f(i'; \wedge; x)^{g_j^{-1}}. \tag{8}$$

Of prime interest for us will be the polynomials $f(i; \wedge; x)$ and $f(i; \vee; x)$ for the case $i(m) = n$. It is obvious that all such polynomials are invariant, and their number is 2^{n-1}. In order to formally conduct induction over the construction, for the sequence $i = (1)$ it would be convenient to assume $i' = i$, $i(m) = 1$, in which case equalities (7) and (8) are not violated.

In this case lemma **6.3.49.** yields the following statement.

6.3.50. Corollary. *Let* $i \in I$ *and* $i(m) < n$.

(1) *If both* $f(i', i(m)+1; \wedge; x)$ *and* $f(i, i(m)+1; \vee; x)$ *trivialize the pair* (a, b), *then* $f(i; \wedge; x)$ *also trivializers the pair* (a, b).

(2) *If both* $f(i, i(m)+1; \vee; x)$ *and* $f(i', i(m)+1; \wedge; x)$ *trivialize* (a, b), *then* $f(i; \vee; x)$ *trivializes* (a, b).

Let us decompose the set I into a union $I_1 \cup \ldots \cup I_n$, where $I_k = \{i \in I | \ i(m) = k\}$. Corollary **6.3.50.** guarantees that if for a certain $m < n$ the polynomials $f(i; \wedge; x)$ and $f(i; \vee; x)$ trivialize the pair (a, b) for all $i \in I_{m+1}$, then it is valid for all $i \in I_m$. By induction we get that if all 2^{n-1} invariant polynomials of I_n trivialize the pair (a, b), then the pair (a, b) is trivialized by the polynomials $f((1); \wedge; x) = f((1); \vee; x) = x$. The theorem is proved.

6.3.51. Corollary. *Let* G *be a finite group of automorphisms of a ring* R, $G \subseteq \mathrm{Aut}R$. *If* R *obeys the maximality (minimality) condition for* G*-invariant left (two-sided) ideals, then* R *also obeys this condition for all* *left (two-sided) ideals.*

Proof. One should remark that the lattice of the left (two-sided) ideals is modular.

6.3.52. Corollary. *Let G be a finite group of automorphisms $G \subseteq \mathrm{Aut}\, R$ of the ring R. Then R obeys the condition of for semiprime ideals iff R obeys the minimality (maximality) condition for invariant semiprime ideals.*

Proof. On a set of semiprime ideals let us determine lattice operations $I \wedge J = I \cap J$ and $I \vee J = \boldsymbol{B}(I + J)$, where $\boldsymbol{B}(I + J)$ is an intersection of all prime ideals containing $I + J$ (i.e., $\boldsymbol{B}(I + J) / I + J$ is a Baer radical of the factor-ring $R / I + J$). Let Ω be a Bullean algebra of subsets of the points of the $Spec\, R$. If we put the union $W(I)$ of all prime ideals containing no I to the semiprime ideal I, then $W(I \wedge J) = W(I) \cap W(J)$; $W(I \vee J) = W(I) \cup W(J)$ and, hence, W is an imbedding of lattices. Since Ω is a distributive lattice, then the lattice of semiprime ideals will also be distributive (and, hence, modular). Now we have to use the Fisher theorem.

6.3.53. Corollary. *Let G be a group of automorphisms $G \subseteq \mathrm{Aut}\, R$ of an invertible order. Then if R^G obeys the condition of minimality (maximality) for semiprime ideals, then the ring R also obeys this condition.*

Proof. If $I \subseteq J$ are semiprime invariant ideals and $I^G = J^G$, then in the factor-ring $\bar{R} = R / I$ we have $\bar{J}^G = 0$, i.e., by the Bergman-Isaaks theorem we have $J^n \subseteq I$, and, hence, $J = I$. Therefore, $I \to I^G$ is a strictly monotonous mapping of the lattice of invariant semiprime ideals of the ring R into the lattice of semiprime ideals of the ring R^G. Now we have to use the preceding corollary.

Let us now consider relations between the Goldie dimensions of the lattices L and L^G under the assumption that L has the biggest element , 1, and the least one, 0.

6.3.54. Definitions. A set of nonzero elements $\{x_1, \ldots, x_n\}$ from L is called *independent* if $x_i \wedge (x_1 \vee \ldots \vee x_{i-1} \vee \vee x_{i+1} \vee \ldots \vee x_n) = 0$ for all i, $1 \leq i \leq n$. The *Goldie dimension of the lattice* L is determined as an exact upper boundary of the powers of independent sets of elements. An element $a \in L$ is called *essential* in L if $a \wedge x \neq 0$ for any nonzero $x \in L$. An element u is called *uniform* if for any nonzero $x, y \leq u$ valid is $x \wedge y \neq 0$.

6.3.55. Theorem (P.Gzesczuk, E.P. Puczylowski). *Let* G *be a finite group of automorphisms of a modular lattice with the least elements. In this case*

$$\text{Gol-dim } L^G \leq \text{Gol-dim } L \leq |G| \cdot \text{Gol-dim } L^G.$$

Before getting down to a direct proof let us consider the properties of the Goldie dimension in modular lattices.

6.3.56. Proposition. *The Goldie dimension of a modular lattice L is equal to a finite number* n *iff L contains an independent set* $\{a_1,.., a_n\}$ *of uniform elements, such that* $a_1 \vee .. \vee a_n$ *is an essential element from L.*

Proof. If the Goldie dimension is $n < \infty$, then we can find an independent family $\{a_1,.., a_n\}$. If $a_1 \vee .. \vee a_n$ is not an essential element, then there is a nonzero element b, such that $(a_1 \vee .. \vee a_n) \wedge b = 0$. By the rule of substitution we see that $\{a_1,.., a_n, b\}$ is an independent family, which contradicts the choice of n. Then, if, for instance, a_1 is not uniform, then we can find nonzero elements $a'_1, a''_2 \leq a_1$, such that $a'_1 \wedge a''_1 = 0$, and again the rule of substitution for modular lattices shows that $\{a'_1, a''_1, a_2,..., a_n\}$ is an independent set.

Let us, inversely, assume that $\{a_1,.., a_n\}$ is a homogenous set consisting of homogenous elements and such that $a = a_1 \vee .. \vee a_n$ is an element essential in L. By induction over n let us show that $\text{Gol-dim } L = n$.

If $n = 1$, we have nothing to prove. Let $n > 1$ and let $\{b_1,.. b_{n+1}\}$ be an independent set.. In this case $a \wedge b_i \neq 0$, $1 \leq i \leq n+1$ and, hence we can suppose that $b_i \leq a$. Let, then, $c = a_2 \vee .. \vee a_n$. Then, by the inductive proposition, the Goldie dimension of the lattice $[0, c]$ is equal to $n-1$. In particular, not more than $n-1$ of elements from $\{b_i\}$ can have a nonzero "intersection" with c. Let $b_i \wedge c \neq 0$ at $1 \leq i \leq k$, $b_i \wedge c = 0$ at $k < i \leq n+1$. Let us set $f_i = b_i \wedge c$ at $i \leq k$ and $f_i = (b_{n+1} \vee b_i) \wedge c$ at $k < i \leq n$. In order to get a contradiction, it suffices to show that $\{f_1,.., f_n\}$ is an independent set.

If $1 \leq i \leq k$, then

$$f_i \wedge (f_1 \vee .. \vee f_{i-1} \vee f_{i+1} \vee .. \vee f_n) \leq$$

$$\leq b_i \wedge (b_1 \vee .. \vee b_{i-1} \vee b_{i+1} \vee .. \vee b_n \vee b_{n+1}) = 0.$$

If $k < i \leq n$, then, accounting for the modularity, we have:

$$f_i \wedge (f_1 \vee \ldots \vee f_{i-1} \vee f_{i+1} \vee \ldots \vee f_n) \leq$$
$$\leq (b_{n+1} \vee b_i) \wedge (b_1 \vee \ldots \vee b_{i-1} \vee b_{i+1} \vee \ldots \vee b_n \vee b_{n+1}) =$$
$$= b_{n+1} \vee [(b_{n+1} \vee b_i) \wedge (b_1 \vee \ldots \vee b_{i-1} \vee b_{i+1} \vee ..$$
$$.. \vee b_n) = b_{n+1}.$$

Since $f_i \leq c$ and $b_{n+1} \wedge c = 0$, we get the required equality.

We now have to show that $f_i \neq 0$ at all i, $k < i \leq n$. Let $f_i = (b_{n+1} \vee b_i) \wedge c = 0$. Since $b_{i+1} \wedge b_i = 0$, then, by the rule of substitution, we get $b_{n+1} \wedge (c \vee b_i) = 0$. From this fact and from formulas M2 we have $c = c \vee (b_{n+1} \wedge (c \vee b_i)) = (b_i \vee c) \wedge \wedge (b_{n+1} \vee c)$, and, hence, $[(b_i \vee c) \wedge a_1] \wedge [(b_{n+1} \vee c) \wedge a_1] \leq c \wedge a_1 = 0$. Since a_1 is a uniform element, this is possible only if $(b_i \vee c) \wedge a_1 = 0$ or $(b_{n+1} \vee c) \wedge \wedge a_1 = 0$. As far as $b_i \wedge c = b_{n+1} \wedge c = 0$, then the rule of substitution results in a contradiction with the fact that the element $c \vee a_1$ is essential. The lemma is proved.

6.3.57. Lemma. *For any* $a \in L$ *the following inequality holds:* $Gol-\dim L \leq Gol-\dim[0, a] + Gol-\dim[a, 1]$.

Proof. Let us first note that if $\{a, x_1, \ldots, x_n\}$ is an independent set in L, then $\{a \vee x_1, \ldots, a \vee x_n\}$ is an independent set in $[a, 1]$. Indeed, using the modularity, we see that for any i, $1 \leq i \leq n$, we have:

$$(a \vee x_i) \wedge (a \vee x_1 \vee \ldots \vee a \vee x_{i-1} \vee a \vee x_{i+1} \vee \ldots) =$$
$$= a \vee ((a \vee x_i) \wedge (x_1 \vee \ldots \vee x_{i-1} \vee x_{i+1} \vee \ldots)) = a$$

Then, the inequality in the formulation of the lemma gets trivial provided one of the dimensions on the right is infinite. Let us, therefore, assume that they are both finite. Let us choose the biggest by the number of elements family $\{x_1, .. x_n\}$, for which $\{a, x_1, \ldots, x_n\}$ are independent. Then all x_i will be uniform, $n \leq Gol-\dim[a, 1]$ and $a \vee x_1 \vee .. \vee x_n$ will be essential in $[a, 1]$. As $d = Gol-\dim[0, a] < \infty$, then there is an independent family $\{y_1, \ldots, y_d\}$ of uniform elements from $[0, a]$, in which case $y_1 \vee \ldots \vee y_d$ is essential in $[0, a]$. By proposition **6.3.56** we are now to show that $y_1 \vee \ldots \vee y_d \vee x_1 \vee \ldots \vee x_n$ is an essential element from L, which immediately results from the following lemma.

6.3.58. Lemma. *If* a' *is an essential element in* $[0, a]$ *and* $a \wedge b = 0$, *then* $a' \vee b$ *is an essential element in* $[0, a \vee b]$.

Proof. Let $0 \neq c \leq a \vee b$, but $c \wedge (a' \vee b) = 0$. Then, by the rule of substitution, $a' \wedge (c \vee b) = 0$ and, as a' is essential in $[0, a]$, then $a \wedge (c \vee b) = 0$, i .e., by formula M 2 we get $b = b \vee (a \wedge (c \vee b)) = (b \vee a) \wedge (b \vee c) \geq c$, or $0 = c \wedge (a' \vee, \vee b) = c \neq 0$ which is a contradiction. Lemmas **6.3.57** and **6.3.58** are proved.

6.3.59. Corollary. *If* $a_1, \ldots, a_n \in L$ *and* $a_1 \wedge \ldots \wedge a_n = 0$, *then* $Gol-\dim L \leq \sum_{i=1}^{n} Gol-\dim[a_i, 1]$.

Proof will be carried out by induction over n. At $n = 1$ the proof is obvious. Let $n \geq 2$ and $\bar{a}_1 = a_2 \wedge \ldots \wedge a_n$. According to the induction proposition, we have:

$$Gol\text{-}\dim\ [\bar{a}_1, 1] \leq \sum_{i=2}^{n} Gol\text{-}\dim\ [a_i, 1].$$

Since $a_1 \wedge \bar{a}_1 = 0$, then $[0, a_1] = [a_1 \wedge \bar{a}_1, a_1] \cong [\bar{a}_1, a_1 \vee \bar{a}_1] \subseteq [\bar{a}_1, 1]$. Thus, we get

$$Gol\text{-}\dim\ [0, a_1] \leq \sum_{i=2}^{n} Gol\text{-}\dim\ [a_i, 1]$$

and we have to use lemma **6.3.57**.

The lemmas to follow reduce the proof of theorem **6.3.55** to complete continuous lattices (a complete lattice L is called *continuous* if for any chain C in L and any $a \in L$ the equality $a \wedge (\vee c) = \bigvee_{x \in C} (a \wedge x)$ holds). It is evident that the lattices of the (left) ideals of a ring are complete and continuous, so that these conditions can be postulated from the point of view of applications. The very idea of reducing is based on the transition to the lattice $I(L)$ of the ideals of L.

A nonempty subset I of L is called *an ideal* of L, provided the conditions $x, y \in I$ imply $[0, x \vee y] \subseteq I$. A set $I(L)$ of all the ideals of the lattice L forms a continuous modular lattice with the operations $\wedge S_\alpha = \cap S_\alpha$, and $\vee S_\alpha =$ {an ideal generated by $\cup S_\alpha$}. It is evident that the mapping $p: L \to I(L)$ acting by the formula $p(a) = [0, a]$ is an imbedding of lattices, so that I can be viewed as a sublattice in $I(L)$. It is also evident that the action of groups with L can be naturally extended onto $I(L)$, the elements of $I(L)^G$ being invariant ideals.

6.3.60. Lemma. *If G is a finite group acting on L, then* $I(L^G) \cong I(L)^G$.

Proof. For $I \in I(L^G)$ and $J \in I(L)^G$ let us determine the mappings $g(J) = J \cap L^G$, and let $f(I)$ be an ideal of L generated by I. Then f and g implement the required isomorphism, provided the group G is finite.

6.3.61. Lemma. $Gol-\dim L = Gol-\dim(I(L))$.

Proof. It is obvious that $Gol-\dim L \leq Gol-\dim I(L)$. Inversely, let $\{I_1,..., I_n\}$ be an independent subset in $I(L)$ and let $0 \neq x_i \in I_i$. Then $\{x_i,..., x_n\}$ is an independent set in L. The lemma is proved.

6.3.62. Proof of lemma 6.3.55. In view of the two preceding lemmas we can assume that L is complete and continuous. By the Zorn lemma, one can find a maximal element l in the set $\{x \in L \mid \wedge_G x^g = 0\}$. By corollary **6.3.59**, it suffices to show that $Gol-\dim[l, 1] \leq Gol-\dim L^G$. This equality will be established if we show that for any family of independent elements $\{x_1,..., x_m\} \in [l, 1]$ the elements $\bar{x}_i = \wedge_{g \in G} x_i^g$ will be independent in L^G. According to the choice of l, the elements $\bar{x}_i$ are nonzero. Let $\bar{y}_i = \bar{x}_i \wedge (\bar{x}_1 \vee .. \vee \bar{x}_{i-1} \vee \bar{x}_{i+1} \vee .. \vee \bar{x}_m)$. Since $l < l \vee \bar{x}_i \leq x_i$ and the set $\{x_i,..., x_m\}$ is independent in $[l, 1]$, then the set $\{l \vee \bar{x}_1,..., l \vee \bar{x}_m\}$ will be also independent in $[l, 1]$. Hence, $l = (l \vee \bar{x}_i) \wedge (l \vee \bar{x}_1 \vee .. \vee l \vee \bar{x}_{i-1} \vee l \vee \bar{x}_{i+1} \vee .. \quad .. \vee l \vee \bar{x}_m) \geq y_i$. Since $y_i \in L^G$, then, by the choice of l we get a contradiction $0 = \wedge_{g \in G} l^g \geq y_i$. The theorem is proved.

From this theorem there results a statement that primarily had a much more complex proof.

6.3.63. Corollary. *Let G be a finite group of automorphisms of a ring R. Then, if in R there is an infinite direct sum of nonzero invariant left ideals, then in R there is an infinite direct sum of nonzero invariant left ideals.*

REFERENCES

J.Fisher [45,46];
P.Grzesczuk, E.P. Puczylowski [53];
V.K.Kharchenko [69].

J. Maximal Ring of Quotients. Let us recall some definitions. The left ideal D of a ring R is called *dense* if for any $r \in R$ the right annihilator in the ring R of the left ideal $Dr^{-1} \overset{df}{=} \{x \in R \mid xr \in D\}$ equals zero. A set $\mathfrak{G}$ of all dense left ideals forms an *idempotent filter* (see, for instance, [146]), i.e., the following statements are valid for it:

(1) if $D \in \mathfrak{G}$ and $r \in R$, then $Dr^{-1} \in \mathfrak{G}$;

(2) if L is the left ideal of the ring R and there is a left ideal $D \in \mathfrak{G}$, such that for all $d \in D$ valid is $Ld^{-1} \in \mathfrak{G}$, then $L \in \mathfrak{G}$.

From these statements one can easily deduce that an intersection of a finite set of dense left ideals will be a dense left ideal. A two-sided ideal I will be dense iff its right annihilator is zero, since $Ir^{-1} \supseteq I$. It means that for a semiprime ring we have $F \subseteq \mathfrak{G}$.

A maximal (left) ring of quotients $Q_{max}(R)$ is determined by a familiar scheme using the filter $\mathfrak{G}$:

$$Q_{max}(R) = \varinjlim_{D \in \mathfrak{G}} Hom(D, R).$$

The only thing that should be accounted here for is the correctness of the definition of a product. If $\varphi_1 \in Hom(D_1, R)$, $\varphi_2 \in Hom(D_2, R)$, then a superposition $\varphi_1\varphi_2$ is determined on $(D_2)\varphi_1^{-1}$ by a common formula $(x)\varphi_1\varphi_2 = (x\varphi_1)\varphi_2$. Therefore, one needs an inclusion $(D_2)\varphi_1^{-1} \in \mathfrak{G}$, which immediately results from the fact that the filter is idempotent: $(D_2\varphi_1^{-1})d^{-1} \supseteq D_2(d\varphi_1)^{-1} \in \mathfrak{G}$ at all $d \in D_1$.

As was the case for a Martindale ring of quotients, we have an imbedding $R \subseteq Q_{max}(R)$ by way of putting $r \in R$ in correspondence to a homomorphism $x \to xr$ which is determined on R. This correspondence has a sense if $R \in \mathfrak{G}$, i.e., the right annihilator of R is zero. It is evident that if $R \notin \mathfrak{G}$, then $\mathfrak{G} = \varnothing$, and the whole construction of a maximal ring of quotients has no sense. It is common practice, when considering problems pertaining to a maximal ring of quotients, to view the initial ring as having unity. We will also keep to this assumption.

Since $F \subseteq \mathfrak{G}$ for a semiprime ring R, then $R_F \subseteq Q_{max}(R)$. To this end we have only to remark that if $\varphi \in Hom(I, R)$, where $I \in F$ and $D\varphi = 0$ for a dense left ideal $D \subseteq I$, then $\varphi = 0$. If $\varphi \neq 0$, then $a\varphi \neq 0$ for a certain $a \in I$. Since D is a dense ideal, then $0 \neq (Da^{-1})(a\varphi) = [(Da^{-1})a]\varphi \subseteq D\varphi$, which is the required proof.

Dense ideals have an important characteristic in terms of an injective hull $E(R)$ of a regular left R-module R It should be recalled that the left module E is called *injective* if any homomorphism $\varphi: S \to E$ determined on a submodule S of any module N is extended to a homomorphism $\bar{\varphi}: N \to E$. It is well-known (see, for instance, [91, **4.2**] that any module M has an *injective hull*, i.e., the least injective overmodule $E(M) \supseteq M$. It is also known that a module is always an essential submodule of its injective hull. In this case the injective hull can be characterized as the biggest extension of M, where M is an essential submodule.

6.3.64. Lemma. *A left ideal D is dense iff for any homomorphism of the left R-modules $\varphi: R \to E(R)$ the equality $\varphi(D) = 0$ implies $\varphi = 0$.*

Proof. Let $(Dr^{-1})w = 0$, $w \neq 0$. Then an inclusion $ar \in D$ implies $aw = 0$ and, hence, correctly determined is the homomorphism $\varphi: Rr / Rr \cap D \to R$, which acts by the formula $\varphi(ar + D) = aw$. Since $Rr / Rr \cap D$ is a submodule of the module R / D, then there is an extension $\bar{\varphi}: R / D \to E(R)$. If $\pi: R \to R / D$ is a natural projection, then $\bar{\varphi}(\pi(D)) = 0$ and $\bar{\varphi}(\pi(R)) \supseteq \bar{\varphi}(\pi(Rr)) = Rw \neq 0$.

Inversely, if $\varphi: R \to E(R)$ and $\varphi(a) \neq 0$, then a submodule $R\varphi(a)$ has a nonzero intersection with an essential submodule $R \subseteq E(R)$. Let $\neq t\varphi(a) \in R$. Then we have

$$[(\ker \varphi)(ta)^{-1}]t\varphi(a) = \varphi[(\ker \varphi)(ta)^{-1}ta] \subseteq \varphi(\ker \varphi) = 0.$$

This means that the kernel of φ cannot be a dense ideal. The lemma is proved.

6.3.65. Lemma. *A left ideal D is dense iff its annihilator in*

$E(R)$ is zero.

Proof. If $Dw = 0$, $0 \neq w \in E(R)$, then the homomorphism $r \to rw$ has D in its kernel, i.e., by the preceding lemma, D cannot be dense. Inversely, if D is not dense, then $\varphi(D) = 0$ for a certain nonzero $\varphi: R \to E(R)$. We have $D \cdot \varphi(1) = \varphi(D) = 0$, which is the required proof.

6.3.66. Lemma. *Let D be a left ideal of R. Let us consider a ring $D^{\#}$ obtained by joining unity to D. Then if U is a left ideal of the ring R, which is contained in $D^{\#}$ and is dense in it, then U is dense in R*

Proof. Let us remark that a dense left ideal is essential: if $D \cap Ra = 0$, then $(Da^{-1}) \cdot a \subseteq Ra \cap D$ and, hence, $a = 0$. Moreover, $D^{\#}$ will be an essential left submodule of the left $D^{\#}$-module $E(R)$; if $0 \neq w \in E(R)$, then Dw is a nonzero left R-submodule, and, hence, it has a nonzero intersection with R, and, therefore, with D as well.

Since the injective hull $E(D^{\#})$ contains all the extensions where $D^{\#}$ is an essential submodule, then $E(R) \subseteq E(D^{\#})$.

If now $w \in E(R)$ and $Uw = 0$, then, since U is dense in $D^{\#}$ we have $D^{\#}w = 0$ and, hence, $w = 0$. The lemma is proved.

6.3.67. Lemma. *Any automorphism from $A(R)$ has a unique extension on $Q_{max}(R)$. Here R is a semiprime ring.*

Proof. Let $g \in A(R)$. Let us choose essential two-sided ideals I, J, such that $J \subseteq I^g \subseteq R$.

It should be remarked that if $D \in \mathfrak{S}$, then $D_1 = J(I \cap D)^g \in \mathfrak{S}$. By the preceding lemma, it suffices to show that D_1 is a dense ideal in $J^{\#}$. If $r \in J^{\#}$, then $r = a^g$ for a certain $a \in I^{\#}$, in which case $D_1 r^{-1} \supseteq J((I \cap D)a^{-1})^g$. Therefore, the equality $D_1 r^{-1} \cdot w = 0$ implies $J((I \cap D)a^{-1} \cdot s)^g = 0$, where $s^g = w \in J^{\#}$ and, hence, $(I \cap D)a^{-1} \cdot s = 0$, i.e., since the intersection $I \cap D$ is dense, we get $s = 0$ and $w = 0$.

If now $\varphi \in Hom_R(D, R)$, then an element $\varphi^g \in Hom_R(D_1, R)$ is determined by the formula $x\varphi^g = (x^{g^{-1}}\varphi)^g$. Standard considerations show $\varphi \to \varphi^g$ is the sought extension of g onto $Q_{max}(R)$. The lemma is proved.

6.3.68. Lemma. *Let R be a semiprime ring, $g \in A(R)$, $\hat{g}$ be an extension of g onto $Q_{max}(R)$. Then $\Phi_g = \Phi_{\hat{g}}$ and, in particular,*

$\mathbb{B}(G) = \mathbb{B}(\hat{G})$ *for any group* $G \subseteq \mathbb{A}(R)$.

Proof. If $\varphi \in \Phi_g$, $\xi \in Q_{max}$, then one can find a dense ideal D, such that $D\xi \subseteq R$. For any $d \in D$ we have $(d\xi)\varphi = \varphi d^g \xi^{\hat{g}} = d\varphi\xi^{\hat{g}}$, i.e., $D(\xi\varphi - \varphi\xi^{\hat{g}}) = 0$, which implies $\xi\varphi = \varphi\xi^{\hat{g}}$.

If $a \in \Phi_{\hat{g}}$, then $xa = ax^{\hat{g}}$ for all $x \in Q_{max}$ and, hence, it suffices to show that $a \in R_F$, i.e., to find an ideal $I \in F$, such that $Ia \subseteq R$. Let D be a dense left ideal, such that $Da \subseteq R$ and J is an essential two-sided ideal, such that $J^g \subseteq R$. In this case $DJa = DaJ^g \subseteq R$ and, hence, $I = DJ$ is the sought ideal. The lemma is proved.

6.3.69. Theorem (J.M.Goursaud, J.L.Pascaud, J.Valette). *Let G be an M-group of automorphisms of a semiprime ring R. Then* $Q_{max}(R^G) = Q_{max}(R)^{\hat{G}}$.

We shall begin with auxiliary lemmas.

6.3.70. Lemma. *Let $w_1, \dots, w_n \in E(R)$ and not all w_i are zero. If $g_1, \dots, g_n \in \mathbb{A}(R)$, then there is an element $r \in R$, such that $r^{g_1}, \dots, r^{g_n} \in R$; $r^{g_1}w_1, \dots, r^{g_n}w_n \in R$ and not all elements $r^{g_1}w_1, \dots, r^{g_n}w_n$ are zero. Here R is a semiprime ring.*

Proof. Let $I, J \in F$ and $J \subseteq I^{g_i}, I^{g_i^{-1}}, I^{g_i^{-1}g_j} \subseteq R$ for all i, j. Since J is a dense ideal, then $I^{g_i}w_i \neq 0$ when $w_i \neq 0$. Let us choose an element $r \in I$ in such a way that among the elements $r^{g_1}w_1, \dots, r^{g_n}w_n$ there was the greatest possible number of elements lying in R, but not all these elements were zero.

Let, for instance, $r^{g_1}w_1 \notin R$. Then $r^{g_1}w_1 \neq 0$ and $Ir^{g_1}w_1$ is a nonzero submodule in $E(R)$. It has a nonzero intersection with R, i.e., there is an element $t \in I$, such that $0 \neq tr^{g_1}w_1 \in R$. Let us set $r_1 = t^{g_1^{-1}}r$. In this case $r_1^{g_i}w_i = t^{g_1^{-1}g_i}r^{g_i}w_i \in R$, providing $r^{g_i}w_i \in R$, which contradicts the choice of r. The lemma is proved.

6.3.71. Lemma. *Let R be a semiprime ring and G be a reduced finite group of automorphisms. If D is a dense left ideal in R, then for* *any invariant form* τ *there is a dense ideal* $D_0 \subseteq D$, *such that* $\tau(D_0) \subseteq D$.

Proof. As was the case in lemma **6.3.10**, let us reduce the form τ to

$$\tau = \sum b_i x^{g_i h_i},$$

where $b_i \in B(G)$. Let I_1 be an ideal from F, such that $b_i I_1 \subseteq R$, and let I be an ideal from F, such that $I^{h_i g_i} \subseteq I_1$ and $I^{h_i^{-1} g_i^{-1}} \subseteq I_1$. Then $D_0 = \bigcap_i (ID)^{h_i^{-1} g_i^{-1}}$ is a dense ideal as a finite intersection of dense ideals. It is also evident that $\tau(D_0) \subseteq D$. The lemma is proved.

6.3.72. Proposition. *Let* R *be a semiprime ring and* G *is a certain* M*-group of its automorphisms. If* D *is a dense left ideal in* R, *then* $D \cap R^G$ *is a dense left ideal in* R^G.

Proof. Let s, w be elements from R^G, $w \neq 0$. Let us find a homogenous idempotent e, such that $we \neq 0$. Since for a certain essential ideal I we have Ie, $I\bar{e} \subseteq R$, then we can find an element $i \in I^G$, such that $0 \neq w_1 = w(i\bar{e}) \in R^G$, in which case $w_1 e \in R$ and $w_1 e \neq 0$, since $w_1 = wi(e \dotplus e^{g_1} \dotplus \ldots \dotplus e^{g_k}) = \sum_k (w_1 e)^{g_k} \cdot i$.

Let $\bar{\tau}_e$ be a principal trace form, D_0 be a dense ideal in R, such that $\bar{\tau}_e(D_0) \subseteq D$. We have $(D_0 s^{-1}) \cdot w_1 e \neq 0$, i.e., there is an element $r \in R$, such that $rs \in D_0$ and $rw_1 e \neq 0$. Since $\bar{\tau}_e$ is a principal form and $rw_1 e \neq 0$, then $\bar{\tau}_e(Ir)w_1 = \bar{\tau}_e(Irw_1) \neq 0$, where I is an ideal from F, such that $\bar{\tau}_e(I) \subseteq R$. It is also obvious that $\bar{\tau}_e(Ir) \cdot s = \bar{\tau}_e(Irs) \subseteq \tau(D_0) \subseteq D$ and, hence, $\bar{\tau}_e(Ir) \subseteq (D \cap R^G) s^{-1}$. The proposition is proved.

6.3.73. Lemma. *Let* R *be a semiprime ring and* $0 \neq \varphi \in Hom(R, E(R))$. *Let us assume that an* M*-group* G *acts on* R. *Then for a certain principal trace form* τ *there is an element* $a \in R$, *such that*

$$0 \neq \varphi(\tau(Ia)) \subseteq R,$$

where I *is an essential two-sided ideal, such that* $\tau(I) \subseteq R$.

Proof. By the condition, there is an element $a_1 \in R$, such that

$0 \neq \varphi(a_1) = w \in E(R)$. Since R is an essential submodule in $E(R)$, then we can find an element $a_2 \in R$, such that $0 \neq \varphi(a_1 a_2) \in R$. Let e be a homogenous idempotent less than $e(\varphi(a_1 a_2))$. Let us consider a principal trace form τ_e.

Let p be a point from $U(e)$. Then τ_p is a principal H-trace form, where $H = \{h \in G \mid e^h = e\}$ is a stabilizer of the idempotent e in the group G. By the definition, this form is

$$\tau_p(x) = \sum_{i=1}^{n} (\sum_{j=1}^{m} \beta_j x \beta_j^*)^{h_i}$$

where $\beta_1,\dots, \beta_m$ is the basis of the space B_p, $\beta_1^*,\dots \beta_m^*$ is a basis dual to it, $1 = h_1,\dots, h_n$ are representatives of the cosets of the group of inner automorphisms $(H_p)_{\text{int}}$ in H_p. These conditions on the coefficients and automorphisms can be presented as the truthfulness of the following Horn formulas:

$$\forall x \exists c_1, \dots, c_m \qquad \pi_B(x) = \sum_{j=1}^{n} \pi_C(c_j)\beta^*_j \tag{9}$$

$$\forall c_1,\dots, c_m \ \mathop{\&}_{i=1}^{m} \ (\sum_{j=1}^{n} \pi_C(c_j)\beta_j = 0 \rightarrow \pi_C(c_i) = 0), \ (10)$$

$$\forall \varphi(\forall x \qquad x^{h_i}\varphi = \varphi x^{h_j}) \longrightarrow \varphi = 0, \qquad (i \neq j) \tag{10}$$

Since these formulas set sheaf predicates, we can find a neighborhood $U(f)$ of the point p, such that the form $\tau_f = f\tau_e$ has a presentation

$$\tau_f = \sum_{i=1}^{n} \sum_{j=1}^{m} (b_j x b_j^*)^{h_i},$$

the coefficients of which, as elements of the ring fQ, and the automorphisms $h_1,\dots h_n$, as automorphisms of the ring fQ, obey conditions (9), (10) and (11). Besides, the ring $\bar{f}Q$ decomposes into a direct sum of isomorphic copies

$$\bar{f}Q = fQ \oplus \dots \oplus fQ,$$

and the form $\bar{\tau}_f$ is $\tau_f \oplus ... \oplus \tau_f$.

Let us present the form $\bar{\tau}_f$ as

$$\bar{\tau}_f = \sum_k \sum_j (b_j x b^*_j)^{g_k}, \quad g_k \in G.$$

Let I be an ideal from F, such that $(Ib_j)^{g_k} \subseteq R$, $(b_j I)^{g_k} \subseteq R$, $(fI)^{g_k} \subseteq R$. If x, a are arbitrary elements from I, then

$$\varphi(\tau_f(xa)) = \sum_k \sum_j (b_j x)^{g_k} \varphi(a^{g_k}(b^*_j)^{g_k}). \tag{12}$$

It should be remarked that we can find an element $a_3 \in If$, such that $\varphi(a_3 b^*_j) \neq 0$ for a certain j. In the opposite case by formula (9) we see that $f = c_1 b_1^* + ... + c_n b^*_n$. Choosing $T \in F$ so that $Tc_i \subseteq R$, we get $\varphi(ITf) = 0$ and, hence, $ITf\varphi(a_1 a_2) = \varphi(ITa_1 a_2 f) \subseteq \varphi(ITf) = 0$, which contradicts the choice of $f \leq e \leq e(\varphi(a_1 a_2))$.

Then, by lemma **6.3.70.** we can find an element $a_4 \in I$, such that $a_4^{g_k} \in I$ and $w_{jk} \overset{df}{=} a_4^{g_k} \varphi(a_3^{g_k}(b^*_j)^{g_k}) \in R$, in which case not all elements w_{jk} are zero. Let us set $a = a_4 a_3$. Then equality (12) yields $\varphi(\bar{\tau}_f(Ia)) \subseteq R$. Besides, $a = af$ and, hence, $w_{jk} = f^{g_k} w_{jk}$.

If $\varphi(\bar{\tau}_f(Ia)) = 0$, then (12) turns into an identity with automorphisms. Since formula (11) is true, theorem **2.3.1** yields the following relations

$$\sum_j b_j^{g_k} \otimes \varphi(a^{g_k}(b^*_j)^{g_k}) = 0$$

at every k. Formula (10) shows that such relations are possible only when $f^{g_k} w_{jk} = f^{g_k} \varphi(a^{g_k}(b^*_j)^{g_k}) = 0$ for all j, k (see lemma **1.6.24**). This is a contradiction that proves the lemma.

6.3.74. Proposition. *Let R be a semiprime ring, G be its M-group of automorphisms. If Δ is a dense left ideal in R^G, then $R\Delta$ is a dense left ideal in R.*

Proof. Let $\varphi(R\Delta) = 0$, but φ be a nonzero homomorphism of the left R-modules $\varphi: R \to E(R)$. By the preceding lemma, we can find an element $a \in R$, such that $\tau(a) \in R$ and $0 \neq \varphi(\tau(a)) \in R$ for a trace form τ. Since Δ is dense, the right annihilator of the left ideal $\Delta_1 = \Delta\tau(a)^{-1}$ in R^G is zero. If ρ is the right annihilator of this ideal in R, τ_1 is an arbitrary principal trace form, and $\tau_1(I) \subseteq R$, $I \in F$, then $\Delta_1\tau_1(\rho I) = \tau_1(\Delta_1\rho I) = 0$, i.e., $\tau_1(\rho I) = 0$ and, hence, $\rho = 0$, as τ_1 is arbitrary (see **5.6.8**).

Thus,

$$0 \neq \Delta_1\varphi(\tau(a)) = \varphi(\Delta_1\tau(a)) \subseteq \varphi(\Delta) = 0.$$

This is a contradiction, which proves the proposition.

6.3.75. Proof of theorem **6.3.69** is based on propositions **6.3.74** and **6.3.72**, as the proof of theorem **3.8.2** was based on lemmas **3.8.3** and **3.8.4**, and is carried out by the same scheme.

Let us first show that $Q_{max}(R^G)$ is naturally imbedded into $Q_{max}(R)$. Let $\xi \in Hom(A, R^G)$, where A is a dense left ideal of the fixed ring. Let us determine the correspondence $\xi^h: RA \to R$ by the formula

$$\left(\sum r_\alpha a_\alpha\right)\xi^h = \sum r_\alpha(a_\alpha\xi)$$

Let us show that ξ^h is a mapping. Let

$$V = \{\sum_\alpha r_\alpha(a_\alpha\xi) \mid \sum_\alpha r_\alpha a_\alpha = 0\}.$$

Then V is also a left ideal and for any fixed form τ and for an essential ideal I, such that $\tau(I) \subseteq R$, we have

$$\tau\sum(ir_\alpha(a_\alpha\xi)) = \sum\tau(ir_\alpha)(a_\alpha\xi) = [\sum\tau(ir_\alpha)a_\alpha]\xi =$$
$$[\tau(\sum ir_\alpha a_\alpha)]\xi = 0,$$

where i is an arbitrary element from I. Therefore, by proposition **5.6.8**, we get $V = 0$. It means that ξ^h is a homomorphism of the left R-modules. Its domain of definition is a dense left ideal of the ring R and, hence, ξ^h determines a certain element from $Q_{max}(R)$, which will be also denoted by ξ^h. It is now obvious that the mapping $h: i\xi \to \xi^h$ is an imbedding of $Q_{max}(R^G)$ into $Q_{max}(R)$.

Now we are to show that the image of h coincides with $Q_{max}(R)^{\hat{G}}$. If $\xi \in Q_{max}(R^G)$, $x=\sum r_\alpha a_\alpha$, where $r_\alpha \in I$, $I^{g^{-1}} \subseteq R$, then

$$x(\xi^h)^g = [((\sum r_\alpha a_\alpha)^{g^{-1}})\xi]^g = [\sum r_\alpha^{g^{-1}}(a_\alpha\xi)]^g =$$
$$(\sum r_\alpha a_\alpha)\xi^h = x\xi^h.$$

Therefore, $IA[\xi^h - (\xi^h)^g] = 0$ and, hence, the image h is contained in $Q_{max}(R)^{\hat{G}}$.

Inversely, let $\varphi \in Q_{max}(R)^{\hat{G}}$, $\varphi\colon D \to R$. Let us consider a restriction of φ on a dense ideal $D \cap R^G$ of the ring R^G. Since $(D \cap R^G)\varphi \subseteq Q_{max}(R)^{\hat{G}} \cap R^G$, then a restriction ξ of the mapping φ belongs to $Q_{max}(R^G)$, in which case $\xi^h = \varphi$, since $R(D \cap R^G)$ is a dense ideal, while a difference $\xi^h - \varphi$ maps it into zero. The theorem is proved.

Since there is an imbedding $R_F \subseteq Q_{max}(R)$, in which case $R_F^G = Q_{max}(R)^{\hat{G}} \cap R_F$, then the proved theorem, combined with lemma **5.6.6** and corollary **5.10.5** make it possible to transfer theorem **3.8.2** on semiprime rings.

6.3.76. Theorem. *Let R be a semiprime ring, G be a certain M-group of its automorphisms. In this case the following equalities are valid:*

$$(R_F)^G = (R^G)_F, \qquad Q(R)^G = Q(R^G).$$

Proof. It suffices to remark that R_F can be characterized as a subring in $Q_{max}(R)$ consisting of elements the set of left denominators of which contain a two-sided essential ideal of the ring R, while $Q(R)$ is a subring of elements from R_F which have essential ideals of the right denominators. The theorem is proved.

It should be remarked that theorem **6.3.76** can also be proved without using a complete ring of quotients, by nearly repeating word per word the proof of theorem **3.8.2** with corresponding substitutions in references.

6.3.77. Theorem. *Let R be a regular left self-injective ring. If G*

is a Maschke group of its automorphisms, then R^G *is also regular and left self-injective.*

Proof. Let us recall that regular left self-injective rings can be characterized as semisimple by Jacobson rings which coincide with their left maximal rings of quotients. Now we have to use theorems **6.3.18** and **6.3.68**. The theorem is proved.

REFERENCES

K.I.Beidar [13];
J.M.Goursaud, J.L.Pascaud, J.Valette [50,51];
J.M.Goursaud, J.Osterburg, J.L.Pascaud, J.Valette [52];
D.Handelman, G.Renault [56];
V.K.Kharchenko [69,80];
I.Kitamura [86];
I.Lambek [91];
A.Page [130];
G.Renault [140];
Bo Sandström [146];

6.4 Relations of a Semiprime Ring with a Ring of Constants

By analogy with the preceding paragraph, we shall consider here the ring properties which are transferred from a ring to a ring of constants and vice versa. As usual, let us neglect the cases of inner derivations and limit ourselves with considering rings with a simple characteristics. The key moment in the proofs of this paragraph is, in most cases, the use of different variations of a local Shirshov finality over a ring of constants.

Let us fix notations for a semiprime ring R of a characteristic $p>0$, of its generalized centroid C, and of a restricted differential Lie C-algebra $L\subseteq D(R)$. Let us assume that L is finitely-generated as a right module over C and contains no nonzero inner for Q derivations. By lemma **1.6.21**, the algebra L can be decomposed into a direct sum of cyclic submodules $L=\mu_1 C\oplus \dots \mu_n C$. Let us fix a strongly independent set $\mu_1,\dots,\mu_n$ and a set of all correct words $\emptyset=\Delta_1<\dots<\Delta_m$. By proposition **5.1.2** the algebra L has an essential trace form

$$f(x)=\sum_j c_j x^{\Delta_j},$$

where $c_j\in C$, in which case $\sup_j e(c_j)e(\Delta_j)=1$.

6.4.1. *A ring* R *has an essential Shirshov locally-finite ideal over a ring of constants.*

Proof. Let us consider a sequence of Horn formulas

$$\Phi_n \Leftrightarrow \forall\, u\, \exists\, a, t_1, \dots, t_n, r_1, \dots, r_n \quad \forall\, x \quad ax = \sum_{i=1}^{n} r_i u f(t_i x) \quad \& \quad e(a) = e(x). \tag{1}$$

Viewing f as a strictly sheaf operation on Q, we see, by proposition **4.4.3**, that on every stalk $\Gamma_{\mathbf{p}}$ of a canonical sheaf one of the formulas Φ_n holds. By metatheorem **1.11.13** and corollary **1.11.20**, on the ring Q also valid is one of formulas (1).

Let us set $u = 1$ and choose an essential two-sided ideal I in such a way that $f(t_i I) \subseteq R$, $Ir_i \subseteq R$, $Ia \subseteq R$. Then IaI is locally-finite over R^L, in which case $1 = e(a) = e(IaI)$, i.e., $IaI \in \mathcal{F}$. The theorem is proved.

6.4.2. Theorem. *For any* (R, R^L)*-subbimodule* V *of* Q *there can be found a central idempotent* e *and an essential ideal* I *in* R*, such that* $Ie \subseteq V = Ve$.

Proof. In formula (1) let us choose an element u in V and an ideal $T \in \mathcal{F}$ in such a way that $Tr_i \subseteq R$, $Ta \subseteq R$, $f(t_i T) \subseteq R$. Then $TaT \subseteq V$ and $e(TaT) = e(a) = e(u)$. Let I_0 be a sum of all ideals TaT when u runs through V. Now we have $e \overset{df}{=} e(I_0) = e(V)$ and we can assume $I = I_0 + ann_R I_0$. The theorem is proved.

The next theorem is proved symmetrical.

6.4.3. Theorem *For any* (R^L, R)*-subbimodule* W *of* Q *there is an ideal* $I \in \mathcal{F}$*, such that* $Ie \subseteq W$*, where* $e = e(W)$ *is a support of* W.

A left analog of theorem **6.4.1** can be formulated in a form allowing for the fact that the coefficients $f(t_i x)$ in formula (1) are linear with respect to x over R^L.

6.4.4. Theorem. *A set* W *of elements* a *of a ring* R*, such that the left* R^G*-module* Ra *is embeddible into a finite direct sum of copies of a free module* R^L *contains an essential ideal of the ring* R.

Proof. Let $u = 1$ in the left analog of formula (1), and let I be an essential ideal of the ring R, such that $f(t_i I) \subseteq R$, $1 \le i \le n$. Then formula (1) shows that at $k \in I$ the mapping

$$xka \longrightarrow f(xt_1) \dot{+} \dots \dot{+} f(xt_n)$$

implements the imbedding of Rka into a direct sum $R^L \dot{+} \dots \dot{+} R^L$. Therefore, $IaR \subseteq W$, and this is an essential ideal, since $e(a) = 1$. The theorem is proved.

6.4.5. Lemma. *If A is a nonzero left (right) ideal of a ring R, then $f(A) \neq 0$.*

Proof. If $f(xa) = 0$ at all $x \in R$, then, by theorem **2.3.1**, we have $e(c_j)e(\Delta_j)e(a) = 0$ at all j. Since f is an essential trace form, then $a = 0$, which is the required proof.

6.4.6. Corollary. *If $I \in F$, then I^L is an essential ideal of a ring R^L. Inversely, if ρ is an essential two-sided ideal of R^L, then $R\rho$ (as well as ρR) contains an essential ideal of the ring R.*

Proof. The first part immediately results from corollary **5.1.3**. To prove the second part, it is sufficient, using theorem **6.4.2**, to establish that the support of ρ is equal to unity. If $\rho xb = 0$ for all $x \in R$, then we have an identity $f(\rho xb) = 0$ which affords $\rho f(xb) = 0$. Choosing $I \in F$ in such a way that $f(Ib) \subseteq R$, we get $f(Ib) = 0$ and, hence, $b = 0$, which is the required proof.

6.4.7. Lemma. *Let I be an essential two-sided ideal of a ring R, such that $f(I) \subseteq R$. Then, if A is an essential left ideal in R, then $f(IA)$ is an essential left ideal in R^L.*

Proof. Let W be a nonzero left ideal in R^L. Then $IW \cap IA \neq 0$ and, hence, $0 \neq f(IW \cap IA) \subseteq f(IA) \cap f(I)W \subseteq f(IA) \cap W$, which is the required proof.

6.4.8. Theorem. *A ring R^L is semisimple artinian iff R is semisimple artinian.*

Proof. Let R^L be simple artinian and let W be an essential locally-finite over R^L ideal. Then $W \cap R^L$ is an essential ideal of R^L and, hence, $1 \in W$. Therefore, R is a finitely-generated left module over R^L. This, in particular, implies that it is artinian as an R-module as well.

Let, inversely, $S = f(R)$. Then $S^2 = f(R)f(R) = f(Rf(R)) = f(R) = S$, in which case S is an essential ideal of R^L.

If $I_1 \supset I_2 \supset \dots$ is a strictly descending chain of left ideals in S, such that $SI_n = I_n$, then there is a number k, such that $RI_k = RI_{k+1}$. This affords $f(R)I_k = f(R)I_{k+1}$, which is a contradiction. Thus, S is a semisimple artinian ring and an essential ideal of R^L, i.e., $S = R^L$. The theorem is proved.

6.4.9. Theorem. *A ring* R^L *is Jacobson semisimple iff semisimple is the ring* R.

Proof. Let R be semisimple and W be an essential locally-finite over R^L ideal. Since $W \cap R^L$ is an essential ideal in R^L, then in the case when $J(R^L) \neq 0$ we can find a nonzero element $w \in W$ which annihilates all irreducible left R^L-modules. By theorem **6.4.2**, we have $R^L wR \supseteq Ie$, where $e = e(w)$, $I \in \mathbb{F}$. Let V be an irreducible left R-module and $IwV \neq 0$. Then V is a finitely-generated left R^L-module. If W is its maximal submodule, then V / W is irreducible, i.e., $W \supseteq wV$, which affords $R^L wRV \subseteq W$ or $eIV \subseteq W$. However, $0 \neq IwV \subseteq eIV$ is a nonzero R-submodule of V, i.e., $V = W$, which is a contradiction.

Inversely, let us remark that $J(R) \cap R^L$ is a quasi-regular ideal in R^L. If $a \in J(R)^L$, then there is an $x \in J(R)$, such that $a \bullet x = x \bullet a = 0$. Therefore, at any $\mu \in L$ we have $a \bullet x^{\mu} = -a$ and, hence, $x^{\mu} = (x \bullet a) \bullet x^{\mu} = x \bullet (a \bullet x^{\mu}) = 0$, i.e., $x \in J(R)^L$.

It should be recalled that if $I^{\mu} \subseteq R$, Then $(I^2)^{\mu} \subseteq I$. Choosing as I an essential ideal, such that $I^{\mu_i} \subseteq R$ for all basic derivations and $c_j I \subseteq R$ for all coefficients f, we find $([IJ(R)]^{m+1})^{\Delta_j} \subseteq IJ(R)$ and, hence, $f((IJ(R))^{m+1}) \subseteq J(R) \cap R^L = 0$, i.e., by lemma **6.4.5**, we get $(IJ(R))^{m+1} = 0$ and, consequently, R is semisimple. The theorem is proved.

6.4.10. Theorem. *A ring* R^L *is primitive iff primitive is the ring* R.

Proof. If V is an exact irreducible left R-module, then it is finitely-generated as a left R^L-module, while the factor-module V / W over a maximal R^L-submodule will be exact: if $0 \neq aV \subseteq W$, then $R^L aRV \subseteq W$, but $R^L aR \supseteq I \in \mathbb{F}$ and, hence, $V = IV = W$.

Inversely, let λ be a maximal modular left ideal of the ring R^L. Let $L_0 = R\lambda + \lambda$. Then $L_0 \cap R^L = \lambda$, since in the opposite case $L_0 \supseteq R^L$ and $L_0 \supseteq R R^L \supseteq T \in \mathbb{F}$, i.e., for the ideal $I \in \mathbb{F}$, such that $f(I) \subseteq R$ we have a contradiction:

$$R^L \supset (IT) \subseteq f(IL_0) \subseteq f(I\lambda) = f(I)\lambda \subseteq \lambda.$$

A set $\mathcal{M}$ of all the left ideals M of the ring R contained in $R' = RR^L + R^L$ and such that $M \cap R^L = \lambda$ is nonempty and inductive. Let N be a maximal element of $\mathcal{M}$. Now one can easily see that $V = R'/N$ is the sought exact irreducible module of the ring R. The theorem is proved.

6.4.11. Theorem. *A ring* R^L *is quite primitive iff quite primitive is the initial ring.*

Proof Let H be a sockle of a ring R. Then H is the least ideal and, hence, $f(H) \subseteq R$. Moreover, $H f(H)$ contains the ideal of the ring R and, hence, $H f(H) = H$. If $f(H)$ has no minimal left ideal, then there is a sequence of elements $s_1,\dots, s_n,\dots$ from $f(H)$, such that $f(H)s_n \supset f(H)s_{n+1}$ for all $n = 1, 2,\dots$. By further multiplying from the left by H, we get $H s_n \supseteq H s_{n+1}$. As $s_1 \in H$, then the module $H s_1$ has a finite rank. This means that there is a number n, such that $H s_n = H s_{n+1}$. Applying f, we come to a contradiction.

Inversely, if e is a primitive idempotent of the ring of constants, then the algebra L is induced onto eRe and theorem **2.2.2** shows that derivations $\bar{\mu}_1,\dots, \bar{\mu}_n$ induced by the basis of the algebra L will be strongly independent in $D(eRe)$ (cf. an analogous lemma **6.3.20**). Since $(eRe)^L = eR^Le$ is a sfield, then, by theorem **6.4.8**, the ring eRe is simple artinian.. If e_1 is a primitive idempotent from eRe, then e_1Re_1 is a sfield and, hence, e_1 is the sought primitive idempotent of the ring R. The theorem is proved.

6.4.12. Theorem. *A ring* R^L *will be a left Goldie ring iff such is* R.

Proof. Let R be a left Goldie ring. The maximality condition for annihilators is transferred onto subrings and is, hence, fulfilled in R^L.

If $A \dot{+} B$ is a direct sum of the left indices in R^L and $f(I) \subseteq R$, $I \in F$, then $f(IA \cap IB) \subseteq f(I)A \cap f(I)B = 0$, i.e., $IA \dot{+} IB$ is a direct sum in R, i.e., there are no infinite direct sums of nonzero left ideals in R^L.

Inversely, let us show R to be the nonsingular ring. If $0 \neq Z(R)$ is a singular ideal of R, then $0 \neq f((IZ(R))^{m+1}) \subseteq Z(R)$, where $I \in F$ is such that $I^{\mu_i} \subseteq R$, $c_j I \subseteq R$. Therefore, we can find a nonzero constant $z \in Z(R)$. Let $Az = 0$, where A is an essential left ideal of the ring R. By lemma **6.4.7**, we get a contradiction $0 = f(IAz) = f(IA)z$, since $Z(R^L) = 0$.

The fact that the ring R has finite Goldie dimension results from the following lemma, which is quite analogous to lemma **6.3.27** and follows from theorem **6.4.4**.

6.4.13. Lemma. *If R^L is a left Goldie ring, then the left R^L-module R is embeddible into a finite direct sum of copies of a regular module R^L.*

This lemma yields the following corollary.

6.4.14. Corollary. *If R^L is a left Noetherian ring, then R is also left Noetherian.*

By way of concluding this paragraph, let us dwell on rings of quotients.

6.4.15. Lemma. *If $\mu \in D(R)$ and D is a dense left ideal in R, then there is a dense left ideal $D_0 \subseteq D$, such that $D_0^{\mu} \subseteq D$.*

Proof. Let $I^{\mu} \subseteq R$, $I \in F$. Let us consider a set $A = \{d \in D \mid d^{\mu} \in D\}$, which is a left I-module: $(id)^{\mu} = i^{\mu} d + id^{\mu} \in D$. Let us set $D_0 = IA$. If $d \in D$, $i \in I$ and $d_1 d^{\mu} \in D$, then $(id_1 d)^{\mu} = (id_1)^{\mu} d + (id_1) d^{\mu} \in D$ and, hence, $Ad^{-1} \supseteq I \cdot D(d^{\mu})^{-1}$ and $D_0 d^{-1} \supseteq I \cdot I \cdot D(d^{\mu})^{-1}$ is a dense ideal. Since the filter of dense ideals is idempotent, we see that D_0 is a dense ideal. The lemma is proved.

6.4.16. Corollary. *For any dense ideal D there is a dense ideal D_0, such that $D_0^{\Delta_j} \subseteq D$ for all correct words Δ_j from basic derivations.*

Proof. It is evident by induction over the word length.

6.4.17. Corollary. *If D is a dense left ideal, then $f(D_0) \subseteq D$ for a certain dense left ideal $D_0 \subseteq D$.*

Proof. We should apply the preceding corollary, replacing R with an ideal $I \in F$, such that $c_j I \subseteq R$ for the coefficients of the universal constant f.

6.4.18. Lemma. *If D is a dense left ideal, then $D \cap R^L$ is a* *dense left ideal in* R^L.

Proof. Let D_0 be a dense left ideal, such that $f(D_0) \subseteq D$. Let $s \in R^L$. Then $D_0 s^{-1}$ is a dense left ideal of R. Therefore, for any $a \neq 0$ we have $D_0 s^{-1} \cdot a \neq 0$. If $a \in R^L$, then $0 \neq f(D_0 s^{-1} \cdot a) = f(D_0 s^{-1}) a$. We now have to remark that $f(D_0 s^{-1}) \subseteq f(D_0) s^{-1}$, since the inclusion $b \cdot s \in D_0$ implies $f(b) \cdot s \in f(D_0)$ and $f(b) \in f(D_0) s^{-1}$. The lemma is proved.

6.4.19. Lemma. *Any derivation from* $D(R)$ *has a unique extension onto* $Q_{\max}(R)$.

Proof. Let $\varphi \in Q_{\max}(R)$. Then $D\varphi \subseteq R$ for a certain $D \in \mathfrak{G}$. Let $\mu \in D(R)$ and $I^{\mu} \subseteq R$, $I \in F$. Let us find a dense ideal $D_0 \subseteq D$, such that $D_0^{\mu} \subseteq D$. Then we can determine $\varphi^{\mu} \in Hom_R(ID_0, R)$ by the formula $d\varphi^{\mu} = (d\varphi)^{\mu} - d^{\mu}\varphi$. Now we can easily see that $\varphi \to \varphi^{\mu}$ is the sought extension. Its uniqueness results from the formula $(d\varphi)^{\mu} = d\varphi^{\mu} + d^{\mu}\varphi$, which we used to set the homomorphism φ^{μ}. The lemma is proved.

6.4.20. Lemma. *The extensions* $\bar{\mu}_1, \ldots, \bar{\mu}_n$ *of basic derivations onto* $Q_{\max}(R)$ *will be strongly independent.*

Proof. Let $\sum c_i x^{\bar{\mu}_i} = ax - xa$ for c_i, $a \in Q_{\max}(R)$ and all $x \in Q_{\max}(R)$. Let us find a dense ideal D, such that $Da \subseteq R$, $Dc_i \subseteq R$. Then we get an identity $\sum (d_1 c_i)(xd)^{\mu_i} = (d_1 a)xd - d_1 x(da)$, $d_1, d \in D$. By theorem **2.3.1** we can find $e(\mu_i) \cdot d_1 c_i \otimes d = 0$ at all i. Since $e(D) = 1$, then we get $c_i e(\mu_i) = 0$ and $e(c_i) e(\mu_i) = 0$. We now have to remark that $e(\mu) = e(\bar{\mu})$: if $R^{\mu} e = 0$, then $iq^{\bar{\mu}} e = (iq)^{\mu} e - i^{\mu} eq = 0$. The lemma is proved.

6.4.21. Theorem (J.Piers Dos Santos). *For a maximal ring of quotients the following formula is valid:*

$$Q_{\max}(R^L) = Q_{\max}(R)^L.$$

Let us prove some more auxiliary lemmas.

Let $\varnothing = \Delta_1 < \Delta_2 < \dots < \Delta_m$ be correct words of basic derivations, $1 = e_1, \dots, e_m$ be supports of the operators $\Delta_1, \dots, \Delta_m$. Let us fix a denotation I for an ideal F, such that $Ie_j \subseteq R$, $I^{\Delta_j} \subseteq R$, $1 \le j \le m$. A direct decomposition $I = e_j I \dot{+} (1 - e_j)I$ induces decomposition of the injective hull $E(R) = E(I) = E(e_j I) \dot{+} E((1 - e_j)I)$. This enables one to determine multiplications of the elements of the injective hull by the central idempotents $e_1, \dots, e_m$, setting $e_j w = w_1$, where $w = w_1 + w_2$ is the corresponding decomposition of w into a direct sum. To every word Δ_j let us put into correspondence an abelian group $\Delta_j e_j E(R)$ which is isomorphic to $e_j E(R)$, under the correspondence $\Delta_j w \leftrightarrow w$. Let us set $M_k = \sum_{j=1}^{k} \oplus \Delta_j e_j E(R)$, and introduce on M_k the structure of the left R-module: $r\Delta w = \sum_{\Delta' \bullet \Delta'' = \Delta} \Delta'(r^{\Delta''} w)$. By the Leibnitz formula we have:

$$(r_1 r)\Delta w = \sum \Delta'(r_1 r)^{\Delta''} w = \sum_{\Delta' \circ \Delta'' \circ \Delta''' = \Delta} \Delta' r_1^{\Delta''} r^{\Delta'''} w =$$

$$= \sum_{\Delta' \circ \Delta'' = \Delta} r_1 \Delta' r^{\Delta''} w = r_1(r\Delta w)$$

so that the determined action indeed turns M_k into a left module. In each of these modules let us consider submodules $N_k = \sum_{j=1}^{k} \oplus \Delta_j e_j R$.

6.4.22. Lemma. *N_k is an essential submodule in M_k.*

Proof. Let us carry out induction over k. If $k = 1$, then M_1 can be naturally identified with $E(R)$, in which case N_1 is identified with R and now we have make use of the fact that R is an essential I-submodule in the injective hull $E(R)$.

Let N_{k-1} be an essential submodule in M_{k-1}. Let us consider a submodule $W_k = I\Delta_k e_k$ and note that $W_k \cap M_{k-1} = 0$. Indeed, if

$0 \neq s \in Ie_k$, then $s\Delta_k e_k = \sum \Delta'_k s^{\Delta''_k} \equiv \equiv \Delta_k s(\text{mod } M_{k-1})$. Now, by the inductive supposition, $W_k \dot{+} N_{k-1}$ is an essential submodule in $W_k \dot{+} M_{k-1}$. Since $N_k \supseteq W_k \dot{+} N_{k-1}$, we now have to prove that $W_k \dot{+} M_{k-1}$ is an essential submodule in M_k. For this purpose let us make use of an evident fact: if $U \subset V \subset W$ is a chain of modules, and V/U is an essential submodule, then V is an essential submodule of W. Under our conditions, M_k / M_{k-1} is isomorphic to $e_k E(R)$ at the mapping $\sum_{i=1}^{k} \Delta_i w_i \to w_k$. This mapping transforms W_k into $e_k I$ and, hence, $W_k \dot{+} M_{k-1} / M_{k-1}$ is an essential submodule in $e_k E(R)$. The lemma is proved.

6.4.23. Lemma. *If w is a nonzero element of the injective hull $E(R)$, then there is an element $a \in R$, such that $0 \neq f(a)w \in R$, where $f(x)$ is an essential trace form and $f(a) \in R$.*

Proof. Let $f(x) = \sum x^{\Delta_j} c_j$, and by I let us denote an ideal from F, such that I^{Δ_j}, Ic_j, $Ie_j \subseteq R$, $1 \leq j \leq m$, where, as above, $e_j = e(\Delta_j)$. Since $\sup_{j=1}^{m} e_j e(c_j) = 1$, then there is a number k, such that $e_k e(c_k) w \neq 0$. By corollary **6.4.6**, we get $RI^L \cdot e_k e(c_k) w \neq 0$, i.e., we can find a constant $t \in I^L$, such that $(c_k t) e_k w \neq 0$ (it should be recalled that by the regularity of the center C, we have $e(c_k) = c_k \cdot c'_k$). In M_k let us consider an element $h = \sum \Delta_j (c_j t) e_j w$. By the construction, the k-th component of this sum is nonzero and, hence, h is a nonzero element in M_m. By lemma **4.6.22**, we can find an element $s \in I$, such that $0 \neq sh \in N_m$.

Let us interrelate every element $g = \sum \Delta_j w_i \in M_n$ with a mapping $\hat{g}: I \to E(R)$ acting by the formula $\hat{g}(x) = \sum x^{\Delta_j} w_i$. We have $\hat{sg}(x) = \hat{g}(xs)$ at any $x, s \in I$. Besides, if $g \in N_m$, then the image of $\hat{g}$ will be contained in R. And, finally, the main point: theorem **2.3.1** shows that if $g \in N_m$ and $\hat{g}$ is a zero mapping, then g is a zero element of N_m.

Now we can find an element $s_1 \in I$, such that $\hat{sh}(s_1) \neq 0$, i.e.,

$\hat{h}(s_1 s) \neq 0$, however, at $x \in I$ we have $\hat{h}(x) = f(xt)\dot{w}$. Setting $a = s_1 st$, we get $0 \neq \hat{h}(s_1 s) = f(a)w = \hat{sh}(s_1) \in R$. The lemma is proved.

6.4.24. Proposition. *If* ρ *is a dense left ideal in* R^L, *then* $R\rho$ *is also a dense left ideal in* R.

Proof. Let $R\rho w = 0$ for a certain nonzero $w \in E(R)$. By the preceding lemma, we can find an element $a \in R$, such that $0 \neq f(a)w \in R$ and $f(a) \in R^L$. Since ρ is dense, the right annihilator of the left R^L-ideal $\rho_1 = \rho f(a)^{-1}$ in R^L and, hence, it is zero in R. Now we have

$$0 \neq \rho_1 \cdot [f(a)w] = [\rho_1 f(a)]w \subseteq \rho w = 0.$$

This is a contradiction which proves the proposition.

6.4.25. Now the proof of theorem **6.4.21** can be obtained by nearly word per word repetition of theorem **6.3.58** , using lemma **6.4.18** and proposition **6.4.24.**

6.4.26. Corollary. *The following equalities are valid for a Martindale ring of quotients:*

$$Q(R)^L = Q(R^L); \qquad (R_F)^L = (R^L)_F .$$

The proof immediately results from theorem **6.4.21** by applying corollary **6.4.6** and an imbedding $R_F \subseteq Q_{max}$.

6.4.27. Corollary. *If* R *is a left Goldie ring, then*

$$Q_{cl}(R)^L = Q_{cl}(R^L).$$

Indeed, in this case the maximal ring of quotients coincides with the classical one.

6.4.28. Corollary. *If* R *is a regular left self-injective ring, then* R^L *is also regular and left self-injective.*

REFERENCES:

V.K.Kharchenko [70,77];
J.Piers Dos Santos [105];
A.Z.Popov [137].

6.5 Hopf Algebras

In this paragraph we shall consider a general concept of the action of Hopf algebras embracing both the case of automorphisms and that of derivations. Let us recall some definitions pertaining to the notion of Hopf algebras.

6.5.1. The definition of a *co-algebra* is obtained from the category definition of an algebra by reversing all the arrows. Namely, *co-multiplication* on a linear space H is a linear mapping $\Delta: H \to H \otimes_C H$, where C is a basic field.

We have already observed earlier that the use of the functor $Hom(-, C)$ results in reversing of arrows (**3.3**). This observation prompts that if on a finite-dimension space H a multiplication is given, then on a dual space H^* a co-multiplication is induced, and vice versa, if on H a co-multiplication is set, then a multiplication is induced on H^* (i.e., H^* transforms into an algebra which is not, generally speaking, associative). Now there arises a natural desire to present this multiplication and co-multiplication in a common or alike form. This can be achieved in terms of structural constants. Let a multiplication U and co-multiplication Δ which are not related, be given on H. Let us fix a certain basis $\{h_i, i \in I\}$ of H. In this case the multiplication is uniquely set by a system of structural constants $\{c_{ij}^{(k)} \in C | i, j, k \in I\}$ which determines the products of basic elements

$$U(h_i, h_j) \overset{df}{=} h_i \cdot h_j = \sum_k c_{ij}^{(k)} h_k. \tag{1}$$

Analogously, the co-multiplication on the basic elements can be set by structural constants $\{\delta_{ij}^{(k)} \in C | i, j, k \in I\}$:

$$\Delta(h_k) = \sum_{i,j} \delta_{ij}^{(k)} h_i \otimes h_j. \tag{2}$$

It would be natural to suggest that a co-multiplication U^* and a multiplication Δ^* is set on H^* by the same sets of structure constants and, respectively,

$$U^*(h^*_k) = \sum_{i,\, j} c^{(k)}_{ij} h^*_i \otimes h^*_j \tag{3}$$

$$\Delta^*(h^*_i,\, h^*_j) = \sum_k \delta^{(k)}_{ij}\, h^*_k\,. \tag{4}$$

where, as usual, by $h^*_1,\ldots, h^*_n$ we denote the dual basis of a dual space. This proposition proves to be correct, and thus, formulas (1)-(4) show that in principle the notion of a co-algebra is as natural as that of an algebra.

For the case of an infinite dimension, formulas (3) and (4) remain valid, but in this case there arise some question to be answered. First, the set $\{h^*_i | \, i \in I\}$ does not form the basis of a dual space. Nevertheless, any functional $\alpha \in Hom(H, C)$ is set by its action on basic elements $\{h_i\}$ and, hence α can be identified with an element $\prod_{i \in I} \alpha(h_i) h^*_i$ of the direct product $\prod C h^*_i$, and inversely, any element $\prod \alpha_i h^*_i$ of the direct product determines the functional: $(\prod \alpha_i h^*_i)(h_j) = \alpha_j$. Therefore, any element from H^* is uniquely presented as an infinite linear combination of elements h^*_i. It would be now natural to determine U^* and Δ^* on infinite combinations with respect to linearity

$$U^*(\sum \lambda_k h^*_k) = \sum \lambda_k U^*(h_k) \tag{5}$$

$$\Delta^*(\sum \lambda_i h^*_i, \sum \mu_j h^*_j) = \sum \lambda_i \mu_j \Delta^*(h^*_i,\, h^*_j) \tag{6}$$

and here arises the second question: how to reduce the similar terms in the right-hand parts of formulas (5) and (6). Here of some help is the fact that in formulas (1) and (2) the sums are finite, which implies that for fixed i, j there is only a finite number k, such that $c^{(k)}_{ij} \neq 0$, and for every fixed k there is only a finite number of pairs (i, j), such that $\delta^{(k)}_{ij} \neq 0$. Therefore, the right-hand parts of formulas (5) contain only finite sets of similar terms. And, finally, the last and most tragic for formula (3) peculiarity is that in the right-hand part of (5) there is an infinite direct sum of basic tensors, i.e., it is an element from $(H \otimes H)^*$. This element can be not a finite sum of basic tensors of infinite linear combinations of functionals $\{h^*_i\}$, i.e., it can lie not in $H^* \otimes H^*$ and, as a result, U^* will not be a co-multiplication in H^*.

Therefore, for an infinite-dimension case we can only claim that if H is a co-algebra, then H^* is an algebra, which is sufficient for putting all the notions of the theory of algebras into correspondence with dual notions of the theory of co-algebras. For example, co-commutativity, co-associativity and the existence of co-unity in H mean commutativity, associativity and the existence of unity in H^*, respectively. Let us write these notions in terms of co-multiplication.

Co-commutativity. If $\Delta(h) = \sum h_i^{(1)} \otimes h_i^{(2)}$, then $\sum h_i^{(2)} \otimes h_i^{(1)} = \Delta(h)$.

Co-associativity. If $\Delta(h) = \sum h_i^{(1)} \otimes h_i^{(2)}$, then $\sum \Delta(h_i^{(1)}) \otimes h_i^{(2)} = \sum h_i^{(1)} \otimes \Delta(h_i^{(2)})$ in the tensor product $H \otimes H \otimes H$.

Co-unity is a functional $\varepsilon\colon H \to H$, such that $\sum h_i^{(1)} \varepsilon(h_i^{(2)}) = \sum \varepsilon(h_i^{(1)}) h_i^{(2)} = h$, where, as above, $\Delta(h) = \sum h_i^{(1)} \otimes h_i^{(2)}$.

A bialgebra is an associative algebra with unity, H, on which set are a co-associative multiplication $\Delta\colon H \to H \otimes H$ and a co-unity $\varepsilon\colon H \to H$, which are homomorphisms of C-algebras.

An antipode of a bialgebra H is an antihomomorphism of algebras, $S\colon H \to H$, such that

$$\sum h_i^{(1)} \cdot S(h_i^{(2)}) = \sum S(h_i^{(1)}) \cdot h_i^{(2)} = \varepsilon(h) \cdot 1,$$

where $\Delta(h) = \sum h_i^{(1)} \otimes h_i^{(2)}$, with the dot denoting multiplication in H.

A Hopf algebra is a bialgebra with unity, co-unity and an antipode.

The action of a Hopf algebra H is said to be set on an associative C-algebra R, if to every element $h \in H$ we put in correspondence a linear transformation $r \to r^h$ of the space R into itself, such that the following identities are valid:

$$x^1 = x, \quad x^{(h_1 h_2)} = (x^{h_1})^{h_2}, \quad x^{\alpha h_1 + \beta h_2} = \alpha x^{h_1} + \beta x^{h_2} \tag{7}$$

and, which of prime importance:

$$(xy)^h = \sum x^{h_i^{(1)}} y^{h_i^{(2)}} \tag{8}$$

where $\Delta(h) = \sum h_i^{(1)} \otimes h_i^{(2)}$.

The role of invariants (or constants) under the effect of a Hopf algebra is played by a subalgebra

$$R^H = \{r \in R \mid \forall\, h \in H \qquad r^h = \varepsilon(h)\, r\}$$

Let us consider the most important examples of the action of Hopf algebras.

6.5.2. *Automorphisms*. Let G be a group. Let us consider a group algebra $C[G]$ and determine on it a co-multiplication Δ, a co-unity ε and an antipode S on the basic elements by the formulas

$$\Delta(g) = g \otimes g\, ,\ \varepsilon(g) = 1\, ,\ S(g) = g^{-1} \tag{9}$$

and, by linearity, extend it onto $C[G]$. One can easily see that in this way a group algebra is transformed into a co-commutative Hopf algebra. If this Hopf algebra acts on a certain associative algebra, then by formula (8) we have $(xy)^g = x^g y^g$ and $(x^g)^{g^{-1}} = x$, and, hence, G acts as a group of automorphisms.

Inversely, if G acts on R as a group of automorphisms, then the action of the Hopf algebra $C[G]$ naturally arises by the formula

$$x^{\sum \alpha_i g_i} = \sum \alpha_i x^{g_i} \tag{10}$$

It is now evident that $R^{C[G]} = R^G$, since if $r \in R^G$, then

$$r^{\sum \alpha_i g_i} = (\sum \alpha_i)\, r = \varepsilon(\sum \alpha_i g_i)\, r.$$

6.5.3. *Group grading*. Let us now assume that G is a finite group and consider the action of a dual algebra $C[G]^*$. The basis of this Hopf algebra consists of elements P_g, $g \in G$, and the table of multiplication and co-multiplication is determined by formulas (3) and (4)

$$\Delta(P_g) = \sum_h P_h \otimes P_{h^{-1}g}$$
$$P_g P_h = \delta_{g,h} P_g$$

where $\delta_{g,h}$ is a Kronecker delta, which implies that $\{P_g,\ g \in G\}$ is a set of mutually orthogonal idempotents, the sum of which is equal to unity. In this case a co-unity is determined by the equalities $\varepsilon(P_1) = 1$, $\varepsilon(P_g) = 0$ at

$g \neq 1 \in G$ and an antipode is set by the formula $S(P_g) = P_{g^{-1}}$.

If $C[G]^*$ acts on R, then we have a decomposition of R into a direct sum of subspaces

$$R = \sum P_g(R)$$

in which case

$$P_f[P_g(R) \cdot P_h(R)] = \sum_{v \in G} P_v(P_g(R)) \cdot P_{v^{-1}f}(P_h(R)) =$$

$$= \begin{cases} 0 & g^{-1}f \neq h \\ P_g(R)\,P_h(R) & f = gh \end{cases}$$

i.e., $P_g(R)P_h(R) \subseteq P_{gh}(R)$. Therefore, a grading by the group G arises on R.

Inversely, if R is graded by a finite group

$$R = \sum_{g \in G} \oplus R_g, \qquad R_g R_h \subseteq R_{gh},$$

then the action of the Hopf algebra $C[G]^*$ can be determined by the formula

$$r^{\sum \alpha_g P_g} = \sum_g \alpha_g r_g,$$

where $r = \sum r_g$ is a decomposition into a sum of homogenous components. In particular, P_g is a projector onto a homogenous component R_g and, hence,

$$(xy)^{P_g} = (xy)_g = \sum_{h \in G} x_h y_{h^{-1}g} = \sum_h x^{P_h} y^{P_{h^{-1}g}},$$

i.e., we really obtained the action of Hopf algebra.

And, finally, under such conditions the role of invariants is played by the first component

$$R^{C[G]^*} = \{ x \in R \mid x^h = \varepsilon(h)x \} = R_1.$$

6.5.4. *Derivations.* Let L be a Lie algebra over a field C, $U(L)$

be its universal enveloping. Let us fix a basis $\{l_i\}$ of the algebra L. In this case the basis of $U(L)$ will be all "monotonic" words from $\{l_i\}$, i.e., the words of type $l_{i_1}, l_{i_2} \dots l_{i_k}$, where $i_1 \leq i_2 \leq \dots \leq i_k$. Let us determine the co-multiplication and antipode on $\{l_i\}$

$$\Delta(l_i) = 1 \otimes l_i + l_i \otimes 1\,,\ S(l_i) = -\,l_i\,, \tag{11}$$

and then extend it onto $U(L)$ in such a way that Δ becomes a homomorphism of C-algebras $\Delta\colon U(L) \to U(L) \otimes U(L)$, while S becomes an antiautomorphism. It is evident that the word V is affected by a co-multiplica-tion and antipode by the formulas

$$\Delta(V) = \sum_{V' \circ V'' = V} V' \otimes V'', \qquad S(V) = (-1)^{|V|} \bar{V} \tag{12}$$

where $|V|$ is the length of the word V, while $\bar{V}$ is obtained from V by reading from right to left. The value of the co-unity ε on all non-empty words is considered to be zero, while on the empty word (i.e., on the unity of a universal enveloping) it is considered to be equal to unity.

Now by induction over the length of a word one can easily check that $U(L)$ is transferred into a co-commutative Hopf algebra. If this algebra acts on R, then formulas (11) show that $x \to x^l$ are derivations, if $l \in L$.

Inversely, let us assume that a Lie algebra of derivations L acts on the algebra R. Then the action of words from $U(L)$ is determined by superposition. The Leibnitz formula shows that identity (8) is valid, so that we get the action of the Hopf algebra $U(L)$. It is also evident that

$$R^{U(L)} = \{x \in R \mid \forall\ l \in L \quad x^l = 0\} = R^L.$$

If a characteristic p of the field C is positive, and L is a restricted Lie algebra, then formulas (12) determine the structure of a Hopf algebra also on the p-enveloping $U_p(L) = U(L)/(l^p = l^{[p]})$, provided we assume that V are correct words from $\{l_i\}$. For a finite-dimensional Lie p-algebra its universal p-enveloping $U_p(L)$ will be finite-dimensional and, hence, there arises a dual Hopf algebra $U_p(L)^*$. This algebra is commutative (since $U_p(L)$ is commutative), in which case the Leibnitz formula (12) and equality (4) show that $V_1^* \cdot V_2^* = (V_1 \cdot V_2)^*$. Therefore, $U_p(L)^*$ is, as an associative algebra, isomorphic to a ring of truncated polynomials

$C[x_1,..., x_n]/(x_i^p = 0)$. (Analogously, in the case of an arbitrary field and a Lie algebra L of an arbitrary dimension, the associative algebra $U(L)^*$ will be isomorphic to a ring of polynomials $C[x_1,..., x_n,..])$. The co-multiplication on $U_p(L)^*$ is determined by the table of multiplication of a Lie algebra by formula (3), the antipode is set by the equality $S(x_i) = -x_i$, while the co-unity maps all variables $x_1,..., x_n$ to zero, and $\varepsilon(1) = 1$. The action of the Hopf algebra $U_p(L)^*$ is quite difficult to determine in a compact form, but this is exactly what could be naturally called grading by a restricted Lie algebra L.

6.5.5. *Skew derivations.* Let R be an algebra over a field C, s be its certain automorphism. A linear mapping $\mu: R \to R$ is called an s-*derivation* (or a skew derivation), if for all $x, y \in R$ the following identity holds:

$$(xy)^{\mu} = x^{\mu} y + x^{s} y^{\mu}.$$

The simplest examples of s-derivations are *inner* s-derivations: if $a \in R$, then $\mu: x \to ax - x^{s}a$ is a skew derivation.

A set L_s of all s-derivations forms a linear space:

$$(xy)^{\mu c + \nu d} = (xy)^{\mu} c + (xy)^{\nu} d = x^{\mu c + \nu d} y + x^{s} y^{\mu c + \nu d}.$$

Inner s-derivations form a subspace $in\ L_s$ in L_s.

It is evident that derivations should be considered together with automorphisms. Hence, of importance is the remark that a group of automorphisms $AutR$ acts by conjugations on a set of all skew derivations:

$$(xy)^{g^{-1}\mu g} = x^{g^{-1}\mu g} y + x^{g^{-1}sg} y^{g^{-1}\mu g},$$

i.e., $L_s^g = L_{g^{-1}sg}$. In this case the subspaces of inner skew derivations prove to be, in a certain sense, invariant.

Known is the fact that a set of common derivations L_1 forms a Lie algebra, while when the characteristic is positive, even a restricted Lie algebra. There arises a natural question: what is the algebraic structure of a set of all skew derivations. This structure proves to be very obscure and essentially dependent on mutual actions of the automorphisms lying in the basis of spaces L_s on these very spaces. Let us, for instance assume that $\mu \in L_s$, $\nu \in L_g$, in which case $\mu^g = \mu$, $\nu^s \nu$. Then, one can easily see that the commutator $[\mu, \nu]$ will be an sg-derivation. If $\mu^g = -\mu$, $\nu^s = -\nu$, then a Jordan composition $\mu\nu + \nu\mu$ will be an sg-derivation. One more example: let $\mu, \nu \in L_s$ and let us suppose $\mu^s = \mu + \nu$, $\nu^s = \nu$. Then $2\mu\nu - 2\nu\mu - \nu^2 \in L_{s^2}$.

Algebraic structures of a set of skew derivations can be studied in terms of Hopf algebras. Let us formulate the necessary definitions.

Let G be a group and let us assume that a mapping γ from G onto a set of linear spaces over the basic field C be set. A pair $\Gamma = (G, \gamma)$ will be called *a comb* if the action of the group G is additionally set on the space $\sum_{g \in G}\oplus\, \gamma(g)$, such that $\gamma(s)^g = \gamma(g^{-1} s\, g)$. The group G will be called a *ground* of this comb, while spaces $\gamma(s)$, $s \in G$ will be called its *teeth*. The *ground of the tooth* $\gamma(s)$ will be the automorphism s.

We now see that skew derivations of the algebra R form a comb with a ground $Aut(R)$ and teeth $\gamma(s) = L_s$. There naturally arises a question: if it is possible to consider the comb Γ over C as a comb of skew derivations (not necessarily of all) of an algebra ? The answer is positive. Let us consider a linear space $L = \sum_{s \in G}\oplus\, \gamma(s)$ and its tensor algebra $R = C < L >$. Each of the elements $s \in G$ acts on L and, hence, uniquely extends to an automorphism of R. In the space $\gamma(s)$ let us fix a certain basis $\mu_1, \dots, \mu_n, \dots$ and determine the action $\mu_i(\mu_j) = \delta_{ij}\mu_j$, where δ_{ij} is a Kroneker delta, and $\mu_i(\gamma(h)) = 0$ at $h \neq s$. Since R is a free algebra, then such an action can be extended to s-derivations of R. Therefore, Γ is implemented by skew derivations.

Let us relate a comb Γ to a Hopf algebra $C < \Gamma >$, which will be called a *free hull of the comb* G. As has been noted above, the group G acts on a free algebra $C < L >$, where $L = \sum \gamma(s)$. Hence, we can determine a crossed product $C < \Gamma > \overset{df}{=} C < L > * G$ with a trivial system of factors (or, in other words, a skew group algebra). The basis of this product consists of all possible words $s\mu_1 \dots \mu_n$, where $s \in G$, μ_i are the elements of fixed bases of teeth. A product of such words is determined by a multiple use of the formula $\mu s = s\,\mu^s$. Let us define co-multiplications on $C < G >$. For the words of a unit length let us set

$$\Delta(\mu) = \mu \otimes 1 + s \otimes \mu, \qquad (\mu \in \gamma(s)),$$
$$\Delta(s) = s \otimes s, \qquad (s \in G).$$

Δ is extended onto words of greater length and on linear combinations in such a way that co-multiplication becomes a homomorphism of algebras. Therefore, $C<\Gamma>$ turns into a bialgebra. This bialgebra has a co-unity $\varepsilon: C<\Gamma>\to C$, which is zero on all the words including basic derivations, and is equal to unity on the elements from G.

Let us determine an antipode S on $C<\Gamma>$. Let us set $S(\mu)=-h^{-1}\mu$, $S(h)=h^{-1}$, where $\mu\in\gamma(h)$. Let us extend it to an anti-automorphism $S: C<\Gamma>\to C<\Gamma>$. The fact that ε is a co-unity and S is an antipode is easily checked by induction over the length of a word. Thus, $C<\Gamma>$ turns into a Hopf algebra.

It is absolutely obvious that the action of this Hopf algebra on the algebra R is equivalent to the implementation (possibly, not exact, i.e., with a kernel) of the comb Γ with skew derivations. Inversely, if Γ is a subcomb of a comb of all skew derivations of the algebra R, then on R one can determine the action of a free hull $C<\Gamma>$ by viewing the word $s\mu_1\dots\mu_n$ as a superposition $x^{s\mu_1\dots\mu_n}=(((x^s)^{\mu_1})\dots)^{\mu_n}$.

The algebraic structure of a set of skew derivations can be characterized by the notion of a relatively primitive element of a free hull.

An element $w\in C<\Gamma>$ is called *relatively primitive* if there can be found an element $s\in G$, such that

$$\Delta(w)=w\otimes 1+s\otimes w.$$

Let us assume that G is a comb of all skew derivations of a certain algebra R. If $w=w(s_1,\dots,s_n,\mu_1,\dots,\mu_m)$ is a certain relatively primitive element, then, by the definition, its action on R will be a skew derivation, i.e., there is a $\mu\in L_s$, such that

$$x^{w(s_1,\,\dots,\,s_n,\,\mu_1,\,\dots,\,\mu_n)}=x^{\mu}\qquad (x\in R).$$

This implies that w determines a certain partial operation on G: $w(s_1,\dots,s_n,\mu_1,\dots\mu_n)=\mu$.

Let us consider some examples. Let μ,ν be skew derivations corresponding to automorphisms s and h, respectively. If $\mu^h=\mu$, $\nu^s=\nu$, then the commutator $\mu\nu-\nu\mu$ is a relatively primitive element. If $\mu^h=-\mu$, then μ^2 is a relatively primitive element. If $s=h$ and $\mu^2=\mu+\nu$, $\nu^s=\nu$, then $2\mu\nu-2\nu\mu-\nu^2$ is a relatively primitive element. In more general terms, let us consider a certain tooth $\gamma(s)$ in an arbitrary comb. Then $\gamma(s)^s=\gamma(s^{-1}ss)=\gamma(s)$, i.e., s acts on $\gamma(s)$ as a non-degenerate linear transformation. Let us assume that in $\gamma(s)$ a basis $\mu_1,\dots,\mu_n$ is chosen in such a way that the matrix of transformations is an elementary Jordan matrix with an eigen number α, i.e., $\mu_1^s = \alpha\mu_1 + \mu_2$, $\mu_2^s = \alpha\mu_2 + \mu_3, .. \mu_n^s = \alpha\mu_n$. One can show that if $\alpha \neq \pm 1$, then a subspace $\gamma(s)^2$ stretched in $C < \Gamma >$ onto the products of pairs of elements of the tooth $\gamma(s)$ has no nonzero relatively primitive elements. If $\alpha = -1$, then the dimension of the subspace of relatively primitive elements in $\gamma(s)^2$ is equal to $[\frac{n+1}{2}]$. If $\alpha = 1$ and the characteristic of the main field is zero, then this dimension is equal to $[\frac{n}{2}]$.

6.5.6. *Algebraic dependences between skew derivations.* Here in the spirit of Chapter 2 we shall consider the problem of dependences between skew derivations. This problem is equivalent to studying generalized identities with automorphisms and skew derivations.

Let R be a prime ring. Any its skew derivation has a unique extension up to the skew derivation of a ring of quotients $Q(R)$, so that, as usual, we shall assume that the objects under investigation are determined on $Q(R)$ as well. Moreover, it would be reasonable to consider not only the skew derivations transferring R in R, but also s-derivations from R in $Q(R)$ transferring a certain nonzero ideal of the ring R in R. Let $s \in A(R)$ and let us set $\mathbb{L}_s = \{\mu$ is an s-derivation of $Q | \exists\ I < R,\ I \neq 0,\ I^{\mu} \subseteq R\}$. In this case a set of all inner s-derivations $ad_s Q$ of the ring Q is contained in $\mathbb{L}_s$.

One can also see that $\mathbb{L}_s$ will be a right vector space over a generalized centroid C of the ring R, i.e., $\mu c\colon x \to \mu(x) \cdot c$. A left multiplication by elements from C is related with a right one by the formula $c\mu = \mu c^s + r_{\mu(c)}$, where $r_{\mu(c)}$ is the operator of right multiplication by the element $\mu(c)$.

If a is an invertible element from $Q(R)$, then an inner automorphism $\tilde{a}$ determined by the element a belongs to $A(R)$, in which case $\mathbb{L}_{s\tilde{a}} = \mathbb{L}_s l_{a^{-1}}$, where $l_{a^{-1}}$ is the operator of left multiplication by the element a^{-1}. This means that we should study dependences only between those teeth which correspond to mutually-outer automorphisms (it should be recalled that automorphisms $g, h \in A(R)$ are called *mutually-outer*, provided gh^{-1} is not an inner automorphism of the ring Q). Then the following definition is prompted.

6.5.7. Definition. A set of skew derivations $\{\mu_1, .., \mu_k\}$ is called *reduced* if

(a) different automorphisms corresponding to these derivations are mutually-outer;

(b) derivations from this set corresponding to the same automorphism s are linearly independent by modulo the subspace of inner s-derivations.

Now we can formulate a theorem showing that there are no dependences without superpositions.

6.5.8. Theorem. *If a prime ring* R *obeys a polylinear identity of the type* $F(x_j^{h_i \mu_k}, x_j^{h_i}) = 0$, *where* $F(z_j^{(i,k)}, y_{ij})$ *is a certain generalized polynomial with coefficients from* R_F, *then on* R_F *the following identity is valid:* $F(z_j^{(i,k)}, y_{ij}) = 0$. *Here* $\{\mu_1, \ldots, \mu_n\}$ *is a reduced set of skew derivations,* $\{h_1, \ldots, h_n\}$ *are pair-wise mutually-outer automorphisms.*

In a general case we can, without essential losses, limit ourselves to considering only the case when the right and left actions of C on the spaces L_s coincide, i.e., to considering the case of algebras over the field C. It is associated with the following two facts: (1) if the automorphism s acts identically on C, then the space L_s consists of only inner s-derivations $L_s = ad_s Q$ and (2) if s acts identically on C, but is not inner, then all s-derivations act trivially on C, i.e., $\mu(C) = 0$.

Under these conditions a set of skew derivations forms a comb Γ over C. Any relatively primitive element w of a free hull of this comb determines a certain algebraic dependence between skew derivations

$$w(s_1, \ldots; \mu_1, \ldots) = \mu, \tag{13}$$

where μ is a skew derivation from Γ, the action of which coincides with that of the operator w from a free hull $C < \Gamma >$.

The solution of the problem of describing algebraic dependences implies studying, to a certain extent, relatively primitive elements. In the simplest case, when an automorphism commutes with skew derivations (i.e., basic automorphisms act trivially on all the teeth of a comb), relatively primitive elements are described easily: their linear combinations form a (restricted, provided the characteristic is positive) Lie algebra generated by the elements of the comb. Therefore, the situation differs but slightly from the case of common derivations.

6.5.9. Theorem. *Let a prime ring* R *obey a polynomial identity of the type* $F(x_j^{h_i \Delta_k}) = 0$, *where* $F(z_j^{(i,k)})$ *is a generalized polynomial with* *the coefficients from* $R_F; \Delta_1,.., \Delta_n$ *are mutually different correct words from a reduced set of skew derivations commuting with all the corresponding automorphisms, and* $h_1,.., h_m$ *are mutually outer automorphisms. In this case the identity* $F(z_j^{(i,\ k)}) = 0$ *is valid on* R_F.

No proofs to theorems **6.5.8** and **6.5.9** will be given here, since under general conditions the problem on identities with skew derivations remains open for discussion (as well as because these proofs are only slightly different from that of theorem **2.2.3**).

6.5.10. *Remarks on the structure of arbitrary Hopf algebras.* Let H be a certain Hopf algebra.The values of co-multiplication on an element determines the character of the action of this element on a product of elements of the algebra R. This enables us to find elements of certain types in H. For instance, if $\Delta(h) = h \otimes h$, then under any action of H we have $(xy)^h = x^h y^h$. Besides, using an antipode, we find $hs(h) = \varepsilon(h)$ and, using a co-unity, we get $h\varepsilon(h) = h$, which affords $\varepsilon(h) = 1$ if $h \neq 0$. This means that such an element h always acts as an automorphism. Let

$$G = \{h \in H \mid h \neq 0, \Delta(h) = h \otimes h\}$$

Let us show that G is a subgroup in H. We have already seen that G consists of invertible elements. Now we have to check if it is closed with respect to multiplication and assuming it inverse:

$$\Delta(gh) = \Delta(g)\Delta(h) = (g \otimes g)(h \otimes h) = gh \otimes gh;$$
$$\Delta(1) = \Delta(g)\Delta(g^{-1}) = (g \otimes g)\Delta(g^{-1}),$$

i.e., $\Delta(g^{-1}) = g^{-1} \otimes g^{-1}$.

It should then be remarked that all the elements from G are linearly independent. Let, on the contrary, $g = \alpha_1 g_1 + .. + \alpha_n g_n$, where $\alpha_i \in C$, $g_i, g \in G$ and the elements $g_1,.., g_n$ are linearly independent. In this case

$$g \otimes g = \Delta(g) = \alpha_1 \Delta(g_1) + \ldots + \alpha_n \Delta(g_n) =$$
$$\alpha_1 g_1 \otimes g_1 + \ldots + \alpha_n g_n \otimes g_n,$$

or

$$\sum_{i,\ j} \alpha_i \alpha_j\ g_i \otimes g_j = \sum_i \alpha_i\ g_i \otimes g_i.$$

Hence, we have $\alpha_i \alpha_j = 0$ at $i \neq j$ and $\alpha_i^2 = \alpha_i$. It is possible only one of the coefficients α_i is equal to unity, while the other are zero, i.e., the initial dependence assumes the form $g = g$, which is the required proof.

Therefore, the linear space generated by G is a group algebra of the group G. In this case the structure of the Hopf algebra induced from H to $C[G]$, is the same as on the group algebra $C[G]$ as a Hopf algebra. Thus, any Hopf algebra contains the biggest Hopf subalgebra which is a group algebra.

Analogously, in H we can find "Lie" elements which always act as derivations (in the theory of Hopf algebras such elements are called *primitive*). Let

$$L = \{ l \in H \mid \Delta(l) = l \otimes 1 + 1 \otimes l \}$$

We can easily see that L is a linear space, and if $l_1, l_2 \in L$, then $l_1 l_2 - l_2 l_1 \in L$, i.e., L forms a Lie subalgebra in H. If the basic field has a positive characteristic p, then

$$\Delta(l^p) = (l \otimes 1 + 1 \otimes l)^p = l^p \otimes 1 + 1 \otimes l^p$$

and, hence, L forms a restricted Lie algebra.

Let us show that the subalgebra generated by L in H will be isomorphic to the universal enveloping of L if $p = 0$, and to the universal p-enveloping if $p > 0$. To this end it would suffice to show that different correct words from a certain fixed basis $\{l_1, ..\}$ of the space L will be linearly independent. Let us carry out induction over the length of a correct word. Let all the words less than V are linearly independent and let us assume that $V = \sum \alpha_i V_i$, where V_i are less words. Let us apply to both parts of the latter equality a co-multiplication

$$\Delta(V) = \sum_{V' \circ V'' = V} V' \otimes V'' = \sum_{i,\, V'_i \circ V''_i = V_i} \alpha_i V'_i \otimes V''_i$$

or, cancelling the sums $\sum \alpha_i V_i \otimes 1 + \sum 1 \otimes \alpha_i V$ from the left and from the right, we get

$$\sum_{\substack{V' \circ V'' = V \\ V', V'' \neq \varnothing}} V' \otimes V'' = \sum_{\substack{i,\, V'_i \circ V''_i = V_i \\ V'_i \neq \varnothing \neq V''_i}} \alpha_i V'_i \otimes V'' .$$

Since, by the supposition of induction, all the words less than V are linearly independent, all the terms must be cancelled. Let $V = \mu_1^k \tilde{V}$, where a subword $\tilde{V}$ does not start with μ_1. Then in the left-hand part the tensor $\mu_1^{k-1} \tilde{V} \otimes \mu_1$ will be encountered exactly k times. In the right-hand part tensors of the type $\mu_1^{k-1} \tilde{V} \otimes w$ will appear only under decomposition of words V_i of the type $l_i \mu_1^{k-1}\tilde{V}$, where $l_i < \mu_1$, in which case the corresponding tensor has the form $\mu_1^{k-1}\tilde{V} \otimes l_i$, while the requirement of cancelling all the terms results in

$$\mu_1^{k-1}\tilde{V} \otimes (k\mu_1 - \sum \alpha_i l_i) = 0,$$

which is impossible, as k is an invertible element in the field C, and the elements $\{\mu_1, l_i\}$ are linearly independent.

Therefore, the Hopf algebra H contains a universal enveloping $U(L)$ (or a p-enveloping $U_p(L)$ when the characteristic is positive). It would be also useful here to remark that $U(L)$ will be a Hopf subalgebra in H: if $l \in L$, then $\varepsilon(l) \cdot 1 + \varepsilon(1)l = l$ and, hence, $\varepsilon(l) = 0$ and, moreover, $S(l) \cdot 1 + S(1) \cdot l = \varepsilon(l) = 0$, i.e., $S(l) = -l$.

Let us then pay attention to the fact that the group G acts by conjugations on L and, hence, on the enveloping as well: $\Delta(g^{-1}lg) = (g^{-1} \otimes g^{-1})(1 \otimes l + l \otimes 1)(g \otimes g) = 1 \otimes g^{-1}lg + g^{-1}lg \otimes 1$. This makes it possible to determine a skew group ring $U(L)*G$ on which naturally arises the structure of a Hopf algebra (the Hopf algebra obtained is a so-called *smash-product* $U(L)\# C[G]$ of the Hopf algebras $U(L)$ and $C[G]$). One can easily see that the subalgebra generated in H by the group G and the Lie algebra L will be isomorphic to $U(L)*G$ and, as a Hopf algebra, to the smash-product $U(L)\# C(G)$. Indeed, one only has to check if the products $g_i V_j$ are linearly independent in H, where $\{V_j\}$ is a basis of $U(L)$, $\{g_i\} = G$, with the induction carried out here with respect to the senior word, by analogy with the one given above.

Now there arises a question: to what extent is the "automorpho-differential" part, $U(L)\# C(G)$, is essential in H? This problem plays an important role in the structure theory of Hopf algebra. Thus, a well-known Costant-Swidler theorem shows this part to quite often cover the whole of a Hopf algebra .

6.5.11. Theorem. *Let C be an algebraically closed field of a zero characteristic. If H is a co-commutative Hopf algebra, then $H = U(L)\# C[G]$.*

In the case of not co-commutative Hopf algebras the "automorpho-differential" part can, of course, be absent (or, more exactly, it can be one-dimensional, $L = \{0\}$, $G = \{1\}$). Let us, for instance, consider grading by finite groups $H = C[F]^*$. Let $h = \sum \alpha_f P_f \in G$. Then

$$\sum \alpha_f P_f \otimes \sum \alpha_f P_f = \sum \alpha_f \Delta(P_f)$$

or

$$\sum \alpha_f \alpha_g (P_f \otimes P_g) = \sum_f \alpha_f \sum_h P_{fh^{-1}} \otimes P_h,$$

wherefrom, calculating the coefficient at $P_f \otimes P_g$ in the right-hand part, we find $\alpha_{fg} = \alpha_f \cdot \alpha_g$. This implies that $\alpha: F \to C^*$ is a character of the group F, $\alpha(f) = \alpha_f$. Now one can easily see that G is isomorphic to a group of characters of the group F with the coefficients in C. If, for instance, $F = [F, F]$, then the group of characters is trivial and $G = \{1\}$. Let us now calculate the "differential" part. Let $l = \sum \alpha_f P_f \in L$. Then

$$\sum \alpha_f P_f \otimes 1 + 1 \otimes \sum \alpha_f P_f = \sum \alpha_f P_{fh^{-1}} \otimes P_h.$$

Allowing for the fact that $1 = \sum_h P_h$ and equating the coefficients at basic tensors, we find $\alpha_{fg} = \alpha_f + \alpha_g$, i.e., $\alpha: F \to C^+$ is a homomorphism of groups. Now we see that as a linear space L is isomorphic to the space $Hom_Z(F/[F, F], C^+)$, while the Lie multiplication on L is zero, as $C[F]^*$ is commutative. Therefore, if $F = [F, F]$, then the "differential" part of $H = C[F]^*$ will be also trivial.

Alongside with automorphisms and derivations, in an arbitrary Hopf algebra we can also find a comb of skew derivations. Let $s \in G$. Let us introduce a notation

$$L_s = \{h \in H \mid \Delta(h) = h \otimes 1 + s \otimes h\}$$

We see that L_s is a linear subspace in H. Besides, if $g \in G$, $h \in L_s$, then

$$\Delta(g^{-1}hg) = g^{-1}hg \otimes 1 + g^{-1}sg \otimes g^{-1}hg,$$

i.e., the group G acts by conjugations, permuting the spaces $L_s^g = L_{g^{-1}sg}$ and we, thus, get a comb $\gamma: s \to L_s$. This comb is closed relative partial operations which were discussed in the preceding paragraph. Indeed, if $C < \Gamma >$ is a free hull of the comb Γ, then an identical mapping on Γ is extended to a homomorphism of the algebras $\varphi: C < \Gamma > \to H$. This homomorphism will be also a homomorphism of co-algebras in the meaning that if $v \in C<\Gamma>$ and $\Delta(v) = \sum v_i^{(1)} \otimes v_i^{(2)}$, then $\Delta(\varphi(v)) = \sum \varphi(v_i^{(1)}) \otimes \varphi(v_i^{(2)})$. Besides, the value of an antipode S and a co-unity ε is uniquely determined on Γ: if $h \in L_g$, then $h = h \cdot \varepsilon(1) + g \cdot \varepsilon(h) = \varepsilon(h) \cdot 1 + \varepsilon(g)h$, i.e., $\varepsilon(h) = 0$ and $\varepsilon(h) = hS(1) + gS(h) = S(h) \cdot 1 + S(g) \cdot h$, and, hence, $S(h) = -g^{-1}h$. Therefore, φ is a homomorphism of Hopf algebras, and if w is a relatively primitive element in $C<\Gamma>$, then $\varphi(w)$ is a relatively primitive element in H, i.e., $\varphi(w) \in L_s$ for a suitable $s \in G$. Now the value of the operation w is the element $\varphi(w)$.

From the viewpoint of studying skew derivations of greatest interest is a Hopf subalgebra $H(\Gamma) \overset{df}{=} (\varphi(C<\Gamma>)$ generated in H by the comb Γ. Many interesting problems arise here. For instance, by analogy with the "automorphodifferential" part, one can expect $H(\Gamma)$ to be, in a certain sense, a universal enveloping for Γ. This would, for instance, mean that if for Hopf algebras H, H' the corresponding (complete) combs, Γ, Γ', are isomorphic, then the Hopf subalgebras $H(\Gamma)$ and $H(\Gamma')$ are also isomorphic. As far as is known, these problems are still awaiting investigation.

REFERENCES

G.Bergen, M.Cohen [18];
M.Cohen [29];
M.Cohen, D.Fishman [30];
M.Cohen, S.Montgomery [31];
V.K.Kharchenko, A.Z.Popov [85];
A.Leroy [94,95];
A.Leroy, J.Matzuk [96,97];
M.E.Swidler [148].

REFERENCES

1 **Alev J.** "Sur l'extension $R^G \subset R$". Seminar a'Algebre Dubreil et Malliavin, Lecture Notes in Mathematics.-1983- v.1029 - Springer-Verlag, Berlin.

2. **Almkvist G.** "Commutative and Non-Commutative Invariants of $SL(2,C)$, a Uniform Combinatorial Approach".

3. **Almkvist G., Dicks W., Formanek E.** "Hilbert Series of Fixed Free Algebras Noncommutative Classical Invariant Theory". Journal of Algebra, - 1985 - v.93 - N1 - pp. 189-214.

4. **Amitsur S.** "Generalized Polynomial Identities and Pivotal Monomials" Trans.Amer. Math. Soc. -1965 -v. 114 - pp.210-216

5. **Amitsur S.** "Identities in Rings with Involutions'. Izr. J..Math. -1969 - v. 7 - N 1 - pp.63-68

6. **Amitsur S.** "On Rings of Quotients" Symposia Mathematica. - 1972 - v. 8 - pp.149-164

7. **Amitzsur S.** "Derivations in Simple Rings". Proc.London Math.Soc. - 1957 - v. 71 - pp.87-112

8. **Andrunakievich V.A., Ryabukhin Yu.M.** "Radicals of Algebras and Structural Theory". 'Nauka', Moscow - 1979 - 495 p.

9. **Artin M.** "Galois Theory" Notre Dame - 1942

10. **Artin M., Shelter W.** "Integral Ring Homomorphisms" Advances in Math. - 1981 - v. 39 - pp.289-329

11. **Baer R.** "Algebrai Theorie der Differentierbarren Funktionenköper. I " Sitzungsberichte, Heidelberger Akademia. -1927 - pp.15-32

12. **Barbaumov V.E.** "Algebraic Automorphisms and *PI*-Algebras". Math. Series. - 1975 - V. 97 - N 1 - pp.59-76

13. **Beider K.I.** "A Ring of Invariants under the Action of a Finite Group of Automorphisms of a Ring". Uspekhi Mat. Nauk. - 1977 - v. 32 - N 1(193) - pp.159-160

14. **Beider K.I.** "Rings with Generalized Identities I, II, III" Vestn.MGU, mat.-

mekh. -1977 - N 2 - pp.19-26; 1977 - N 3 - pp.30-37; 1978 N 4 - pp.66-73.

15. **Beider K.I., Mikhalev A.V.** "Orthogonal Completeness and Minimal Prime Ideals". Trudy seminara im. I.G.Petrovskogo. - 1984 - vyp. 10 - pp.227-231.

16. **Beider K.I., Mikhalev A.V.** "Orthogonal Completeness and Algebraic Systems". Uspekhi Mat.Nauk. - 1985 - v.40 - N 6 (246) - pp.79-115.

17. **Beider K.I., Ten V.** "On Local Finiteness of Certain Algerbras" SMZh - 1977 - v.18 - N 4 - pp.934-939.

18. **Bergen J., Cohen M.** "Actions of Commutative Hopf Algebras". Bull.London Math.Soc. - 1986 - v. 18 - N 2 - pp. 159-164

19. **Bergman G.** "A Derivation on a Free Algebra whose Kernel is a Nonfree Subalgebra". Preprint - 1981

20. **Bergman G., Cohn P.** "Symmetric Elemrents in Free Powers of Rings". J.London Math.Soc. - 1969 - v. 44 - pp.1-10

21. **Bergman G., Isaaks I.M.** "Rings with Fixed Point-Free Group Actions" Proc.London Math.Soc. -1973 - v. 27 - N 3 - pp.69-87

22. **Berkson A.** "The u-Algebra of a Restricted Lie Algeb ra is Frobenius". Proc.Amer. Math.Soc. - 1964 - v. 15 - N 1 - pp.14-15

23. **Bokut' L.A.** "Associative Rings, I (Ring Constructions)". Ser. 'Biblioteka kafedry algebry i matematicheskoj logiki NGU' - 1977 - v.18 - pp.1-82

24. **Bokut' L.A.** "Embeddings into Simple Associative Algebras" Algebra i Logika. - 1976 - v. 15 - N 2 - pp.115-246

25. **Burris S., Werner H.** "Sheaf Constructions and their Elementary Properties". Trans.Amer.Math.Soc. - 1979 - v. 248 - N 2 - pp.269-309

26. **Chang C.L., Lee P.H.** "Noetherian Rings with Involutions". Chinese J. Math. -1977 - V.5 - pp.15-19

27. **Childs L.N., DeMeyer F.R.** "On Automorphisms of Separable Algebras". Pacific J.Math. - 1967 - V. 23 - pp.25-34

28. **Cohen M.** "Semiprime Goldie Centralizers". Izrael J.Math. -1975 - v. 20 - pp.37-45; Addenum. -1976 - v. 24 - pp.89-93

29. **Cohen M.** "Smash Product", Inner Action and Quotient Rings". Pacific Journal of Math. -1986 - v. 125 - N 1 - pp.45-66

30. **Cohen M., Fishman D.** "Hopf Algebra Actions". Journal of Algebra. - 1986 - v. 100 - N 2 - pp.363-379

31. **Cohen M., Montgomery S.** "Group Graded Rings, Smash Products and Group Actions". TAMS - 1984 - v. 282 - N 1 - pp.237-257

32. **Cohen M., Montgomery S.** "Semisimple Artinian Rings of Fixed Points". Canad.Math.Bull. -1975 - v. 18 - pp.189-190

33. **Cohen M., Montgomery S.** "Tracelike Functions on Rings with no Nilpotent Elements". Trans. Amer. Math.Soc. - 1982 - v. 273 - N 1 - pp. 131-145

34. **Cohn P.** "Free Rings and their Relations". London Math. Soc., Monograph, N 2, 1971; second edition, N 19, 1985

35. **Cohn P.** "On a Group of Automorphisms of a Free Algebra of Rank 2". Preprint -1978

36. **Curtis C.W., Rainer I.** "Representation Theory of Finite Groups and Associative Algebras". Interscience Publishers (a division of John Wiley & Sons), New-York-London, 1962, 650 p.

37. **Czerniakiewich A.** "Automorphisms of Free Algebras of Rank 2, I, II". TAMS - 1971 - V. 160 - pp. 393-401; Ibid. - 1972 - V. 171 - pp.309-315

38. **Dicks W., Formanek E.** "Poinçare Series and Problem of S.Montgomery" Linear and Multilinear Algebra. - 1982 - v. 12 - pp.21-30

39. **Diendonne J.** "La Théorie de Galois des Anneaux Simples et Semisimples" Comment. Math. Helv. -1948 - v. 21 - pp.154-184

40. **Dieudonne J., Carrol J..** "Invariant Theory, Old and New". Academic Press, New-York-London, 1971

41. **"Dnestrovskaya Tenrad' "** ("Dnestrovskaya Notebook"). IM, Novosibirsk - 1982 - 72 p.

42. **Elizarov V.P.** "Strong Pretorsions and Strong Filters, Modules and Rings of Quotients". SMZh - 1973 - v. 14 - N 3 - p.549-559

43. **Faith C.** "Galois Subrings of Ore Domains are Ore Domains". Bull. Amer.

Math. Soc. - 1972 - V.78 - pp.1077-1080

44. **Farkas D., Snider R.** "Noetherian Fixed Rings". Pacific J. of Math. - 1977 - V. 69 - N 2 - pp.347-353

45. **Fisher J.** "Chain Conditions for Modular Lattices with Finite Group Actions". Canadian J. - 1979 - V. 31 - pp.558-564

46. **Fisher J.** "Semiprime Ideals in Rings with Finite Group Actions". Preprint.

47. **Fisher J., Montgomery S.** "Invariants of Finite Cyclic Grouops Acting on Generic Matrices". Preprint

48. **Fisher J., Osterburg J.** "Semiprime Ideals in Ringswith Finite Group Actions". Jour. of Algebra - 1978 - V. 50 - pp.488-502

49. **Fisher J., Osterburg J.**. 'Finite group actions on noncommutative rings: a survey since 1970'. In: "Ring Theory and Algebra III", Lecture Notes in Math., Dekker, New-York, 1980, pp. 357-393

50. **Goursaud J.M, Pascaud J.L., Valette J.** "Sur la Travaus de V.K.Kharchenko" Lecture Notes - 1982 - N 924 - pp.322-355

51. **Goursaud J.M., Pascaud J.L., Valette J.** "Actions de Groups et Contextes de Morita " Comm.in algebra. -1983 - v. 11 - N 8 - pp.2069-2105

52. **Goursaud J.M., Ostenburg J., Pascaud J.L., Valette J.** "Points Fixes Anneaux Reguliers Autoinjectifs a Gauche" Comm.in algebra - 1981 - v. 9 - N 3 - pp.13543-1394

53. **Grezesczuk P., Puczylowski E.R.** "Goldie Dimension and Chain Conditions for Modular Lattices with Finite Group Actions". Can.Math.Bull. - 1986 - v. 29 - N 3 - pp.274-280

54. **Guralnick R.** "Invarinats of Finite Linear Groups on Relatively Free Algebras". Linear Algebra and its applications - 1985 - v.72 - pp.85-92

55. **Hacque M.** "Théorie de Galois des Anneaux Presque-Simples". Jour. of Algebra - 1987 - V.108 - N 2 - pp. 534-577

56. **Handelman D., Renualt G.** "Actions of Finite Groups on Self-Injective Rings". Pacific J. of Math. - 1980 - V. 89 - N 1 - pp.69-80

57. **Herstein I.** "Noncommutative Rings". Carus Mathematical Monographs, N 15, Amer. Math. Soc., 1968

58. **Higman G.** "Ordering by Divisibility in Abstract Algebras". PLMS - 1952 - V. 2 - N 7 - pp.326-336

59. **Higman G.** "Groups and Rings which Have Automorphisms with no Nontrivial Fixed Elements". J. L.M.S. - 1957 - V. 31 - pp.321-334

60. **Hochschield G.** "Automorphisms of Simple Algrebras". Trans. A.M.S. - 1950 - V. 69 - pp.292-301

61. **Hochschield G.** "Double Vector Spaces over Division Rings". Amer. J. Math. - 1949 - V. 71 - pp.443-460

62. **Jacobson N.** "A note on Division Rings ". Amer.J.Math. - 1947 - v. 69 - pp.27-36

63. **Jacobson N.** "Abstract Derivations and Lie Algebras" Trans.Amer.Math.Soc. - 1937 - v. 42 - pp.206-224

64. **Jacobson N.** "The Fundamental Theorem of the Galois Theory for Quasifields" Ann.of Math. -1940 - v. 41 - pp.1-7

65. **Jacobson N.** "Structure of Rings". Amer. Math.Soc.Colloquium. Publ., Providence, 1964

66. **Kanzaky T.** "On Galois Extension of Rings". Nagoya Math.J. - 1966 - v. 27 - N 1 - pp.43-50

67. **Kartan G.** "Theorie de Galois pour les Corps non Commutatifs" Ann.Ecol.Norm. -1947 - v. 64 - pp.59-77

68. **Kharchenko V.K.** "Galois Subrings of Simple Rings". Matem. Zametki - 1975 - V. 17 - N 6 - pp.887-892

69. **Kharchenko V.K.** "Galos Extensions and Rings of Quotients" . Algebra i Logika - 1974 - V. 13 - N 4 - pp.460-484

70. **Kharchenko V.K.** "New Structure Theory and Derivations of Semiprime Rings". Mathematisches Forschungsinstitut Obervolfach - Tagungstericht, 18 - 1986 - p.7

71. **Kharchenko V.K.** " Differential Identities of Semiprime Rings". Algebra i Logika - 1979 - V. 18 - N 1 - pp.86-119

72. **Kharchenko V.K.** " Galois Extensions of Radical Algebras". Matem Sb. - 1976 - V. 101(143) - N 4(12) - pp.500-507

73. **Kharchenko V.K.** " Ring Identities with Automorphisms". SMZh - 1976 - . 17 - N 2 - pp.446-467

74. **Kharchenko V.K.** "A Note on Central Polynomials". Matem Zametki - 1979 - V. 26 - N 3 - pp.345-346

75 **Kharchenko V.K.** "Actions of Groups of Lie Algebras on Noncommutative Rings". Uspekhi Mat. Nauk. - 1980 - V. 35 - N 2 - pp.66-90

76. **Kharchenko V.K.** "Centralizors of Finite-Dimensional Algebras". Algebra i Logika - 1981 - V. 20 - N 2 - pp.231-247

77. **Kharchenko V.K.** "Constants of Derivations of Prime Rings". Izv. AN, ser. Matem. - 1981 - V. 45 - N 2 - pp.435-461

78. **Kharchenko V.K.** "Differential Identities of Prime Rings". Algebra i Logika - 1978 - V. 17 - N 2 - pp.220-238

79. **Kharchenko V.K.** "Fixed Elements Relative a Finite Group Acting on a Semiprime Ring". Algebra i Logika - 1975 - V. 14 - N 3 - pp.328-344

80. **Kharchenko V.K.** "Galois Theory of Semiprime Rings". Algebra i Logika - 1977 - V. 16 - N 3 - pp.313-263

81. **Kharchenko V.K.** "Generalized Identities with Automorphisms". Algebra i Logika - 1975 - V. 14 - N 2 - pp.215-237

82. **Kharchenko V.K.** "Noncommutative Invariants of Finite Groups and Noetherian Varieties". Jour. of Pure and Applied Algebra - 1984 - V. 31 - N 1-3 - pp.83-90

83. **Kharchenko V.K.** "On Algebras of Invariants of Free Algebras". Algebra i Logika - 1978 - V. 17 - N 4 - pp.478-487

84. **Kharchenko V.K.** "Generalized Identities with Automorphisms of Associative Rings with Unity". Algebra i Logika - 1975 - V. 14 - N 6 - pp.681 - 696

85. **Kharchenko V.K., Popov A.Z.** "Skew Derivations of Prime Rings". Trudy Instituta Matematiki SO AN SSSR - 1989, V. 16, pp. 183-196

86. **Kitamura Y.** "Note on the Maximal Quotient Ring of a Galois Subring".

J.Okayama Univ. -1976 - v. 19 - pp.55-60

87. **Kolotov A.T.** "On Free Subalgebras Of Free Associative Algebras". SMZh. - 1978 - v. 19 - N 2 - pp.328-335

88. **Koryukin A.N.** "On Noncommutative Invariants of Reductive Groups". Algebra i logika. - 1984 - v. 23 - N 4 - pp.419-429

89. **Kreimer H.** "A Galois Theory for Noncommutative Rings". Trans. Amer. Math.Soc. -1967 - v. 127 - N 1 - pp.127 - N 1 - pp.29-41; 42-49

90. **L'vov I.V., Kharchenko V.K.** "Normal Elements of an Algebra of General Matrices are Central". Sib.Mat. Zhurnal. - 1982 - V. 23 - N 1 - pp.193-195

91. **Lambek J.** "Lectures on Rings and Modules". Blaisdell Publ. Comp., Waltman, Massachussetts-Toronto-London, 1966

92. **Lane D.R.** "Free Algebras of Rank 2 and their Automorphisms". Ph.D.thesis, London University. - 1976

93. **Lansky C.** "Differential Identities in Prime Rings with Involutions". Trans.Amer.Math.Soc. -1985 - v. 291 - N 2 - pp.765-787

94. **Leroy A.** "(S)-derivations Algebriques sur les Cops Gauches et sur les Anneaux Premiers". -1986 - v. 14 - N 8.

95. **Leroy A.** "Derevees Logarithmiques pour une S -derivation Algebrique". Comm.in Algebra. -1985 - v. 13 - N 1 - pp.85-100

96. **Leroy A., Matzuk J.** "Derivations et Automorphismes Algebriques d'anneaux Premiers". Comm.in Algebra. - 1985 - v. 13 - N 6 - pp.1245-1266

97. **Leroy A., Matzuk J.** "Quelques Remarques a Propos der S-derivations". Comm.in Algebra. - 1985 - v. 13 - N 6 - pp.1229-1244

98. **Levitsky J.** "On Automorphisms of Certain Rings". Ann.Math. - 1935 - v.36 - pp.984-992

99. **Lorenz M., Montgomery S., Small L.W.** "Prime Ideals in Fixed Rings. II". Comm. in Algebra. - 1982 - v. 10 - N 5 - pp.449-455

100. **Lyubetsky V.A., Gordon E.I.** "Imbedding of sheaves in a Geiting-sign Universum". Dep.VINITI, N 4782-82 - 1982 - Moscow - pp.1-29

101. Makar-Limanov L.G. "Automorphisms of Free Algebras with Two Generators". Funrz. analiz i ego prilozh. - 1970 - V. 4 - pp.107-108

102. Mal'tsev A.I. "Algebraic Systems". Nauka, Moscow. - 1970 - 392 p.

103. Mal'tsev A.I. "On a General Method of Obtaining Local Thorems of the Group Theory". Uchen. zap. Ivanovsk. ped. inst. - 1941 - V.1 - N 1 - pp.3-9

104. Martindale W.S. "Fixed Rings of Automorphisms and the Jacobson Radical". J.London Math.Soc. - 1978 - V.17 - N 2 - pp.42-46

105. Martindale W.S. "Prime Rings Satisfying a Generalized Polynomial Identity". J.Algrebra. - 1969 - V. 12 - N 4 - pp.576-584

106. Martindale W.S., Montgomery S. "Fixed Elements of Jordan Automorphisms of Associative Rings". Pacific Journal of Math. - 1977 - V, 72 - N 1 - pp.181-196

107. Martindale W.S., Montgomery S. "The Normal Closure of Coproducts of Domains". Journal of Algebra. - 1983 - V. 82 - N 1 - pp.1-17

108. Miyashita Y. "Finite Outer Galois Theory of Non-Commutative Rings". J.Fac. Sci. Hokkaido Univ. -1966 - V.19 - N 3 - pp.114-134

109. Molien T. "Über die Invarianten der Linearen Substitutionsgruppe". Sitzugsber Konig.Preuss. Akad. Wiss. - 1897 - pp.1152-1156

110. Montgomery S. "Group Actions on Rings: Some Classical Problems". Proc. of NATO A.S.I. in Ring Thoery". - 1983 - Antwerpen.

111. Montgomery S. "Outer Automorphisms of Semiprime Rings". J.London Matrh.Soc. - 1978 - V.18 - N 2 - pp.209-221

112. Montgomery S. "*X*-inner Automorphisms of Filtered Algebras". Proc. Amer. Math. Soc. - 1981 - v.83 - N 2 - pp.263-268; 1983 - V.87 - N 4 - pp.569-575

113. Montgomery S. "Fixed Rings of Finite Automorphism Grouops of Associative Rings". Lect.Notes in Math. - Berlin, Heidelberg, New York - Springer-Verlag - 1980

114. Montgomery S. "Prime Ideals in Fixed Rings" Comm. in Algrebra - 1981 - V.19 - N 4 - pp.423-449

115. Montgomery S. "The Jaconsin Radical and Fixed Rings of Automorphisms".

Comm. in Algebra - 1976 - V.4 - N 5 - pp.459-466

116. Montgomery S., Passman D. "Prime Ideals in Fixed Rings of Free Algebras". Comm. in Algebra - 1987 - V.15 - N 11 - pp. 2209-2234

117. Montgomery S., Passman D.S. "*X*-inner Automorphisms of Group Rings". Houston J.Math. - 1981 - V.7 - N 3 - pp.395-402; Ibid. - 1982 - V.8 - N 45 - pp. 537-544

118. Montgomery S., Passman D.S. "Galois Theory of Prime Rings". Journal of Pure and Applied Algebra - 1984 - V.31 - N 1-3 - pp.139-184

119. Montgomery S., Small L. "Fixed Rings of Noetherian Rings". Bull. L. M. S. - 1981 - V.13 - pp.33-38

120. Montgomery S., Small L. "Some Remarks on Affine Rings". Proc. AMS - 1988

121. Montgomery S., Small L.W. "Integrality and Prime Ideals in Fixed Rings of P.I.Rings". Jour. of Pure and Appl. Algebra - 1984 - V.31 - N 1-3 - pp. 185-190

122. Moors R. "Théorème Foundamental de la Théorie de Galkois Finite pour Certains Anneaux Non-Commutatifs". Bull.de la Société Roy des Sci. de Liège. - 1970 - V.39 - N 11, 12 - pp.541-550

123. Nagarajan K. "Groups Acting on Noetherian Rings". Neuw. Archief. voor Wiskunde - 1968 - V.16 - pp.25-29

124. Nakayama T. "Galois Theory of Simple Rings". Trans. Amer.Math.Soc. - 1952 - V.73 - pp.276-292

125. Nakayama T., Azumaya G. "On Irreducible Rings ". Ann. of Math. - 1947 - V.48 - N.2 - pp.949-965

126. Noether E. "Nichtkommutative Algebra". Math.Z. - 1933 - V.37 - pp.514-541\

127. Ostenburg J. "Copmpletely Outer Galois Theory of Perfect Rings". Pacific J.Math. - 1975 - V.56 - N 1 - pp.215-220

128. Ostenburg J. "Fixed Rings of Simple Rings". Comm.in Algrebra - 1978 - V.6 - N 17 - pp.1741-1750

129. Ostenburg J. "The Influence of the Algebra of the Group". Comm. in

Algrebra - 1979 - V.7 - pp.1377-1396

130. **Page A.** "Actions de Groupes". Lect. Notes in Math. -1979 - N 740 - pp.9-24

131. **Paré R., Schelter W.** "Finite Extensions are Integral". J. of Algebra - 1978 - V. 53 - pp. 477-479

132. **Pascaud J.-L.** "Actions de Groupes et Traces". Comm. in Algebra - 1982 - V.10 - N 10 - pp.1101-1117

133. **Pascaud J.-L.** "Two Results on Fixed Rings". Proc. Amer.Math.Soc. -1981 - V.82 - N 4 - pp.517-520

134. **Passman D.S.** "Infinite Crossed Products". Pure and Appl. Math. - 1989 - V. 135 - 468 p.

135. **Passman D.S.** "It's Essentially Maschke's Theorem". Rocky Mountain Jour. of Math. - 1983 - V.13 - N 1 - pp.37-54

136. **Piers Dos Santos.** "Derivations des Anneaux Semi-Premiers I". Comm. in Algebra - 1986 - V.14 - N 8 - pp.1523-1559

137. **Popov A.Z.** "On Differentiation of Prime Rings". Algebra i Logika - 1983 - V. 22 - N 1 - pp.79-92

138. **Posner E.** "Derivations in Prime Rings". Proc. Amer. Math.Soc. -1957 - V.8 - N 6 - pp.1093-1100

139. **Quinn D.** "Integrality over Fixed Rings".

140. **Renault G.** "Action de Groupes et Anneaux Regulars Injectifs". Lect. Notes in Math. -1979 - N 734 - pp.236-248

141. **Rosengebr A., Zelinsky D.** "Galois Theory of Continuous Transformation Rings". TAMS - 1955 - V.79 - pp.429-452

142. **Rowen L.** "Generalized Polynomisal Identities". J.Algebra - 1975 - V.34 - N 3 - pp.458-480

143. **Shirshov A.I.** "Subalgebras of Free Lie Algebras". Matem Sb. - 1953 - V. 33(75) - pp. 441-453

144. **Shirshov A.I.** "On Rings with Identities". Matem Sb. - 1957 - V. 43(85) - pp.277-283

145. Slin'ko A.M. "Notes on Radicals and Detrivations of Rings". SMZh - 1972 - V.13 - pp.1395-1397

146. Stendström Bo. "Rings of Quotients". - Springer-Verlag - Berlin, Heidelberg, New York - N 237 - 1971

147. Sundström T. "Groups of Automorphisms of Simple Rings". J. of Algebra - 1974 - V. 29 - pp.555-566

148. Swidler M.E. "Cocommutative Hopf Algebras with Antipode". Bull. Amer.Math. Soc. - 1967 - V. 73 - pp.196-228

149. Tominaga H. "Note on Galois Subgroups of Prime Goldie Rings". Math. J. Okayama Univ. - 1973 - V. 16 - pp.115-116

150. Tominaga H., Nakayama T. "Galois Theory of Simple Rings". Okayama Math. Lecturesa - 1970 - Okayama Univ.

151. Veil H. "The Classical Groups, their Invariants and Representations". Institute for Advanced Study", 1939

152. Villmayor O.E., Zelinsky D. "Galois Theory with an Infinite Number of Idempotents" Nagoya Math.J. - 1969 - v. 35 - pp.83-98

153. Weisfeld M. "On Derivations in Division Rings". Pacific J. Math., 1960 - v. 10 - pp. 335 - 343

154. Wolf M.C. "Symmetric Functions of Noncommuting Elements". Duke Math. Journal. - 1936 - v. 2 - pp.626-637

155. Yakovlev A.V. "Galois Theory of Sheaves of Sets". Trudy LOMI - 1978 - V. 148 - pp.253-266

156. Zalessky A.E., Neroslavsky O.M. "On simple Noether Rings". Izvestiya AN BSSR - 1975 - v. 5 - pp.38-42

157. Zalessky A.E., Neroslavsky O.M. "There exists a simplke Noetherian Ring with Zero Divisors but without Idempotents" Comm. in Algebra. - 1977 - v. 5 - pp.231-245

INDEX